Historiography of the History of Science in Islamicate Societies

This book presents eight papers about important historiographical issues as debated in the history of science in Islamicate societies, the history of science and philosophy of medieval Latin Europe and the history of mathematics as an academic discipline. Six papers deal with themes about the sciences in Islamicate societies from the ninth to the seventeenth centuries, among them novelty, context and decline. Two others discuss the historiographical practices of historians of mathematics and other disciplines in the nineteenth and twentieth centuries.

The central argument of the collected papers is that in addition and beyond the study of scientific texts and instruments historians of science in Islamicate societies need to pay attention to cultural, material and social aspects that shaped the scientific activities of the authors and makers of such texts and instruments. It is pointed out that the diachronic, de-contextualized comparison between methods and results of scholars from different centuries, regions and cultures often leads to serious distortions of the historical record and is responsible for the long-term neglect of scholarly activities after the so-called "Golden Age".

The book will appeal in particular to teachers of history of science in Islamicate societies, to graduate students interested in issues of methodology and to historians of science grappling with the unresolved problems of how to think and write about the sciences in concrete societies of the past instead of subsuming all extant texts, instruments, maps and other objects related to the sciences under macro-level concepts like Islam or Latin Europe.

Sonja Brentjes is an historian of science with specialization in Islamicate societies, the late medieval Mediterranean and early modern Catholic and Protestant Europe. Currently, she is a visiting scholar at the Max Planck Institute for the History of Science in Berlin. Her latest books include: *Teaching and Learning the Sciences in Islamicate Societies, 800–1700* (2018) and *The Routledge Handbook of the Sciences in Islamicate Societies: Practices from the 2nd/8th to the 13th/19th Centuries* (2023), co-edited with Peter Barker (associate editor) and Rana Brentjes (assistant editor).

Also in the Variorum Collected Studies series

AVERIL CAMERON
From the Later Roman Empire to Late Antiquity and Beyond (CS1113)

ROBERTO TOTTOLI
Studies in Islamic Traditions and Literature (CS1112)

BRIAN CROKE
Engaging with the Past, c.250-c.650 (CS1111)

FELICE LIFSHITZ
Reading Gender
Studies on Medieval Manuscripts and Medievalist Movies (CS1110)

DOROTHEA MCEWAN
Studies on Aby Warburg, Fritz Saxl and Gertrud Bing (CS1109)

GAD G. GILBAR
Trade and Enterprise
The Muslim Tujjar in the Ottoman Empire and Qajar Iran, 1860–1914 (CS1108)

JEAN-CLAUDE HOCQUET
Le marchand et les poids et mesures (CS1107)

CHRISTOPHER ALLMAND
Aspects of War in the Late Middle Ages (CS1106)

PHILIP BUTTERWORTH
Staging, Playing, Pyrotechnics and Magic: Conventions of Performance in Early English Theatre
Shifting Paradigms in Early English Drama Studies (CS1105)

ALEXANDER CALLANDER MURRAY
The Merovingians:
Kingship, Institutions, Law, and History (CS1104)

CRAIG KALLENDORF
The Virgilian Tradition II
Books and Their Readers in the Renaissance (CS1103)

www.routledge.com/Variorum-Collected-Studies/book-series/VARIORUM

Historiography of the History of Science in Islamicate Societies

Practices, Concepts, Questions

Sonja Brentjes

VARIORUM COLLECTED STUDIES

Routledge
Taylor & Francis Group
LONDON AND NEW YORK

First published 2024
by Routledge
4 Park Square, Milton Park, Abingdon, Oxon OX14 4RN

and by Routledge
605 Third Avenue, New York, NY 10158

Routledge is an imprint of the Taylor & Francis Group, an informa business

British Library Cataloguing-in-Publication Data
A catalogue record for this book is available from the British Library

Library of Congress Cataloging-in-Publication Data
Names: Brentjes, Sonja, author
Title: Historiography of the history of science in Islamicate societies : practices, concepts, questions / Sonja Brentjes.
Description: Abingdon, Oxon ; New York, NY : Routledge, [2024] | Series: Variorum collected studies | Includes bibliographical references and index.
Identifiers: LCCN 2023008050 (print) | LCCN 2023008051 (ebook) | ISBN 9781032445052 (hardback) | ISBN 9781032445069 (paperback) | ISBN 9781003372493 (ebook)
Subjects: LCSH: Science—Islamic countries—Historiography. | Islam and science—Historiography. Classification: LCC Q127.I742 B74 2024 (print) | LCC Q127.I742 (ebook) | DDC 509.17/671—dc23/eng/20230322
LC record available at https://lccn.loc.gov/2023008050
LC ebook record available at https://lccn.loc.gov/2023008051

ISBN: 978-1-032-44505-2 (hbk)
ISBN: 978-1-032-44506-9 (pbk)
ISBN: 978-1-003-37249-3 (ebk)

DOI: 10.4324/9781003372493

Typeset in Times New Roman
by Apex CoVantage, LLC

VARIORUM COLLECTED STUDIES SERIES CS1114

CONTENTS

	Acknowledgements	vii
	Introduction	1
1	Novelty as cultural value: places, forms and norms of the claims to novelty in Islamic societies. English version of: 'La Nouveauté comme valeur culturelle,' in Sarah Carvallo, Sophie Roux (eds.), *Du nouveau dans les sciences*, Recherches sur la philosophie et le langage 24, Grenoble: Université Pierre Mendès France, 37–70.	4
2	Reflections on the role of the exact sciences in Islamic culture and education between the twelfth and the fifteenth centuries, published in Mohammed Abattouy (ed.), *Etudes d'histoire des sciences arabes*, Casablanca: Fondation du Roi Abdul-Aziz, 2007, 15–33.	28
3	What could it mean to contextualize the sciences in Islamic societies of the past?, published in Mohammed Abattouy (ed.), *Les sciences dans les sociétés islamiques. Approches historiques et perspectives d'avenir*, Casablanca: Fondation du Roi Abdul-Aziz, 2007, 15–41.	42
4	The mathematical sciences in Safavid Iran: questions and perspectives, published in D. Hermann, F. Speziale (eds.), *Muslim Cultures in the Indo-Iranian World during the Early-Modern and Modern Periods*, Berlin: Klaus Schwarz Verlag, Tehran: Institut Français de Recherche en Iran, 2010, 325–402.	62
5	The prison of categories – 'decline' and its company, published in F. Opwis, and D. Reisman (eds.), *Islamic philosophy, science, culture, and religion: studies in honor of Dimitri Gutas*. Leiden: Brill, 2012, 131–56.	129

6 Towards a new approach to medieval cross-cultural exchanges, co-authored with Fidora, A., & Tischler, M. M., published in *Journal of Transcultural Medieval Studies*, *1*(1) (2014), 9–50. 151

7 Practicing history of mathematics in Islamicate societies in 19th-Century Germany and France, published in V. R. Remmert, M. R. Schneider, & H. Kragh Sørensen (eds.), *Historiography of Mathematics in the 19th and 20th Centuries*, Cham: Birkhäuser, (2016), 25–52. 185

8 Was there a shift from faith-neutral to faith-based scholarly communities in Islamic societies from the classical to the post-classical period?, English version of 'Sans Islam dan kerjasama lintas agama,' in Symseuddein Arif (ed.), *Islamic Science. Paradigma, Fakta dan Agenda*, Jakarta: Institute for the Study of Islamic Thought and Civilizations, (INSISTS), 2016, 138–55. 213

Index 228

ACKNOWLEDGEMENTS

I thank my coauthors Alexander Fidora and Matthias Tischler (article 6) for their agreement to republish our joint paper in this volume. A special thanks goes to all publishers who accepted the republication of my articles from books and journals produced in their edition houses: Birkhäuser/Springer, Brepols, Brill, De Gruyter, Fondation du Roi Abdul-Aziz, IFRI, INSISTS and Université Pierre Mendès France.

INTRODUCTION

This volume unites eight articles on historiographical issues related to the history of science in Islamicate societies and their neighbours between the ninth and the seventeenth centuries. These papers address major historiographical concepts and practices that shaped research, engagement, and interpretation with the sciences in past Islamicate societies by historians of science and historians alike since the nineteenth century. These concepts include above all "decline", "novelty", "marginalization", and "the Golden Age". The positions taken towards these four concepts shared two main historiographical consequences, which profoundly shaped the historical practices in the field – the long-term, far-reaching neglect of scholarly work after the so-called "Golden Age" and the interpretive view that the so-called "ancient sciences" were rejected by large parts of the educated elites, in particular the scholars active in the religious disciplines. In 1987, A. I. Sabra raised the first major challenge against aspects of the historical practices characteristic for historians of science studying Islamicate societies, focusing on concepts such as "marginalization", criticizing the treatment of the sciences in Islamicate societies as an intermediary stage between ancient Greek and medieval Latin texts and instruments and doubting the view of a radical exclusion of the sciences from the madrasa as a religious school. Preceding this call to focus in the study and interpretation of the sciences on their conditions within the Islamicate world itself, history of science at large turned to the extensive study of the contexts in which scientific works were undertaken overcoming the previously dominating dichotomic juxtaposition of internal and external factors for their development. Historians of science in Islamicate societies had rejected this move with the argument, explicitly made in particular by Willy Hartner, that first the texts needed to be published and analysed before any evaluation of their political background could and should be dealt with. Hence, article 3 (2007) focuses on the question what it could mean to contextualize the sciences within Islamicate societies. Article 2 (2007) raises the question where the "exact sciences" were placed in Islamicate societies after "the Golden Age". Written long before their dates of publication, they signal one aspect of my long-term turn to studying the life cycles of the sciences in Islamicate societies from the ninth to the late seventeenth centuries. Article 4 (2010) on the mathematical sciences in Safavid Iran exemplifies

DOI: 10.4324/9781003372493-1

how to translate Sabra's methodological demands to the study of the sciences in a concrete Islamicate society after "the Golden Age". It argues that any comparison with developments in other countries can only be done fairly and fruitfully when the sociocultural conditions are included (i.e., a formal comparison between the mathematical sciences in an early modern Islamicate society and a Christian country in early modern Europe needs to consider the differences between their institutional and further sociocultural conditions to avoid distortions).

"Novelty" and "decline" are two closely interconnected historiographical concepts in debates about the scientific achievements by scholars from Islamicate societies. In article 5 (2012), I reject "decline" as a meaningful historiographical category while accepting it as a possible phenomenon under concrete historical conditions, which need to be proven on a solid basis of sound methodological definitions. In article 1 (2006), I discuss the concept of "novelty" in its various Arabic and Persian designations, which show that the existing constraints were time and again circumvented or at least modified. I argue that "novelty" is not to be limited to new scientific results but embodies large sociocultural developments which on their own shape the opportunities to think and work of the members of a given specific scholarly community. Article 6 (2014), a joint work with Fidora and Tischler, transgresses, as does article 7 (2016), the boundaries of Islamicate societies and takes Christian societies in Europe into consideration. Each of them also goes in another manner beyond the other writings collected in this volume. The joint paper with Alexander Fidora and Matthias Tischler calls for a substantial shift in programmatic objectives and methodology when studying cross-cultural exchanges between different cultures and their communities of knowledge. It provides a long-term, extensive survey of the methodological and interpretive positions taken by the leading scholars of the twentieth century who studied Latin and vernacular as well as Hebrew translations of Arabic and Greek scholarly texts. Article 7 on German and French nineteenth-century practices in history-writing about the mathematical sciences in Islamicate societies approaches the issue of contextualized studies from the perspective, not of past scientists, but of past historians of mathematics. It raises the increasingly important point that current interpretations of contributions of previous historians of science (including mathematics, medicine, philosophy or the so-called occult sciences) will distort the works of older generations of academics if they do not contextualize their works. Article 8 (2016), the last one in this volume, was published in an Indonesian translation. It originated from a conference on convivencia, where I raised the question whether Bernard Goldstein's observations that the sciences in the classical period were a faith-neutral domain of intellectual activities characterized by cross-denominational cooperation needs to be modified. Furthermore, I discuss whether the conditions under which the sciences were practiced in the post-classical period differed in these two respects significantly from those in the earlier centuries. Taken together, these eight papers strongly argued for a fundamental shift in the perspectives, approaches and practices of contemporary historians of science in Islamicate societies. Many colleagues in the field have by now accepted the necessity for such a shift and offer their own views on the conditions under which scholars in Islamicate societies produced their knowledge. Their work adds new insights, which rectify the most important distortion inflicted on the history of science in Islamicate societies: the claim that there were no noteworthy scientific activities in Islamicate societies after "the Golden Age". The writings assembled in this volume have pushed this shift by showing which other questions need to be raised for understanding scientific activities in Islamicate societies in addition and beyond their authors' sources, methods and results and which kind of material can be used to answer them in a meaningful manner.

1

NOVELTY AS CULTURAL VALUE

Places, forms and norms of the claims to novelty in Islamic societies[1]

When the question of the Latin Middle Ages is raised, the idea according to which the scientific approach in the sciences was fundamentally different from that of the new sciences which began to emerge since the sixteenth century has undoubtfully lost its authority. When, however, the Islamic societies are concerned, this question remains stubbornly asked. One often names the originality, the novelty, the methodological openness, criticism and self-criticism, rational proofs, shared experiences, technological innovations, social flexibility and a new understanding of God as the most important factors that conditioned the break with the old and the emergence of the new in the Protestant and Catholic societies in Europe during the sixteenth and seventeenth centuries.[2] The historians of science of other periods and regions continue to believe that those factors were almost completely absent in Islamic societies and that, as a consequence, the new sciences could not develop. Another widespread opinion is that even the ancient sciences had disappeared in Islamic societies after 1100, chased out by the religious stigmatization of *bidʿa* (innovation), the denunciation of *falsafa* (philosophy) as heretical by the imminent Muslim theologian and Sufi Abū Ḥamīd al-Ghazālī (d. 1111), and the exclusion of the ancient sciences from the *madrasa*, the main institution of Sunni education supported by governments since the second half of the eleventh century.[3]

1 I thank Sophie Roux, David A. King and Ulrich Rebstock for their help and their precious comments. Mark Naimark translated the English version into French. Sarah Carvallo and Sophie Roux provided the final check of the French text.

2 P. E. Rossi, *The Birth of Modern Science*, Oxford, Blackwell Publishers, 2000, 3–5.

3 See, for instance, T. E. Huff, *The Rise of Early Modern Science: Islam, China and the West*, Cambridge, Cambridge University Press, 1993 or H. F. Cohen, *The Scientific Revolution: A Historio-*

 DOI: 10.4324/9781003372493-2

Since several decades, historians of science in Islamic societies have proven that these pretended obstacles to the scientific life in Islamic societies did not exist, given that the production of instruments and scientific texts continued in those societies until the beginning of the twentieth century. They demonstrated that between 750 and 1550 numerous scholars produced original and novel results in mathematics, that is, in geometry, number theory, astronomy and the theory of music, but also in a significant number of subdisciplines such as algebra, optics, surveying, arithmetic or the study of magic squares.[4] They have shown that important elements of the new astronomy of Copernicus resulted from centuries of critical debate on planetary theory by the scholars of Islamic societies.[5] Although these proofs received only a limited audience among the historians of science and the public in general, it is not the place to repeat this type of work in this paper. On the other hand, the history of the mathematical sciences in Islamic societies has not yet investigated systematically the values that the scholars associated with their work and the forms in which those values were expressed.

In this paper, I will treat the value given to novelty by mathematicians in Islamic societies until 1700. This time limit was chosen in order to exclude the substantial modifications which diverse Islamic societies underwent both in regard to internal factors as well as in their relations to Protestant and Catholic societies in Europe since the early eighteenth century. Except for the first part, I will focus on

---

graphical Inquiry, Chicago: The University of Chicago Press, 1994. The view of those authors is not only characterized by a great lack of familiarity with the original sources, but at times even by a surprising lack of familiarity with the most recent research results in this domain. Hence, they usually do not understand the scientific content of the models or theories of authors from Islamic societies.

4 J. P. Hogendijk, "Greek and Arabic Constructions of the Regular Heptagon," *Archive for History of Exact Sciences* 30 (1984), 179–330; D. A. King, *Islamic Mathematical Astronomy*, London, Variorum Reprints, 1986; King, *Astronomy in the Service of Islam*, Aldershot, Brookfield, Variorum, 1993; King, "On the Role of the Muezzin and the Muwaqqit in Medieval Islamic Society," in *Tradition, Transmission, Transformation*, S. Livesey, F. J. Ragep and S. P. Ragep (eds.), Leiden, New York and Köln, E. J. Brill, 1996, 285–346; King, *In Synchrony with the Heavens: Studies in Astronomical Timekeeping and Istrumentation in Islamic Civilization*, 2 vols., Leiden, E. J. Brill, 2004, 2005; F. J. Ragep, "Freeing Astronomy from Philosophy: An Aspect of Islamic Influence on Science," *Osiris* 16 (2001), 49–71; Sabra, *The Optics of Ibn al-Haytham*, London, The Warburg Institute, 1989; G. Saliba, *A History of Arabic Astronomy: Planetary Theory during the Golden Age of Islam*, New York: New York University Press, 1994; J. Samsó, *Islamic Astronomy and Medieval Spain*, Aldershot, Brookfield, Variorum, 1994; J. Sesiano, *Un traité médiéval sur les carrés magiques. De l'arrangement harmonieux des nombres*, Lausanne, Presses polytechniques et universitaires romandes, 1996.

5 A. Dallal, *An Islamic Response to Greek Astronomy. Kitāb Taʿdīl Hayʾat al-Aflāk of Ṣadr al-Sharīʿa*, edited with Translation and Commentary, Leiden, E. J. Brill, 1995; E. S. Kennedy and S. Ghanem, *The Life and Work of Ibn al-Shatir*, Aleppo, Aleppo University, 1976; F. J. Ragep, *Naṣīr al-Dīn al-Ṭūsī's "Memoir on Astronomy" (al-Tadhkira fī ʿilm al-hayʾa)*, 2 vols., New York-Berlin *et al.*, Springer Verlag, 1993; Ragep, "ʿAlī al-Qushjī and Regiomontanus: Eccentric Transformations and Copernican Revolutions," *Journal for the History of Astronomy* 36 (2005), 359–371; G. Saliba, "A Sixteenth-Century Arabic Critique of Ptolemaic Astronomy: The Work of Shams al-Dīn al-Khafrī," *Journal for the History of Astronomy* 25 (1994), 15–38.

developments from the eleventh to the fourteenth centuries. To nourish the discussion, I will rely on sources that historians of mathematics in Islamic societies often neglect. Among those sources one finds titles and prefaces of mathematical works, anecdotes and stories about the practitioners of these sciences, and finally remarks found in mathematical texts and the margins of their manuscripts. The norms of the new sciences in Protestant and Catholic Europe inspired me to look for scholarly claims and their implementation, their intentions and their goals in well-defined sites: the titles, headings and introductions of scientific works. In Persian and Arabic scientific works, their introductions and prefaces played indeed such a role, but the titles, the headings and the certificates of transmission, which had a similar function like the privileges in printed editions at the beginning of the Western early modern period, received a different treatment.[6] Indeed, they rarely express claims to novelty. Like any other human activity, the cultural modes of scientific diffusion differ from society to society and experience their own and specific histories. Thus, the expression and valorization of novelty in the sciences in Islamic societies deserve their own historical analysis. Usually, those formal elements are only reported in anecdotal manner. Most often, the claims to novelty are ignored. They are seen as truisms without interest to the historian of science. If they discuss such claims, they evaluate their veracity in a qualitative manner by determining the degree of innovation achieved by a historical author.[7] I will take a different approach by offering a systematic study of the modes and forms in which novelty or theoretical and technical innovation were claimed. I will ask which value was ascribed to them and how those modes and forms evolved over time. Furthermore, I will discuss other objectives than novelty or innovation that scholars tried to tackle and which importance they accorded to those respective goals.

The missionaries of the early modern period and the question of *bidʿa*

Catholic missionaries from Italy, Spain, Portugal, and France sent to Iran since the early seventeenth century, above all Gabriel du Chinon (d. 1668) and Raphael du Mans (d. 1690), reported about the exchange of mathematical and astrological knowledge as a means to push the Safavid court to accept their religious goals. There exists little proof that their confirmations mean more than occasional discussions about mathematical or astrological themes, but they compiled dictionaries in which they assembled a significant number of scholarly terms, a good part of which are related to mathematics or astrology. These dictionaries provide us with information going beyond the immediate

6 I render by "certificate of transmission" the term *ijāza,* which describes a written permission given by a professor to his student to transmit the received teaching to other students.

7 L. J. Berggren and J. P. Hogendijk, *The Fragments of Abū Sahl al-Kūhī's Lost Geometrical Works in the Writings of al-Sijzī*, Utrecht, Universiteit Utrecht, Department of Mathematics, Preprint 1226 (2002), 4–18.

preoccupation of the missionaries.[8] They reflect – through the chosen Persian words as equivalents for the Latin or Italian terms – the knowledge, the interest and the values of their translators. Depending on the ethnic, religious and cultural origin of the translators, the translations assumedly express the attitudes of different parts of the Safavid society.[9] Even in such a complex and heterogenous context, in some of the dictionaries it is possible to find words that go straight to heart of the matter discussed in this paper: did the scholars of Islamic societies, in particular in Safavid Iran, in the early modern period oppose innovation in general or within the mathematical sciences in particular? An Italian-Persian-Turkish dictionary with some Armenian notes serves as an example. It was compiled in Isfahan in the seventeenth century. I will investigate the Italian terms for invention, to invent and inventor together with their Persian translations *ikhtirāʿ* (*inuentione*), *ikhtirāʿ konande* (*inuentare*), *az nav peydā konande* (*inuentare*) and *mukhtariʿ* (*inuentore*).[10] Certainly, the terms used in this dictionary do not have the same meaning today as in the seventeenth century, even when one continues to use and translate them in the same manner. That is why it makes sense to try to figure out what *ikhtirāʿ* may have meant then. The substantive *ikhtirāʿ* is not found in classical Arabic dictionaries, but the verb *ikhtiraʿ* is present in the great dictionaries like the *Ṣiḥāḥ* of al-Jawharī (d. 1007), the *Abwāb* of al-Ṣafhānī (d. 1262) or the *Qāmūs* of al-Fīrūzābādī (d. 1409). In the *Ṣiḥāḥ*, it has already the sense of: *to newly or the first time create, invent, conceive, innovate, do, act, produce, cause to be or to exist, to bring into existence, something that had never been or existed before and which was not similar to anything that was pre-existent.*[11] At the beginning of the early modern period, the substantive *ikhtirāʿ* had come into use. It can be found, for instance, in a Persian text composed in the Safavid empire and in another Persian text apparently written in India. The title of the Safavid text is *Dar sharḥ-i ikhtirāʿ ālati keh navvāb-i Īrān-madārī namūdeh and* (On the Explication of an Instrument, Which His Excellence, the Pinnacle of Iran, Has Made). This text was probably composed in the middle of the seventeenth century.[12] While the word *ikhtirāʿ* appears in the title, in the text, the production of

8 An example is the dictionary used for this paper: MS Vatican City, Biblioteca Apostolica, Borg. Pers. 14.

9 The translators were most likely Armenians of Catholic or Orthodox education. It is possible that some Iranian Shiites, who had secretly converted to Catholicism, also participated in some of those projects. The social origins of the translators may also have differed considerably.

10 MS Borg. Pers. 14, ff. 88a,1–4; 131a, 20 + 22; 131b,2; 183b, 18–19; 185a, 10–11.

11 Summarized and reordered after E. W. Lane, *An Arabic-English Lexicon*, Beirut, Imprimérie libanaise, 1968, Part II, 724.

12 This text is attributed to Muḥammad Ṭāhir whom M. T. Bahār identified as the author of a history of ʿAbbās II, to which this commentary was added in the margins (Bahār, *Sabk-i shenāsi-yi tārīkh-i taṭavvur-i nathri fārsī*, Tehrān, Intisharāt zawār, s.d., vol. 3, 277; see the remark between parentheses and footnote 1). I thank B. and F. Hāshimī, Isfahan for their help with the translation and interpretation of the expression *navvāb-i Īrān-mādarī*.

an instrument is described by *vaḍʿ*. The author highlights that in the eyes of his compatriots the word *vaḍʿ* is linked to *gharīb* and *nadīr* – the strange, the curious and the rare. He comments that this appears less wondrous if one practices in the astral sciences, is experienced in the geometrical disciplines, is able to put ideas together and is supported by (wise) advice.[13] Thus, *ikhtirāʿ* and *vaḍʿ* are both used in this text to characterize a recent human invention, which at first glance seems to be out of human reach, but which, when given a closer look, can be realized by some men with rare gifts.

The Indian text is called *ʿUqūl-i ʿashara* written in 1084 h/1673 by Muḥammad Baranī Ummī Qāqshāl.[14] The emphasis of the ninth chapter is on novelties, inventions and wonders: *ʿaql-i nahom dar waḍʿ va-khtirāʿ o-ʿajāʾib* and one finds there the same ambivalence as in the Safavid text. The author describes *ikhtirāʿ* as heroic human or semi-divine deeds.[15] Nonetheless, these novelties, inventions and wonders are results of *ʿaql* (reason or ratio). They complete the long list of things accessible by human reason such as the sphere of the stars, the astrolabe, the sphere of the earth, medicine, mountains, animals, plants, stones, seas, or time and space.[16] What is more, *ikhtirāʿ* does not only concern the production of an instrument, but numerous other inventions: the introduction of the monarchy, the production of iron from stone, the making of weapons, the fabrication of silk, the application of fur to cloth, or the invention of the wheel of navigation, the invention of writing, the invention of composing tunes and so on. Although similar enumerations are found in certain historical or literary works, this list does not belong to the general tradition of classifying the arts and the sciences. Moreover, from the point of view of this Indo-Persian author, *ikhtirāʿ* shares with the mathematical and rational sciences an anchorage in reason, belonging thus to a different genre. The mathematical and rational disciplines demand an education, attention and the submission to masters and to authoritative texts. Thanks to this obeisance, ordinary men can acquire knowledge and produce objects that are useful to their contemporaries and their successors. In contrast, the activities worthy to be called *ikhtirāʿ* come most often from unique acts of extraordinary men, capable of forming an entire culture.

The study of other branches such as calligraphy or painting indicates that this restriction does not concern all sectors. In the arts of the sixteenth century, it is possible to claim invention and novelty, and that not merely by cultural heroes, but by ordinary men. When certain inventions such as the art of writing or even algebra are ascribed to cultural heroes, in particular ʿAlī, the son-in-law of the Prophet, then other inventions such as specific forms of calligraphy are proudly attributed to well-known artists. Around 1015 h/1596–7, the Safavid calligrapher and painter Qāḍī Aḥmad b. Mīr-Munshī (d. ca 1015 h/1606)

13 Bahār, *Sabk-i shenāsi-yi tārīkh-i taṭavvur-i nathri fārsī*, vol. 3, 278.

14 MS Vienna, ÖNB, Mixt. 271, f 376b,8.

15 Ibid., ff. 378a,4–379b,1.

16 Ibid., ff. 6b, 61b, 87a, 240b, 348b, 354a, 372b, 380b.

wrote a treatise, in which he promised to examine "the first appearance of the qalam and the invention of writing, tracing the origin of the latter to His Holiness the Shah (i.e., ʿAlī b. Abī Ṭālib)."[17] Furthermore, he wishes to present "the biographies of each of the masters, artists and all men of talent who are connected with this glorious company and excellent class, or with books and libraries."[18]

Can one conclude from those elements that at the beginning of the early modern period, authors in Islamic societies felt free to write books about innovations in whatever domain of human activity? If that was the case, did this attitude find its origin in their own societies? There are at least three reasons to hesitate giving an unreserved, affirmative response to these questions.

1) As far as we know, outside the arts and the production of instruments and beyond a circle of extraordinary men, few authors explicitly made claims to novelty or innovation.
2) The Europeanists have often declared that during the sixteenth and seventeenth centuries with respect to novelty a new attitude emerged in Catholic and Protestant Europe. In support of their thesis, they have compiled lists of titles containing the adjective *novus*, *neuf* or *new*, discovered the formulation of an ideal of scientific progress in the works of Francis Bacon and Blaise Pascal and studied the *Querelle des Anciens et Modernes*.[19] Because the elements I just presented emerged in the same period, when the exchange between different European and Asian countries increased considerably and Arabic texts show again the word novelty into their titles, one can legitimately ask whether the introduction of terms expressing novelty in Iran and India in the early modern period did not result equally from those cultural exchanges.
3) The entry in the mentioned Italian-Persian-Turkish dictionary, which follows that of *inventione* is *inventione malo*. Its interpretation leads us to our last, but not least reason to hesitate.[20] The Persian and Turkish translations of the term are identical: the Arabic word *bidʿa*. *Bidʿa* has a colorful history. After a *ḥadīth* ascribed to Muḥammad:

> the worst things are those that are novelties, every novelty is an innovation, every innovation is an error and every error leads to Hell-fire.

Bidʿa was, as the *ḥadīth* attributed to the Prophet indicates and as the Italian-Persian-Turkish dictionary confirms, effectively depreciated. Combined with an accusation of apostasy it could become a powerful weapon of denunciation

17 Translated from V. Minorsky, *Calligraphy and Painters, a Treatise by Qadi Ahmad, Son of Mir Munshi, Translated from the Persian*, Washington, DC, White Lotus Press, 1959, 44.

18 Ibid.

19 For the treatment of novelty in the classical epoch, see the introduction to this volume, pp. 8–16.

20 MS Borg. Pers. 14, f. 131b,1 + 3. In this case, the inventor is called *inventore di cosa mala*.

and critique. It was often employed against aspects of cult and ritual, such as the introduction of the Prophet's birthday as a holy day or the installation of eunuchs as guardians of the two holy places in Mecca and Medina. It could also be directed against changes in legal custom, such as the direction of prayer in a mosque aligned to an incorrect astronomical position, or against intellectual norms, such as the use of logic within jurisprudence. Due to this fact, writing about novelty and innovation had the potential to arouse accusations of *bidʿa*.

Does the inherited power of the term *bidʿa* imply that the scholars in the mathematical sciences were indeed attacked, judged and condemned because of this motif when they mentioned their new results? Did it hinder the production of novelty or innovation? Based solely on the *ḥadīth* evoked above, Toby Huff has drawn the conclusion that in medieval Islamic societies scientific innovation was neither permitted institutionally nor tolerated culturally.[21] In contrast, ʿAbd al-Hamid Sabra emphasizes that

> the concept of innovation in intellectual endeavor found expression in terms like *istikhrāj* or *istinbāṭ* (discovery), which denoted accomplishments that went beyond merely elucidating, emending, or completing an earlier contribution to knowledge; and a critical attitude clearly revealed itself in the not infrequent composition of *shukūk* (*aporia*, *dubitationes*), a form of argument in which difficulties or objections were raised against ancient authorities.[22]

In a sense, both analyses are complementary: Huff wished to discuss the attitude of groups within society or of society at large towards the sciences and their relationship to innovation, while Sabra focused on the language of scientists expressing what they considered as novel. Their different points lead to two questions: was it possible to make claims to novelty in a society which culturally denounced innovation and which forms did such claims take? Was there a relation – direct or indirect – between *bidʿa* in the domain of religion and the attitudes towards innovation that we can find in scientific treatises?

These questions have not yet been made the object of an in-depth examination by historians of science in Islamic societies, who usually do not pay attention to them or consider them as simple questions of rhetoric. In order to answer the questions, I will proceed in three steps. First, I examine the status of the ancient authorities who sometimes are opposed to in certain intellectual practices called *shakk* (doubt), *radd* (refutation) and *mushkila* (difficulty). Then, I consider the forms of titles of scientific treatises. Finally, I analyse some introductions, marginal notes and colophons.

21 Huff, *The Rise of Early Modern Science*, 234.
22 Sabra, *The Optics of Ibn al-Haytham*, 133.

A critical attitude towards the ancients

It is well known that major ancient authorities such as Aristotle, Euclid, Ptolemy and Galen received written critique by scholars of the Islamic world. In principle, two types of criticism can be found. One rejects the ancient authorities by pointing to their religious beliefs and their belonging to a different culture. The other refers to theories, methods, claims, parameters or precise. They highlight the lacunas – real or perceived – in observational results in order to engage in an intellectual debate with the ancients. Most of the critiques of the second type present themselves in one of the three already-mentioned literary forms. In the *shakk,* one raises doubts about the opinions of authors, in particular ancient ones. In the *radd,* the opinions of any opponent, whether ancient or modern, are refuted. In the *mushkila,* the difficulties, raised or named by an ancient author, are resolved. These criticisms are expressed in the texts, but the authorial intentions appear emblematically in specific formulations in the titles.

1) Commentators from late Antiquity may have set the precedence for works labeled *Kitāb shukūk ʿalā Uqlīdis* or *Baṭlāmiyūs* or *Arisṭūṭālīs* (Book of Doubts Against Euclid or Ptolemy or Aristotle). The first known author of such a text in Islamic times was the Christian Qusṭā b. Lūqā from Baalbek (d. ca. 910). His critique of Euclid is lost. A critique of Aristotle's theories of vision and comets was written in the late ninth or early tenth century by ʿAlī al-Hāshimī (fl. ca. 890): *Taʿdīd shukūk talzamu maqālat Arisṭūṭālīs fī baṣar wa-taʿdīd shukūk fī kawākib al-dhanab* (Enumeration of the Doubts in Aristotle's *Book on Vision* and Enumeration of Doubts on Comets).[23] The two best-known authors of *shukūk* texts are the Muslim Ibn al-Haytham (d. ca. 1041) and the Jew Maimonides (d. 1204).[24] The former wrote *dubitationes* against Euclid and Ptolemy, the latter against Aristotle and Galen.

2) The genre of *radd* is mostly confined to religious and philosophical texts, but at least once it also was applied to the discussion about the universe and the earth. The writer who did this was al-Ḥasan b. Mūsā al-Nawbakhtī (d. ca. 920), a theologian, philosopher and astronomer. His theological and philosophical interests may have contributed to his choice of this kind of genre of criticism. Of two today lost works, one was directed against Ptolemy: *Kitāb al-radd ʿalā Baṭlāmiyūs fī hayʾat al-falak wa-l-arḍ* (Book on the Refutation of Ptolemy About the Form of the Universe and the Earth); the other was opposed – on the basis of Aristotle's natural philosophy – against Neoplatonic beliefs that the universe was a rational living being *Ḥujaj ṭabīʿiyya mustakhraja min kutub Arisṭāṭālīs fī l-radd ʿalā man zaʿama anna l-falak ḥayy nāṭiq* (Natural Philosophical Arguments Taken From Aristotle's Books Against Him Who Claims That the Universe Is Living and Rational).

23 I corrected the date according to more recent literature and completed the title.

24 Ibn al-Haytham, *Kitāb fī ḥall al-shukūk fī Kitāb Uqlīdis*, Fuat Sezgin, Frankfurt am Main, Institute for the History of Arabic-Islamic Science at the Johann Wolfgang Goethe University, 1985; Ibn al-Haytham, *Al-Shuklūk ʿalā Baṭlāmiyūs*, A. I. Sabra and N. Shehaby (eds.), Cairo, Dār al-Kutub, 1971; Maimonides, Commentary on Hippocrates' Aphorisms; Fuṣūl Mūsā; Y. T. Langermann, "Criticism of Authority in the Writings of Moses Maimonides and Fakhr al-Dīn al-Rāzī," *Early Modern Science and Medicine* 7.3 (2002), 255–275, in particular 256–265.

3) Certain authors affirmed to have solved difficulties left in the works of other authors or in their own earlier works. This is in particular the case of Badr al-Dīn al-Fārisī in his *Maʾārij al-fikr al-wahīj fī ḥall mushkilāt al-zīj* (The Scales of the Ardent Thinking About the Solution of the Difficulties of the Zīj); Naṣīr al-Dīn al-Ṭūsī in his work *Ḥall mushkilāt al-Muʿīniyya* (Solution of the Difficulties of the Muʿīniyya); or Niẓām al-Dīn al-Nīsābūrī in his treatise *Ḥall mushkilāt al-Majisṭī* (Solution of the Difficulties of the *Almagest*).

Beyond these more standardized forms of expressing doubt, dissent and superiority with regard to the work of predecessors or contemporaries from Antiquity and Islamic societies, titles of mathematical and astronomical works contain a variety of individual expressions pointing to errors, mistakes and other shortcomings. Aḥmad b. Muḥammad b. al-Sarī (d. 1154) attacked sharply Ibn al-Haytham, Jābir b. Ibrāhīm al-Ṣābīʾ, al-Fārābī, ʿAbd al-Raḥmān al-Ṣūfī, al-Bīrūnī, Aristotle and Ptolemy for their errors, miscalculations, and imaginations (*ghalaṭ*, *khaṭāʾ*, *wahm*). He felt himself capable to correct all of them and promised unnamed readers filled with doubts in respect to XIV,12 of Euclid's *Elements* to deliver a sound explanation. Al-Samawʾal al-Maghribī (d. ca. 1175), in turn, accused geometers of glossing over things and astronomers of being insufficient and erroneous when predicting heavenly events. ʿAbd al-Laṭīf al-Baghdādī (1162–1231), well-known for his knowledge in mathematics, medicine, philology, philosophy and history, reprimanded Ibn al-Haytham for his statements about space with the term *tahāfut* alluding to al-Ghazālī's famous critique of philosophy *Tahāfut qawl Ibn al-Haytham fī makān* (The Incoherence of Ibn al-Haytham's Discourse on Space).

All these titles indicate that self-confidence was widespread among scholars of the ancient sciences in Islamic societies and that they did not hesitate to take on everybody whom they thought merited critique, whether famous or not. Based on this insight, we might be tempted to deduce that novelty had found a place in the titles of scientific works. Several words were indeed used to express, that is, to announce, the novelty of a result. Some of them were already mentioned: *ikhtirāʿ* (apparatus, invention), *istikhrāj* (determination, discovery), *instinbāṭ* (discovery), *ibtikār* (invention), *ḥadūtha* (being new or young), *ḥadatha* (to take place), *aḥdatha* (to create, to produce), *istaḥdatha* (newly introduce, renew, create), and forms of *jadda* (being new, having recently taken place, being important, being serious). Such terms are, however, rarely found in titles of scientific works with the exception of a few astronomical or philosophical texts: Ibn Kammūna (d. 1284), a Jewish scholar; Ibn al-Shāṭir (d. 1375), an astronomer and the main *muwaqqit* (timekeeper) at the Umayyad Mosque in Damascus; or the Timurid prince Ulugh Beg (d. 1449) and his scholarly collaborators used the word *jadīd* (new) in the titles of their works.[25] In contrast, one finds such words inscribed on scientific instruments. Some of these inscriptions clearly indicate that the object

25 Ibn Kammūn, *al-Jadīd fī l-ḥikma*; Ibn al-Shāṭir, *Kitāb zīj al-jadīd*; Ulugh Beg, *Zīj-i jadīd-i sulṭānī*. The last work also was assigned other titles. A new and stimulating discussion on the meaning of the title of Ibn Kammūna's work and its content is found in Y. T. Langermann, "Ibn Kammūna

was not a mere variant of an already produced or fabricated instrument, but to actually highlight its novelty. On an astronomical instrument constructed by the astrolabe maker Hibat Allāh al-Ḥusayn in the twelfth century one can read, for instance, several inscriptions concerning the form of the astrolabe, one stating that he had fabricated (*ṣanaʿa*) the instrument as the astronomer, mathematician and ambassador Abū Jaʿfar al-Khāzin had invented (*istanbaṭa*) it.[26] The manufacturers of instruments still claimed novelty in the fourteenth century. In general, one does not expect claims of novelty in that period, neither in actual implementations nor in the cultural rhetoric. Such instruments indicate that our expectations should be put in question and modified. When Ibn al-Shāṭir, for instance, produced his *Ṣandūq al-yawāqīt* (Box of Sapphires) – a compound astronomical instrument containing a compass needle – one exemplar today lost carried an inscription indicating the title, the intended audience, the scholar who had produced (*ṣanaʿa*) and invented (*ibtakara*) it, namely Ibn al-Shāṭir, as well as the date of its production.[27] The professional astronomers (*muwaqqit*s) repeated these claims to novelty in their treatises on instruments. Occasionally, they even admit that they were not the first who had conceived a new instrument particularly complete, elaborate or marvellous. This is the case, for instance, of Ibn al-Sarrāj's (fourteenth century) (re-)invention (*istinbāṭ*) of the universal astrolabe first constructed in the eleventh century by the Toledan astronomer ʿAlī b. Khalaf al-Shakkāz.[28]

Titles

In order to understand the vocabulary used in the titles, one has to know the rhetorical codes employed to set them in scene, although they did not evolve before the ninth century and changed in a significant manner in the following three to four centuries. During the first centuries of flourishment of the sciences in Islamic societies the texts were rarely, if at all, endowed with titles. A well-known example is al-Khwārizmī's book on algebra. In the following centuries, titles become more the norm, but take on a special form inspired by the flowery literary titles widespread in other domains of knowledge.

1) Assuming that at least some of the titles in extant manuscripts with scientific works had been chosen by the authors of those texts, we need to conclude that until the tenth century, simple, direct and sometimes explanatory titles dominated in the mathematical sciences. This applies to *Tarkīb al-aflāk* (The Composition of the Orbs/the Spheres) by the astronomer-astrologer Yaʿqūb

and the 'New Wisdom' of the Thirteenth Century," *Arabic Sciences and Philosophy* 15 (2005), 277–327.

26 D. A. King, "New Light on the *Zīj al-Ṣafāʾiḥ* of Abū Jaʿfar al-Khāzin," *Centaurus* 23 (1979), 105–117, above all 109–110.

27 L. Janin and D. A. King, "Ibn al-Shāṭir's Ṣandūq al-Yawāqīt: An Astronomical Compendium," *Journal for the History of Arabic Science* 1 (1977), 187–256, in particular 190, plate 6.

28 King, *Islamic Astronomical Instruments*, IX, plate 1.

b. Ṭarīq (d. ca. 796) cited by the polymath Abū l-Rayḥān al-Bīrūnī (d. after 1053); to *Kitāb ittifāq al-falāsifa wa-khtilāfihim fī ḥuẓūẓ al-kawākib* (The Book of the Agreement Between the Philosophers and their Disagreement About the Lots of the Stars) by the astronomer-astrologer ʿUmar b. Farrukhān al-Ṭabarī (d. ca. 815) cited by the book trader and bibliophile Ibn al-Nadīm (d. 995); to *Ziyāadāat fī l-maqāla al-khāmisa min kitāab Uqlīdis* (Addition to the Fifth Fascicle of the Book of Euclid) by the courtier, astronomer and mathematician al-ʿAbbās b. Saʿīd al-Jawharī (first half of the ninth century), of which one finds today at least five copies; *Ḥisāb al-aqalīm al-sab*ʿa (The Calculation of the Seven Climates) by the astronomer and engineer (or geometer) Aḥmad b. Muḥammad al-Farghānī (ninth century); or the *Risāla fī anna quṭr al-murabbaʿ lā yushāriku min al–ḍilʿ min ghayr al-handasa* (Treatise on [the Claim] that [the Fact that] the Diagonal of the Square Is Not Commensurable With the Side [Does Not Belong to Geometry]) by the physician and philosopher Muḥammad b. Zakariyyāʾ al-Rāzī (865–925 or 935).

2) A more literary style appears at the beginning of the tenth century in geographical and historical writings, but it remains still quite sober in comparison to the later flowery style. Aḥmad b. Rusta modestly claims to have written a *Kitāb al-aʿlāq al-nafīsa* (Book on the Exquisite Preciousness), which deals with geography and history. One of the first mathematicians to follow this lead was Abū Kāmil al-Miṣrī (ca. 850-ca. 930), who called one of his works on number theory and algebra *Ṭarāʾif al-ḥisāb* (The Rarities of Arithmetic). The next author who is known to have chosen some qualifier outside the field of dry scientific speaking was the mathematician, astronomer, librarian and courtier Abū Jaʿfar al-Khāzin (d. between 961 and 971) mentioned above. He called one of his works *al-Ālāt al-ʿajība al-raṣadiyya* (The Marvelous/Astounding Observational Instruments). But it took more than another fifty years before rhymed titles began to be used by scholars of the ancient sciences. It was one of the most famous, most creative and most productive Muslim writers about the mathematical and historical sciences who adopted such a style for numerous of his works – Abū l-Rayḥān al-Bīrūnī. He called his work on chronology and calendars *al-Āthār al-bāqiya ʿan al-qurūn al-khāliya* (The Remaining Monuments of the Previous Centuries). His introduction into astrology written for a young woman carries the title *Kitāb al-tafhīm li-awāʾil ṣināʿat al-tanjīm* (The Book of Understanding for the Beginnings of the Art of Astrology). Abū l-Rayḥān titled his work on projections *Risāla fī tasṭīḥ al-ṣuwar wa-tabṭīḥ al-kuwar* (Treatise on the Projection of Constellations and the Representation of the Sphere on a Plane). The title of his famous work on geodesy follows the same style: *Kitāb taḥdīd nihāyât al-amākin li-taṣḥīḥ masāfāt al-masākin* (Book of the Determination of the Limits of Places for the Correction of the Distances of Inhabited Localities). The elements available to me in catalogues seem to indicate that al-Bīrūnī was one of the first Muslim scientists who gave rhymed titles to their works.[29]

29 Al-Bīrūnī was one of the most innovative scholars of all Islamic societies. His innovations are much more substantial than the possible fact that he was the first Muslim scientist to have used rhymed titles. This is in particular the case in his works on mathematics, astronomy and history.

A consequence of rhymed literary titles was the influx of non-technical terminology, a fact the standard translations of al-Bīrūnī's titles do not reveal. The title of his work on projections says literally: Treatise on the flattening of images and the throwing of spheres to the ground. Although *tasṭīḥ*, used in the first part of the title, was well established as a technical term for projection in the time when al-Bīrūnī wrote, *tabṭīḥ*, used in the second part of the title, never acquired this status. According to the previously prevalent norms, the second part of the title is completely superfluous. Over the following centuries, the preference for literary and rhymed titles for any type of book grew. Although short and straightforward titles continued to be given to mathematical and astronomical texts, rhymed and often longwinded titles became more and more fashionable. In the thirteenth century, many authors in the mathematical sciences chose fully or partially rhymed titles for their works, the famous as well as the mediocre. Flowery titles had become the standard.

3) The imagery used in rhymed or flowery titles is manifold.[30] Titles of books on natural history and to some extent geography often combined expressions of wonder, excitement, strangeness and rarity with creation and being. The earliest such title known to us is the *ʿAjāʾib al-makhlūqāt wa-gharāʾib al mawjūdāt* (The Marvels of the Creatures and the Rarities of the Beings) by Muḥammad b. Maḥmūd al-Ṭūsī, a writer on geography and cosmography of the twelfth century. A more famous and more influential book with the same title was written in the next century by Zakariyyāʾ b. Muḥammad al-Qazwīnī (1203–1283). He was a geographer, astronomer, cosmographer, jurist and imam who worked in Damascus, Hilla and Wasit. In the mathematical sciences, the authors used similar cultural markers in the titles evoking the brilliant, the splendid and the marvelous. A good number of works with such sparkling titles possess a content that we describe as partially or fully new. This is the case, for instance, in works such as *al-Badīʿ fī l-ḥisāb* (The Marvelous on Arithmetic) by Muḥammad al-Karajī (d. ca. 1030), *al-Bāhir fī ʿilm al-ḥisāb* (The Shining on the Science of Arithmetic) by al-Samawʾal al-Maghribī, mentioned earlier, or *Nihāyat al-idrâāk fī dirāyat al-aflāk* (Limit of Perception in the Understanding of the Spheres) by Quṭb al-Dīn al-Shīrāzī (1236–1311) contain profound novelties in algebra and planetary theories. When in Islamic societies the nature of texts and scientific disciplines was clearly marked out and distinct from other domains of inquiry and other literary genres, certain forms – in particular flowery or rhymed titles and diverse types of poetry – were thus largely shared across all the fields of knowledge and this since the eleventh century. Nonetheless, not all flowery or rhymed titles emphasize extraordinary qualities as the above-cited examples. Thus, one can conclude that, even when novelty and innovation are not directly claimed in such titles, the selection of nouns which refer to the

However, as I emphasized at the beginning of this paper my goal here is to analyze the modes and forms of claims to novelty, not the innovations themselves.

30 I do not know any study as to whether the imagery used for scientific texts differed markedly from that employed for poetry, medicine, law, philosophy, *ḥadīth*, *kalām* or history.

extraordinary may have constituted a cultural alternative to announce the pride and ambition of an author, without risking to enter the minefield of *bidʿa*.

Introductions, marginal notes and colophons

The introductions, the marginal notes and the concluding formulations at the end of a text, as well as inscriptions engraved on scientific instruments also present significant elements to better understand the attitudes and beliefs of scholars in Islamic societies: did they refer to creating new ideas, methods, theories or practices or to conserving what their predecessors had believed or cherished? Once more, an evolution with respect to those features took place between the ninth and the fourteenth centuries.

1) In the ninth century, the standard scientific text has no introduction. If at all, it opens with a short *basmallāh* reciting: "In the Name of God, the Most Gracious and the Most Merciful" and then plunges into the subject matter. A good example is Thābit b. Qurra's (d. 901) "On the Measurement of the Parabola". The Arabic text as edited by Rushdi Rashed starts with *bismillāhi l-raḥmāni l-raḥīm*. Then a title follows whose form makes clear that it is a later addition: *Fī misāḥat al-mujassamāt al-mukāfiyya li-Thābit b.* Qurra (On the Measurement of the Parabolic Bodies by Thābit b. Qurra). Between parentheses, the editor identified the subsequent text as "Definitions" using the only much later employed term *taʿrīfa*. Then the first definition begins: "The solid figures which I call parabolic are of two kinds . . ."[31]

One finds, however, exceptions to this general rule, for instance, the introductions to Muḥammad b. Mūsā al-Khwārizmī's *Algebra* and to the translation of Apollonius's *Conics*. Al-Khwārizmī's *Algebra*, for instance, carries a preface longer than a page. It starts with the *basmallāh* followed by the brief remark that "this book was composed by Muḥammad b. Mūsā al-Khwārizmī who opened it by saying . . ."[32] This part is obviously a later addition. But what follows may well be from al-Khwārizmī himself, although the religious utterances are very untypical for mathematical treatises of the time. In this long introduction, al-Khwārizmī formulates a kind of general characteristics of the meaning of research. According to him, ancient scholars and nations strove with much pain to unveil the secrets of knowledge and its obscure parts. They wrote one book after the other for those who came after them in the hope that their heirs would accumulate and memorize knowledge and acknowledge their predecessors' share in truth finding. The deeds of contemporary scholars al-Khwārizmī sees in three points:

31 R. Rashed, *Les Mathématiques infinitésimales du IXe au XIe siècle*, 3 vols, London, Al-Furqan Islamic Heritage Foundation, 1416/1996, vol. 1, p. 321, French translation p. 320.

32 *Kitāb al-jabr wa-l-muqābala*, Taqdīm wa-taʿlīq ʿAlī Muṣṭafā Musharrafa wa-Muḥammad Murshī Aḥmad, al-Qāhira, Dār al-Kitāb al-ʿarabī li-l-ṭibāʿa wa-l-nashr, 1968, 15.

> The man, who comes to what was not discovered before him, leaves it for him who comes after him. The man, who comments on what the ancients left as incomprehensible, clarifies its way, simplifies its approach, and gets near to its method. The man, who finds a gap in some books, but does not rectify it, does not assist in its need, and does not allay the doubt through its partner, did not search for it and cannot proudly (claim) it as his own deed.[33]

2) In the tenth century, longer introductions, explaining the purposes of an author and what he set out to achieve, became more widespread as witnessed by the texts written by Thābit b. Qurra's grandson Ibrāhīm b. Sinān and by Abū Sahl Wayjan b. Rustam al-Kūhī. Part of this process was a trend to situate one's own work within one's own research and within the achievements of earlier scholars. In difference to al-Khwārizmī, the chosen approach did not focus any longer on a general outline of the purpose of research, but emphasized specific subjects, specific reasons for writing a text and the specific merits of the author and his achievement.

In the introduction to his treatise on the measurement of the parabola, Ibrāhīm b. Sinān narrates the very personal story of the text. After he had already written a work on this subject, he had changed his mind and a theorem. Unfortunately, both versions got lost over time. Thus, he felt compelled to rewrite what he had done and to explain to his readers why they, perhaps, might encounter variants of his book that differed from this new version. He added that his grandfather Thābit b. Qurra and (his contemporary) al-Māhānī had already written works on this very same subject.[34] The historical placement given by Ibrāhīm b. Sinān is still very crude and imprecise. Neither titles are mentioned nor judgments are expressed.

In al-Kūhī's work on the determination of the volume of the paraboloid, the narration is more complex. The author begins with characterizing his personal competence and methods of working: careful, meticulous, exhausting all possibilities and personal skills. Then follows a proud comparison of the author's achievement vis-à-vis the ancients as well as the moderns. The author's rank is elevated by pointing to the fame of the ancients in geometry and the relative inferiority of contemporary scholarship when compared with the ancients. As the third element, al-Kūhī then introduces himself as a motivated scholar, encouraged by his success in solving newly invented problems. In a sense, al-Kūhī presents himself as a tenth-century Captain Kirk – to boldly go where no man went before. Al-Kūhī keeps to this theme in the future course of his introduction. He reminds his readers that he has only one predecessor, Thābit b. Qurra, who wrote a well-known and famous book about the subject. But in al-Kūhī's eyes it is deeply flawed – too long,

33 Ibid.

34 Rashed, *Les Mathématiques infinitésimales*, vol. 1, 719, French translation p. 718. Ibn al-Haytham employed an analogous strategy in his *Fī uṣūl al-misāḥa* (On the Elements of Surveying). Rashed, *Les Mathématiques infinitésimales*, vol. 3, 539, French translation p. 538.

too cumbersome, too many propositions and lemmas to achieve a single goal, too many fields of mathematics involved in the procedure: arithmetic, geometry and others. But worst of all, it was almost incomprehensible, while Archimedes's difficult book *On the Sphere and the Cylinder* with its many purposes appeared to be much easier to understand. The author's own work caused by these flaws of Thābit's treatise is presented as superior since he found an accessible method and dispensed of all lemmas. In his self-portrayal, al-Kūhī insists that he would never have bothered with discussing the works of his predecessors if his own research had not forced him to deal with a particular mathematical problem discussed by them. He presents a double explanation for this novel attitude towards previously generated knowledge. On the one hand, his self-confidence is strong enough to focus only on what he himself is capable of doing. Hence "it is not his custom" to bother with what others did before him. On the other hand, "the ways of this science" of geometry "are manifold and vast" and hence everybody can find his own way.[35]

Al-Kūhī's words suggest that a scholar did not need to claim novelty nor superiority. A veritable, competent and innovative scholar does the opposite – he remains silent about his rivals, praises himself for his novelties and lets his results speak for themselves.[36] Certain mathematicians of the tenth century, among them al-Kūhī, did not merely express their personal ambitions and discuss their results but situated the desire to produce new mathematical results in the larger social and cultural contexts of their time. In that period, such explanations were undeniably integrated in a rhetoric that insisted on the importance of a courtly patron for the work of the scholar. As a consequence, honoring above all their royal benefactors with tremendous praise, the scholars do not claim any longer that the new developments of their research manifested a general tendency towards innovation in the arts, technology, literature and the sciences. This general climate of innovation, al-Kūhī wrote, resulted from the preoccupation of the Buyid sovereign with justice, propriety, the quality of life and security. Thanks to those conditions of life, the scholars could successfully focus on addressing problems, which the ancients could not solve.[37]

3) In the eleventh century, the introductions to mathematical texts expanded their length and assumed a more standardized structure. At the beginning stands a much longer expression of religious sentiments or formulas. Then follow the explanation of the purpose of the work, the reason why it was written, possibly its

35 Rashed, *Les Mathématiques infinitésimales*, vol. 1, 851 and 853; French translation pp. 850 and 852.

36 Such an attitude was apparently current in al-Kūhī's period. David King informed me that the astronomer Ibn Yūnus started his astronomical handbook by indicating the famous *Mumtaḥan Zīj* (The Verified *Zīj*) prepared for the Abbasid caliph al-Ma'mūn (r. 813–833) had been composed by a group of scholars, while his work had been done by a superior scholar, working alone, that his by himself.

37 Rashed, *Les Mathématiques infinitésimales*, vol. 3, 765 and 767; French translation 764 and 766.

title and its author, and a description of its chapters and subchapters. Variations of this scheme are possible and manifold, but the convergence towards such a unified structure is undeniable. An enumeration of the components of such a new kind of standardized introduction can be found, for example, in the book on astronomy and astrology called *Kifāyat al-taʿlīm fī ṣināʿat al-tanjīm wa-maʾrifat al-hayʾa wa-l-aḥkām* by Abū l-Maḥāmid Muḥammad al-Ghaznawī.[38]

A further question needs to be raised here: do such tendencies towards normalization express new, more severe attitudes with regard to novelty and innovation or did they even contribute to forming them? At a first glance, this does not seem to be the case. The authors of mathematical texts of the eleventh century continue to claim real or pretended discoveries. An anonymous author of a text on magic squares and their mathematical foundations, for example, leaves no doubt that its author wished above all for his work to be seen as the result of his own reflections, his ingenuity and his search for new results. The reflections and the ingenuity revealed to him the singularities of his field and permitted him to find marvelous and subtle (insights). The imagination finally led him to innovation.[39]

Comparisons of examples from different mathematical fields

In order to understand, which attitudes were expressed in the introductions to mathematical texts in the eleventh century, I now discuss two types of introductions, which are quite different, but conform nonetheless to the newly formalized norms. The first type appears in ʿAlī b. al-Khiḍr b. al-Ḥasan al-Qurashī's (d. 1067–8) book on arithmetic and the calculation of inheritance shares *Kitāb al-Tadhkira bi-uṣūl al-ḥisāb wa-l-farāʾiḍ wa-ʿawlihā wa-taṣḥīḥihā* (Book on the Memoirs on the Elements of Arithmetic and the Parts of Inheritance and the Augmentation [of the Denominators] and Their Correction). The second type appears in some of the works by the mathematician Ibn al-Haytham. They lead me to examine Ibn al-Sarī's commentaries on Ibn al-Haytham's works.

In his book about arithmetic and the calculation of inheritance shares, after some pious sentences, al-Qurashī explains: 1) why he wrote the text, 2) in which style he composed the work, 3) in what parts he divided it, 4) what the useful aspects of those parts were for various groups of professionals, 5) who the predecessors in the field were, 6) what the fundamentals of arithmetic were, 7) what

38 "Every author has to introduce at the beginning of his work eight things. A) Precise determination to which kind of science his work belongs because every science is a search and a result. B) Naming of his book by something which corresponds to each of its purposes. C) Mentioning of his own name for the <student>, so that he knows whether he is the author or not. D) Explanation of the purpose of the text. H) Explanation of the usefulness of the text. W) Mentioning of the style of the sequence of his words. Z) Mentioning of the time of study. H.) Indication of the parts of the book." Ms Berlin, Staatsbibliothek, Preussischer Kulturbesitz, Wetzstein 1154, f 1b,14–19.

39 Sesiano, *Les carrés magiques dans les pays islamiques*, Lausanne, Presses polytechniques et universitaires romandes, 2004, 13.

type of proof would be applied and 8) which sources were used. The introduction ends with a short self-portrayal and the demand that possible errors should not be censured. The author of this book is not, as my previous examples, a representative of theoretical mathematics. He was rather a legal scholar who also covered *ʿilm al-farāʾiḍ*, the branch of jurisprudence where knowledge of arithmetic, algebra and occasionally geometry was needed or considered useful. As a result, he places himself first and foremost in a lineage of legal scholars who wrote about the calculation of inheritance shares. While acknowledging that the calculators, beginners, jurists, secretaries and secretaries in the department of certificates can draw profit from his book, it is with respect to those who deal with the determination of inheritance shares that al-Qurashī mentions a few chosen predecessors, among them Muḥammad b. Mūsā al-Khwārizmī. In contrast to al-Khwārizmī's view of the places of predecessors and successors in the sciences, al-Qurashī emphasizes the insight into tradition as the major goal when reading those works and hence their preservation:

> The calculator of inheritance shares needs it (i.e., arithmetic) too as a methodological tool that simplifies the access for those, who wish to gain insight into the books of the predecessors, the calculators and the teachers. Muḥammad b. ʿAbdallāh al-Baṣrī called Ibn Labbān composed an introduction and good summary of greatest completeness and usefulness. Ayyūb b. Sulaymān al-Baṣrī also mentioned in his book something beautiful about the knowledge of calculating inheritance shares and commented on it in respect to the rules for dividing inheritance shares. Among those who expressed opinions about legacies and testaments, nobody surpassed Abū Bakr Muḥammad b. Mūsā al-Khwārizmī.[40]

This linkage to a special branch of jurisprudence and its representatives is combined with a reference to Euclid and three major Islamic representatives of arithmetic and algebra – Abū Yūsuf al-Ūqlīdisī, Abū Kāmil, Abū l-Wafāʾ – and again al-Khwārizmī. It is here that the author stresses his own efforts in two directions – the continuous practice of learning and finding new ways that give fast results and the reproduction of Euclidean mathematics in the author's own words as best as he can:

> I will explain this on the basis of the statements of the predecessors about that what the honorable Euclid has said about (this matter) –

40 Translated from U. Rebstock, *Al-Tadhkira bi-uṣūl al-ḥisāb wa-l-farāʾiḍ* (Buch über die Grundlagen der Arithmetik und der Erbteilung) von ʿAlī b. al-Khiḍr b. al-Ḥasan al-Qurashī (st. 1067), Frankfurt am Main, Institut für Geschichte der Arabisch-Islamischen Wissenschaften, 2001, 18. The *kunyā* of al-Khwārizmī is, however, not the ordinary one, which is Abū Jaʿfar. It belongs to another al-Khwārizmī who wrote in the tenth century an encyclopedia on all the fields of knowledge. Abū Bakr as *kunyā* of al-Khwārizmī was repeatedly used by authors in the Yemen. See King, "A Medieval Arabic Report on Algebra Before al-Khwārizmī," 26–28.

> insofar as it belongs to the figures from the *Elements* and to that what he determined by numbers – with my (own) words and as best as I can. I also will mention, which notions I took over from Abū Yūsuf al-Ūqlīdisī, Abū Kāmil Shujāʿ b. Aslam, Abū l-Wafāʾ, Muḥammad b. Muḥammad b. Lakhmī al-Muhandis, and Abū Bakr Muḥammad b. Mūsā al-Khwārizmī. They are the masters of this art insofar as I can judge it. I never stopped trying to find new, fast answers. I remained a student on the path of redemption.[41]

Other authors of books on the calculus of inheritance shares formulated similar positions towards the appropriateness of preserving the tradition of their field. This attitude did not prevent them from introducing new names, concepts, methods and examples. It was, however, an attitude better in tune with the mother discipline of jurisprudence than with theoretical mathematics. As the example of al-Qurashī demonstrates, professionals who wished to apply mathematical knowledge outside of mathematics proper faced a different set of demands and had to satisfy a different clientele. The presentation of their attitudes incorporated elements not found in texts on theoretical mathematics. Al-Qurashī's introduction also shows that the novelty could appear in different meanings and adopt diverse values. In his case, the focus was on fast replies to known problems. In difference to al-Khwārizmī, for al-Qurashī novelty was only to be justified when it derived from ancient or modern predecessors. It was this continuity and this anchorage in tradition that assured al-Qursahī's innovations a place of respect.

While al-Qurasī was a legal scholar who claimed reading not only the books of his own discipline, but also the works written by mathematicians, Ibn al-Haytham was a mathematician, astronomer, philosopher, physician and engineer who, in his younger years, also wrote about religious and theological themes as well as the arithmetic of commerce. The introductions to his mathematical writings differ substantially from that by al-Qurashī. They are often much shorter, integrate religious formulas into the text rather than placing them before and after it, and do not give so much bibliographical information than comments on why the author came to think about the subject matter of his work. Within this explanatory presentation, Ibn al-Haytham situates himself in the field and determines his own value and success. He takes up problems formulated, but not solved, by the ancients. He admires their marvelous properties and strange notions. Finally, he unveils his solutions to his sovereign in order to induce him to acquire new knowledge, to ponder it and then to use it for pointing out the virtue of the science of geometry and its hidden notions. One cannot miss the parallelism to the language used in natural history. Wondrous, astonishing and strange things happen in both kinds of sciences. The

41 Translated from Rebstock, *Al-Tadhkira bi-uṣūl al-ḥisāb wa-l-farāʾiḍ*, 18. Once more, the *kunyā* of the mathematician, astronomer, geographer and historian of the ninth century is the same as that of the secretary of the tenth century.

desire to unveil the hidden truth is another element shared by authors of both disciplines. Ibn al-Haytham may have taken recourse to this kind of language in order to allude to shared ideals between geometry and the religious beliefs of the Fatimid caliph to whom he dedicated his treatise *Fī l-hilāliyāt* (On the Lunules).[42] A similar use of religious rhetoric and imagery can be detected in al-Khwārizmī's *Algebra*, but this time referring to tenets dear to Caliph al-Maʾmūn, such as the debate about God's throne and the divine properties.[43] Striving for virtue and serving the community by useful achievements is a standard maxim invoked by authors of biographical dictionaries in most of their entries. The only specific element in Ibn al-Haytham's application of this rhetoric is its extension from scholars to the discipline he served.

In a second treatise on the lunules, characterized as exhaustive, Ibn al-Haytham applies a rhetorical figure analogous to the one used by al-Qurashī. He claims that he had written an earlier text on the squaring of the lunule for some of his friends, for whom, because he was in a hurry, he solved the problems by particular methods. Since then, he has had time to think more closely about the issue and he had a (novel) idea how to solve the problem (in a better manner) by "scientific" methods. Applying these methods, he managed to find new kinds of lunules and their squaring. The very interesting contrasting of particular and scientific methods cannot be followed up here.

In a third extant text on a similar problem, the squaring of the circle, Ibn al-Haytham attacks ancient and modern philosophers for having discussed in vain the question as to whether it is possible to find by an exact method a polygon with the same area as a circle. Ibn al-Haytham was convinced that he had found what the ancients could not achieve and expressed his pride, again using a language, which alludes to religion and turns the scientist into a relative of a prophet: he has thought about the problem in a profound manner; its solution was revealed to him; he showed that the belief (*iʿtiqād*) of those who believed that a circle cannot be squared is erroneous.[44]

42 Rashed, *Les Mathématiques infinitésimales*, vol. 2, 71; French translation, p. 70. Rashed claims that the anonymous addressee of Ibn al-Haytham's dedication was a man of letters or of science, of an elevated class, but not that in power, undoubtedly a friend, because in the second treatise on the lunules, Ibn al-Haytham wrote that he had redacted an older text on this subject after the demand of one of his "brethren" (*ikhwān*), a term, scholars most often used for describing their colleagues (Ibid., 32). Nonetheless, the formulas employed by Ibn al-Haytham to describe the addressee of this text do not correspond to the informality of the second text. In the dedication written by a mathematician in the period of Ibn al-Haytham and dedicated to the Grand vizier al-Muʾayyad al-Manṣūr Fakhr al-Mulk Abū Ghālib, the formulas are essentially identical with those of Ibn al-Haytham. Moreover, it appears more probable that Ibn al-Haytham's treatise was dedicated to a person of the Fatimid court or its entourage than to a scholarly colleague. On this point, see S. Shalhūb, *Al-Kāfī fī l-ḥisāab li-Abī Bakr Muḥammad b. al-Ḥasan al-Karajī*, Ḥalab, Jāmiʿat Ḥalab, Maʿhad al-turāth al-ʿilmī al-ʿarabī, 1406/1986, 35.

43 *Kitāb al-jabr wa-l-muqābala*, 16.

44 Rashed, *Les Mathématiques infinitésimales*, vol. 2, 93 and 95; French translation pp. 92 and 94.

The scribe who copied Ibn al-Haytham's text was himself a person interested in geometry. He searched for critical comments and found them in works by an author he could not name with certainty, but thought he may have been the physician ʿAlī b. Riḍwān (d. 1067). Leaving aside that the first part of the author's critique implies that his mathematical knowledge was not very firm, the second part of his objection is of greater interest for my subject. Here, the author attacks what Ibn al-Haytham had declared as being the purpose of his treatise: to show that it is possible to square the circle, but not to find the square itself whose area equals the area of a given circle. In a sense, the author's criticism here is principally similar to the critique by twentieth-century constructivist mathematicians at infinite methods and existence theorems. The medieval author namely admonished his predecessor for the uselessness of his result: the proof of the possible existence of a square whose area equals that of a given circle without giving a method for constructing the square adds nothing new to the knowledge of the ancients. And as if this attack was not strong enough, the author summarizes his epistemological position as follows:

> The inaccessible is that whose knowledge is impossible to us. Our belief that its knowledge is possible is of no use. If we wish to make clear (what we have to say) about these notions, (we say that) they can be divided into three parts, namely: the notion is known, meaning that it is established by proving it, i.e., that it is known; or its knowledge is impossible, meaning that it cannot be established by proof that it is known; nor (can it be proven) that its knowledge is impossible. (An example is the knowledge of) the chord of the ninth of a circle or the knowledge of the chord of a degree. (Things) similar to the two are many in this part.[45]

Whoever authored this addition to Ibn al-Haytham's treatise, he was not the only reader of Ibn al-Haytham's mathematical works who submitted them to severe objections. Better known is Ibn al-Sarī's (d. 1154) series of small works all aiming at elucidating the errors of Ibn al-Haytham. In his text titled *About the Explanation of Abū ʿAlī b. al-Haytham's Error in the First Theorem of Book X of Euclid's 'Elements'*, Ibn al-Sarī wrote a different introduction than those discussed so far. The difference is partly caused by the single focus of his text and partly by the fact that Ibn al-Sarī does not merely situate his own work in a field of predecessors and contemporaries but wishes to expose substantial errors of a celebrated, first-class scholar of the previous century. Ibn al-Sarī first outlines in detail what Ibn al-Haytham had written about, what he had hoped to achieve and the way in which he talked about his results. He emphasizes that Ibn al-Haytham disagreed with the view of numerous of his predecessors, that he believed that mathematical

45 Rashed, *Les Mathématiques infinitésimales*, vol. 2, 101; French translation p. 100; modified translation.

concepts should be formulated in accordance with what a mathematician needs for a concrete mathematical purpose, that their validity should be proven and that they then could be possibly generalized, hence producing new knowledge of different sorts.[46] Then Ibn al-Sarī comes forward with his own story. Its principal element resembles strongly Ibn al-Haytham's approach to self-representation: Ibn al-Sarī meditated about the aim of his predecessor, investigated it and discovered mistakes. In contrast to the generalities found in the remarks about the history of a certain mathematical problem investigated by the authors I discussed previously, Ibn al-Sarī now enumerates in detail the errors he has found in Ibn al-Haytham's text:

> First, in the understanding of the sense of universal and particular; second, in the understanding of the sense of Euclid's theorem and of the theorem in which he applied this theorem (i.e., Book X, theorem 1) and his belief that his theorem can replace Euclid's theorem; and third, the demand to limit oneself to the subject of this theorem as exposed by Euclid, in the only book of Euclid, by mentioning thus the need for this theorem and by neglecting all the others.[47]

Finally, Ibn al-Sarī erects defenses for himself in order to escape any future attack by pointing to his duty to expose those errors to the future student. Nobody who studies should doubt that Ibn al-Haytham's purpose was wrong, that Euclid's theorem was not a universal, but only a particular truth, but nonetheless the tool for solving all geometrical problems in the field of non-homogenous areas and volumes. Hence, Ibn al-Sarī turns out to be a conservative in content, but a rebel in form. He includes in his rebellious stance Ibn al-Haytham's vision of history by pointing to those authors and their writings whom Ibn al-Haytham either did not mention at all or misrepresented according to Ibn al-Sarī.[48]

Over time, several alternative styles emerged. Among them the three principal ones were: 1) the aim to satisfy a request by a friend, colleague, student or sovereign; 2) to offer a certain knowledge to a patron and 3) to fulfill the desire to follow the route shown by one's elders. These types evolved progressively into the norm. The texts in which an author explicitly claimed novelty slowly moved to the backstage.

Conclusions

During the four or five centuries in which the sciences flourished in Islamic societies the field of discoveries, inventions and modifications of ancient knowledge was not limited to mathematics. An attitude aiming to implement new objectives and to explore new territories of knowledge already appeared in the ninth century.

46 Rashed, *Les Mathématiques infinitésimales*, vol. 2, 503; French translation p. 502.
47 Ibid., 503; modified translation.
48 Ibid., 505; French translation p. 504.

It was most often expressed in philosophical writings. In the tenth century, the experts of the mathematical sciences were convinced of their value. The specialists of geometry, arithmetic, number theory, algebra, astronomy and astrology trusted the certitude of their tools, their methods and their theories. They were ready to defy the ancient authors and to propose their own works as innovations.

There was, however, no obligatory link between the rhetoric of novelty and the existence of novel mathematical results. Authors of claims to novelty could, after all, overestimate their own contributions. In contrast, authors who did not claim novelty could in fact present discoveries, because they were very sure of their own capabilities or because they did not think that innovation was the aim of their text. The rhetoric of novelty was part of a larger culture of rejection, argumentation and critique. This culture continued to inform the scientific, medical and philosophical rhetoric, even when the rhetoric of novelty became less prominent. Equally, the spaces in which the rhetoric of novelty could be expressed varied.

In difference to Latin, Italian, French, Dutch, German and English texts, which in the sixteenth and seventeenth centuries asked for novelty from their titles and cover pages onward, in Islamic societies, innovations, inventions and discoveries were most often evoked inside a text and sometimes on an instrument or in a map. Titles were rarely considered a convenient place for such claims, in particular after the emergence of rhymed titles in other fields and their adoption in the sciences. Introductions and remarks inside a text appear to have been the privileged sites for such claims. Although claims of novelty cover only a very small part of the scientific discourse in later Islamic societies, this does not mean that the study of the sciences was abandoned in those societies nor that one stopped altogether to search for novelty. Despite its importance, *bidʿa* had not the power to extinguish the need and the will to change.

Bibliography

Sources

al-Qurashī, ʿAlī b. al-Khiḍr, *Al-Tadhkira bi-uṣūl al-ḥisāb wa-l-farāʾiḍ* (Buch über die Grundlagen der Arithmetik und der Erbteilung), U. Rebstock (ed.), Frankfurt am Main: Institut für Geschichte der Arabisch-Islamischen Wissenschaften, 2001.

Bahār, M. T., *Sabk-i shenāsi-yi tārīkh-i taṭavvur-i nathri fārsī*, Vol. 3, Tehrān: Intisharāt zawār, s.d.

Dallal, A., *An Islamic Response to Greek Astronomy. Kitāb Taʿdīl Hayʾat al-Aflāk of Ṣadr al-Sharīʿa*, edited with Translation and Commentary, Leiden: E. J. Brill, 1995.

Ibn al-Haytham, *Al-Shukūk ʿalā Baṭlāmiyūs*, A. I. Sabra and N. Shehaby (eds.), Cairo: Dār al-Kutub, 1971.

Ibn al-Haytham, *Kitāb fī ḥall al-shukūk fī Kitāb Uqlīdis*, Fuat Sezgin, Frankfurt am Main: Institute for the History of Arabic-Islamic Science at the Johann Wolfgang Goethe University, 1985.

Ibn al-Haytham, *The Optics of Ibn al-Haytham*, A. I. Sabra (ed.), 3 Vols. London: Warburg Institute, 1989.

Kitāb al-jabr wa-l-muqābala, Taqdīm wa-ta'līq 'Alī Muṣṭafā Musharrafa wa-Muḥammad Murshī Aḥmad, al-Qāhira, Dār al-Kitāb al-'arabī li-l-ṭibā'a wa-l-nashr, 1968.
MS Berlin, Staatsbibliothek Berlin, Preussischer Kulturbesitz, Wetzstein 1154.
MS Vatican City, Biblioteca Apostolica, Borg. Pers. 14.
MS Vienna, ÖNB, Mixt. 271.
Rashed, R., *Les Mathématiques infinitésimales du IXe au XIe siècle*, Vol. I, London: Al-Furqan Islamic Heritage Foundation, 1416/1996.
Rashed, R., *Mathématiques infinitésimales du IXe au XIe siècle*, Vol. II, London: al-Furqān Islamic Heritage Foundation, 1993.
Rashed, R. *Mathématiques infinitésimales du IXe au XIe siècle*, Vol. III, London: al-Furqān Islamic Heritage Foundation, 1421/2000.

Secondary literature

Berggren, L. J. and Hogendijk, J. P., *The Fragments of Abū Sahl al-Kūhī's Lost Geometrical Works in the Writings of al-Sijzī*, Utrecht, University Utrecht, Department of Mathematics, Preprint no. 1226, February 2002, 4–18.
Cohen, H. F., *The Scientific Revolution: A Historiographical Inquiry*, Chicago: The University of Chicago Press, 1994.
Hogendijk, J. P., "Greek and Arabic Constructions of the Regular Heptagon," *Archive for History of Exact Sciences* 30 (1984), 179–330.
Huff, T. E., *The Rise of Early Modern Science: Islam, China and the West*, Cambridge: Cambridge University Press, 1993.
Janin, L. and King, D. A., "Ibn al-Shāṭir's Ṣandūq al-Yawāqīt: An Astronomical Compendium," *Journal for the History of Arabic Science* 1 (1977), 187–256.
Kennedy, E. S. and Ghanem, S., *The Life and Work of Ibn al-Shatir*, Aleppo: Aleppo University, 1976.
King, D. A., "The Astronomical Instruments of Ibn al-Sarrāj: A Brief Survey," in King, D. A. *Islamic Astronomical Instruments*, Aldershot: Brookfield, Variorum, 1987, IX.
King, D. A., *Astronomy in the Service of Islam*, Aldershot: Brookfield, Variorum, 1993.
King, D. A., *In Synchrony with the Heavens: Studies in Astronomical Timekeeping and Istrumentation in Islamic Civilization*, 2 Vols., Leiden: E. J. Brill, 2004, 2005.
King, D. A., *Islamic Mathematical Astronomy*, London: Variorum Reprints, 1986.
King, D. A., "A Medieval Arabic Report on Algebra before al-Khwārizmī," *al-Masāq* 1 (1988), 25–32.
King, D. A., "New Light on the *Zīj al-Ṣafā'iḥ* of Abū Ja'far al-Khāzin," *Centaurus* 23 (1979), 105–117.
King, D. A., "On the Role of the Muezzin and the Muwaqqit in Medieval Islamic Society," in *Tradition, Transmission, Transformation*, S. Livesey, F. J. Ragep and S. P. Ragep (eds.), Leiden, New York and Köln: E. J. Brill, 1996, 285–346.
Lane, E. W., *An Arabic-English Lexicon*, 8 Parts, Beirut: Imprimérie libanaise, 1968.
Langermann, Y. T., "Criticism of Authority in the Writings of Moses Maimonides and Fakhr al-Dīn al-Rāzī," *Early Modern Science and Medicine* 7.3 (2002), 255–275.
Langermann, Y. T., "Ibn Kammūna and the 'New Wisdom' of the Thirteenth Century," *Arabic Sciences and Philosophy* 15 (2005), 277–327.
Minorsky, V., *Calligraphy and Painters, a Treatise by Qadi Ahmad, Son of Mir Munshi, Translated from the Persian*, Washington, DC: White Lotus Press, 1959.

Ragep, F. J., "ʿAlī al-Qushjī and Regiomontanus: Eccentric Transformations and Copernican Revolutions," *Journal for the History of Astronomy* 36 (2005), 359–371.
Ragep, F. J., "Freeing Astronomy from Philosophy: An Aspect of Islamic Influence on Science," *Osiris* 16 (2001), 49–71.
Ragep, F. J., *Naṣīr al-Dīn al-Ṭūsī's "Memoir on Astronomy" (al-Tadhkira fī ʿilm al-hayʾa)*, 2 vols., New York-Berlin *et al.*: Springer Verlag, 1993.
Rossi, P. E., *The Birth of Modern Science*, Oxford: Blackwell Publishers, 2000.
Sabra, A. I., *Optics, Astronomy and Logic: Studies in Arabic Science and Philosophy*, Aldershot: Brookfield, Variorum, 1994.
Saliba, G., *A History of Arabic Astronomy: Planetary Theory During the Golden Age of Islam*, New York: New York University Press, 1994.
Saliba, G., "A Sixteenth-Century Arabic Critique of Ptolemaic Astronomy: The Work of Shams al-Dīn al-Khafrī," *Journal for the History of Astronomy* 25 (1994), 15–38.
Samsó, H., *Islamic Astronomy and Medieval Spain*, Aldershot: Brookfield, Variorum, 1994.
Sesiano, J., *Les carrés magiques dans les pays islamiques*, Lausanne: Presses polytechniques et universitaires romandes, 2004.
Sesiano, J., "Une compilation du XIIe siècle sur quelques propriétés des nombres naturels," *SCIAMVS* 4 (2003), 137–189.
Sesiano, J., *Un traité médiéval sur les carrés magiques. De l'arrangement harmonieux des nombres*, Lausanne: Presses polytechniques et universitaires romandes, 1996.
Shalhūb, S., *Al-Kāfī fī l-ḥisāab li-Abī Bakr Muḥammad b. al-Ḥasan al-Karajī*, Ḥalab: Jāmiʿat Ḥalab, Maʿhad al-turāth al-ʿilmī al-ʿarabī, 1406/1986.

2

REFLECTIONS ON THE ROLE OF THE EXACT SCIENCES IN ISLAMIC CULTURE AND EDUCATION BETWEEN THE TWELFTH AND THE FIFTEENTH CENTURIES

In this article, I wish to challenge the dominant current picture of the place of the exact sciences in Islamic culture and education. I claim that central parts of it need to be revised, because they are based on methodologically doubtful investigations, a too narrow base of sources with regard to the exact sciences, and a set of assumptions which are the result of historically grown prejudices. To substantiate this claim, I will first briefly characterize the main aspects of the current view which need to be newly investigated. Then I will present a few examples concerning the relation between the production of mathematical and astronomical manuscripts and the *madāris*. The center of this paper will be a discussion of a history of education in Damascus written by what is commonly thought of as a religious or legal scholar of the late fifteenth and early sixteenth centuries, the book *al-Dāris fī taʾrīkh al-madāris* by ʿAbd al-Qādir b. M. al-Nuʿaymī (845/1441–927/1520).

Five claims constitute the dominant current picture of the place of the exact sciences in Islamic culture and education:

1 There was a permanent hostility of religious and legal scholars towards all ancient sciences (viz., philosophy, logic, mathematics, astronomy, astrology, medicine, and the occult sciences). This hostility increased after the eleventh century as a result of the profound criticisms of the ancient sciences by the outstanding religious scholar al-Ghazālī (d. 505/1111) and the Seljuq vizier Niẓām al-Mulk's (d. 485/1092) elevation of the *madrasa* to the core institution of Sunni education against Shiʿi Islam and other tendencies declared to be heresies.

 DOI: 10.4324/9781003372493-3

2 The *madāris* developed a regular curriculum from which the ancient sciences were excluded. This curriculum encompassed the study of the religious sciences such as the reading and interpretation of the Qur'an or the traditions of the Prophet, grammar and literature as ancillary sciences, and law according to one of the four major teachings of Sunni law as the ultimate goal of education. If contemporary scholars concede that some of the exact sciences have been taught by scholars connected with a *madrasa*, they often claim that such a teaching focused on practical aspects.

3 There was a sharp decline of productivity in all ancient sciences in Islam after the thirteenth century until they died out some centuries later. This decline was caused by the hostile attitude towards the ancient sciences and by their exclusion from the *madrasa*.

4 The ancient ideal of scientific, above all philosophical, inquiry to nourish the rational soul so that it can gain perfection and thus salvation was replaced by the ideal of a pious Muslim as embodied in the *sīrat Muḥammad*, the life of the Prophet. This shifted the focus of study from philosophy to religious matters.

5 The main historical significance of the pursuit of the ancient sciences in Islam was the preservation of Greek knowledge, texts, instruments, and practices and its partial transmission to Catholic Christianity in Western Europe.[1]

The points I wish to challenge in this paper are the first and the second claim. To this end, I am going to raise three questions. After an introduction, in which I will present some examples which – in my view – have the power to challenge the dominant current picture of the place of the ancient sciences in Islam not only locally, but globally. I will ask first whether the ancient sciences indeed were ignored by scholars connected with the *madāris*. I will discuss this question with information given by al-Nuʿaymī about the fields of education and practice pursued by the scholars of Damascus and by looking at the terminology he used to describe educational patterns in the religious, legal, philological, and ancient sciences. My second question will address the participation of scholars active in the field of the ancient sciences in important social rituals and the stance taken by the community in this regard. Third, I will study stories of conflicts between scholars and rulers and within the scholarly community of Damascus. The aim of this study is to check whether those conflicts in which scholars active in the ancient sciences were involved revolved mainly around their pursuing the ancient sciences. The second point of relevance in this study of conflicts is to challenge our own methodological approach to Muslim biographical literature which too often

1 The perhaps most influential works to shape this current picture are those by Ignaz Goldziher 1915, Martin Plessner 1931, Gustav E. von Grunebaum 1964, Franz Rosenthal 1965, and George Makdisi 1981. A neat recent outline of this picture is to be found in Gerhard Endress 1987. Aspects of this picture have been criticized by Abd al-Hamid Sabra 1987, 1996, Jamil Ragep 1993, Michael Chamberlain 1992, or Dimitri Gutas 1998.

neglects to question the meaning and context of the variety of narratives of one and the same conflict extant in the sources.

1 Introduction

In September 9, 1215, a great-great-great-grandson of Niẓām al Mulk finished copying a collection of mathematical texts by al-Sijzī (fl. late fourth/late tenth century) and Ibn al-Haytham (d. 440/1048). He undertook this copying at the Niẓāmiyya *madrasa* in Baghdad, the most important *madrasa* of the Shāfiʿī *madhhab* in town. The *madrasa* was the first Niẓām al-Mulk had founded. The copied texts mostly treated higher theoretical geometry. Their being copied at the very cornerstone of the claimed anti- and non-scientific system of Islamic education needs at least a local interpretation if it is seen as the famous exception that proves the rule. Everyone, however, who has read colophons, manuscript dedications, ownership stamps, historical chronicles, biographical dictionaries, or travel accounts knows that it is easy to compose a long list of other examples of a similar kind. The following six examples are given to substantiate this point, but they are by no means exhaustive:

1 Ibn Sīnā's (d. 428/1037) theoretical introduction to his famous medical encyclopedia *al-Qanūn fī l-ṭibb* was copied at the Niẓāmiyya *madrasa* in Baghdad April 13, 1283.
2 At the same *madrasa*, Naṣīr al-Dīn al-Ṭūsī's (597/1200–673/1273) text on theoretical astronomy, *al-Tadhkira fī ʿilm alhayʾa*, was copied in 1283.
3 Al-Kindī's (d. ca. 257/870) book on optics was copied in 1491 at the Kāniliyya *madrasa* in Cairo.
4 In 1675, Shams al-Dīn al-Khafrī's (d. after 932/1525) commentary on Naṣīr al-Dīn al-Ṭūsī's *al-Tadhkira* was copied by M. Riḍā Shīrāzī, a student of the Ismaʿiliyya *madrasa* in Shīrāz for Mīrzā Ibrāhīm, his teacher and the school's headmaster.
5 Muḥammad b. Ibrāḥīm al-Ījī (d. after 783/1381), a student of the eminent *mutakallim* ʿAḍud al-Dīn al-Ījī (d. 756/1355), characterized "the study of all natural and supernatural phenomena, the data of physics, geography, and metaphysics" as the proper object of history.[2] Among his sources, he cited the works *al-Qānūn al-masʿūdī, Nihāyat al-idrāk*, and *al-Tuḥfa al-shāhiyya* by Abū l-Rayḥān al-Bīrūnī (d. after 444/1053) and Quṭb al-Dīn al-Shīrāzī (d. 711/1311).
6 Muḥibb al-Dīn M. b. Muḥammad known as Ibn al-ʿAṭṭār (fl. ca. 830/1426–874/1469), a Shāfiʿī jurist and member of then Wafāʾīyya Sufi order, studied astronomy with Egyptian scholars working as *muwaqqitūn* at mosques,

2 Rosenthal 1952, p. 202.

> among them Jamāl al-Dīn al-Māridīnī (d. 809/1406) and Shihāb al-Dīn M. al-Majdī (d. 850/1446). He later also taught astronomy, composed texts on it, and devoted about 30 years of his life to this science. His teacher Jamāl al-Dīn al-Māridīnī taught Euclid's *Elements* at the Azhar Mosque, another of the most famous educational institutes in Sunni Islam.[3]

Al-Kindī, Ibn Sīnā, al-Bīrūnī, Ibn al-Haytham, Naṣīr al-Dīn alṬūsī and Quṭb al-Dīn al-Shīrāzī were all first-class representatives of the ancient sciences in Islam. Their works, copied and taught in *madāris* or mosques and used in modern historiography, are among the best products of theoretical research and writing of Muslim philosophers and scientists. The copying of their works took place between the thirteenth and seventeenth centuries in towns located in such varied regions as Egypt, Syria, Iraq, and Iran under the rule of a variety of Muslim dynasties of different ethnic origin, social structure, and cultural upbringing such as the Ilkhanids, Ayyubids, Mamluks, Safavids, and Ottomans. The people and institutions involved in this copying were connected with various Muslim religious and social groups such as a Sunni law school, a Shiʿi *madrasa*, and a Sufi order. These aspects make the examples highly interesting. They imply that the dominant current picture of the place of the ancient sciences in Islam is not only locally, but globally distorted.

To arrive at more balanced and more nuanced answers to the question of the place of the ancient sciences in Islam we urgently need studies of local conditions over a limited period; that is, we need micro-histories of the exact sciences at Muslim courts, in towns, at *madāris*, in Sufi monasteries, private circles, etc. There are different types of historical sources available for such studies, among them biographical dictionaries and topographical histories of towns and regions. Although the concepts and methods of critical historical analysis developed by medieval Muslim scholars differ greatly from current approaches, as convincingly shown by Franz Rosenthal in his book *A History of Muslim Historiography*, the medieval historical writings contain a great wealth of detail and information relevant for a better understanding of the place of the exact sciences in Muslim societies. The methods and concepts applied by medieval Muslim historians themselves are an important clue to such an understanding. They document that the historians followed values and norms which shaped their stories and made them select data and evaluate them. Excavating the particular values and norms which governed the presentation of the lives, deeds, and connections of scholars active in the fields of the ancient sciences is one of the tasks we have to face.

3 Ragep 1993, vol. 1, p. 77; Rosenthal 1952, p. 244; Arberry ed. 1955–64, no. 4138; ʿAẓīmu'd-dīn *et al.* eds. 1937, no. 2451 and p. 94.

2 Damascene *madāris* and the ancient sciences

The book I have chosen to illustrate the claims of this paper is ʿAbd al-Qādir al-Nuʿaymī's *al-Dāris fī taʾrīkh al-madāris*, a local history of the institutes of education in Damascus between 1100 and 1500. Al-Nuʿaymī divided his book in twelve chapters and an appendix treating eight types of educational institutes:

1 the Houses for the (studies) of the Qurʾan,
2 the Houses for the traditions of the Prophet,
3 the *madāris* for studies of Shāfiʿī, Ḥanafī, Ḥanbalī, and Mālikī law and for medical training,
4 the *khānqāh* (inns),
5 the *ribāṭ* (Sufi monasteries),
6 the *turbāt* and *qubbāt* (tombs and domes),
7 the *jāmiʿāt* (Friday mosques) and *masājid* (ordinary mosques).

Each chapter treats the individual establishments of one type. The description of each establishment is subdivided in at least two main parts – data about the topography of the institution, its founder and his or her family, and the stipulations of the deed of foundations if known to al-Nuʿaymī; and a list of professors and, occasionally, other staff with information on educational background, professional career and duties, personal relations, social and other conflicts, and, again, occasionally, special events in individual biographies.

According to al-Nuʿaymī logic, arithmetic, algebra, and some parts of the other ancient sciences were taught at Shāfiʿī and Ḥanafī *madāris*, houses for the traditions of the Prophet, and at the Umayyad Mosque. M. b. ʿAlIī b. Ibrāhīm known as al-Fakhr al-Miṣrī (691/1291–751/1350), the later secretary of the Mamluk Quṭlūbīk, for instance, studied "arithmetic with al-Nuʿmān and logic with a whole group of scholars, the most famous among them were Shaykh Rāḍī l-Dīn al-Manṭiqī (646/1248–732/1332) and Shaykh ʿAlāʾ al-Dīn al-Qūnawī (668/1269–729/1329)."[4] Shams al-Dīn Abū ʿAbdallāh M. b. M. known as Ibn Muʾadhdhin al-Zanjīlī al-Ḥanafī (d. 819/1418) studied inheritance mathematics at the Umayyad Mosque.[5] Ṣadr al-Dīn Abū ʿAbdallāh M. b. ʿUmar known as Ibn al-Wakīl (665/1267–716/1315) taught at the Ashrafiyya house for the traditions of the Prophet, mixing those traditions with words "from a multitude of sciences, among them medicine, philosophy, and *kalām* ('theology'); but they do not belong to this science, but to the sciences of the ancients."[6] Physicians teaching at medical *madāris* also studied philosophical, mathematical, and astronomical texts, but al-Nuʿaymī refrained from stating precisely where they did so.[7] He related,

4 al-Nuʿaymī 1988, vol. 1, p. 247; see also p. 245.
5 *Ibid.*, vol. 1, p. 529.
6 *Ibid.*, vol. 1, p. 28.
7 *Ibid.*, vol. 2, pp. 127–138.

however, that several of these physicians not only studied parts of the ancient sciences, but busied themselves also with the traditions of the Prophet, Arabic language and literature, poetry, or legal doctrine.[8] Amīn al-Dīn b. Dā'ūd (665/1267–732/1332) studied medicine and some of the traditions of the Prophet with 'lmād al-Dīn al-Dunayṣarī (606/1209–686/1287), the donor of the medical *madrasa* named after him.[9] Jamāl al-Dīn Aḥmad b. 'Abdallāh al-Dimashqī (d. 694/1293) not only held a position at the medical *madrasa* al-Dakhwāriyya donated by the physician Mudhahhab al-Dīn 'Abd al-Raḥīm known as al-Dakhwār (565/1 170–628/1231), but also one at the Shāfi'ī *madrasa* al-Farūkhshāhiyya.[10]

The terminology used by al-Nu'aymī to describe educational patterns and teaching profiles is the same for all disciplines whether traditional or rational. Special knowledge in arithmetic, algebra, astronomy, logic, and medicine is praised with the same intensity as that in law, the traditions, or exegetic studies or that in grammar, literature, or poetry. The same holds true if remarkable ignorance was displayed in the one or the other field. Shihāb al-Dīn b. Taymiyya 'Abd al-Ḥalīm al-Ḥarranī (627/1230–682/1283), the father of the famous Ḥanbalī judge Taqī l-Dīn Aḥmad b. 'Abd al-Ḥalīm b. Taymiyya, who taught at the Sukriyya house for the traditions of the Prophet and at the Umayyad Mosque, is said to have had "a long hand in inheritance mathematics, arithmetic, and *al-hay'a*" (cosmology, astronomy, and geography).[11] Taqī l-Dīn (661/1263–728/1328) himself "fully mastered inheritance mathematics, arithmetic, algebra, as well as other sciences. He occupied himself with *kalām*, surpassed in it its representatives, and proved their leaders wrong."[12] Zayn al-Dīn b. Talḥa (d.8311/1430)

> studied inheritance mathematics and arithmetic, was excellent in both, . . . wrote beautiful books on that, read on it many (lectures), . . . but did not take a renumeration for his teachings in inheritance mathematics and arithmetic, while he accepted payment for algebra . . . At the end of his life, he wrote legal opinions on problems in inheritance mathematics and arithmetic and took fees as other followers of this discipline[13]

Aḥmad b. 'Abdallāh al-Dimashqī mentioned above taught at the Farūkhshāhiyya *madrasa* of the Shāfi'ī *madhhab*, "made himself well known in law, gave legal opinions, worked as a tutor, and was excellent in medicine. Because of his being more advanced in the medical art than others, he led the scholars of the Dakhwāriyya (medical *madrasa*). He treated the sick in the al-Nūrī hospital according to the methods of the physicians . . . He had much intellect and bad

8 *Ibid.*, vol. 2, pp. 128, 130–134.
9 *Ibid.*, vol. 2, pp. 132–133.
10 *Ibid.*, vol. 2, p. 131.
11 *Ibid.*, vol. 1, p. 75.
12 *Ibid.*, vol. 1, p. 76.
13 *Ibid.*, vol. 1, pp. 88–89.

knowledge in many disciplines."[14] Amīn al-Dīn b. Dā'ūd is called *mudarris al-tibb*.[15] Shams al-Dīn M. b. a. Bakr known as al-Aykī (627/1230–697/1 298), the first teacher at the Ghazāliyya *madrasa*, named after al-Ghazālī who had taught in Damascus for some time, then head of the scholars of Cairo, "was one of the most able persons to solve problems and to explain riddles, above all in the sciences of the fundaments of religion and law, logic, and the ancient sciences."[16] Al-Shihāb al-Adhraʿī (d. 783/1381), however, the first teacher at the Bahā'iyya house for the traditions of the Prophet, had "no knowledge in inheritance mathematics whatsoever. After he had made several mistakes (in this field), a jurist, called al-Nawī, put them together in a collection."[17]

3 Public rituals and conflicts

An important social ritual reflecting the degree of public reputation was the burial procedure after the death of a scholar. Several teachers of the religious sciences or law, but also those of medicine and other fields, were honored by public praying ceremonies, poems, and laudations. Some of them were buried at famous places in graveyards, that is, beside famous scholars and other men or women of influence. When the aforementioned Shams al-Dīn al-Aykī had died, "many people came to his burial, among them the main judge. He was buried at the graveyard of the Sufis."[18] For the physician Badr al-Dīn b. Tarkhān (641/1243–711/1311), a special tomb with cupola was built.[19] The Ḥanafī jurist Majd al-Dīn ʿAbd al-Wahhāb b. Sakhnūn (619/1222–694/1295) who taught at the Damāghiyya *madrasa* for Shāfiʿī and Ḥanafi law, preached the sermon on Fridays, and was the head of the physicians. "He was an excellent physician with great skills. When he died a public prayer was held for him at the Friday mosque of the Ṣāḥiliyya (quarter of Damascus)."[20]

Sayf al-Din al-Āmidī (551/1156–631/1233), a Shāfiʿī scholar famous for his knowledge in *kalām*, logic, and other ancient sciences, built for himself at Mount Qāsiyūn near Damascus a tomb with cupola. Sibṭ b. al-Jawzī (d. 654/1256) reported that he ordered the bones of his dead cat to be sent from Ḥamā to be buried in his tomb. Sibṭ b. al-Jawzī took this to prove the softness of al-Āmidī's heart.[21] Despite al-Āmidī's conflict with the Ayyubid rulers of Damascus and of Egypt, al-Ashraf (d. 635/1238) and al-Kāmil (d. 635/1238), which allegedly caused his

14 *Ibid.*, vol. 2, p. 131.

15 *Ibid.*, vol. 2, p. 132. Compare George Makdisi 1981, p. 153 who claims that the use of the title *mudarris* without a complement applied exclusively to a law-teacher, whereas the title *shaykh* was generally used for teachers of other disciplines.

16 *Ibid.*, vol. 1, p. 422, see also vol. 2, p. 160.

17 *Ibid.*, vol. 1, pp. 57–58.

18 *Ibid.*, vol. l, p. 422.

19 *Ibid.*, col. 2, p. 132.

20 *Ibid.*, vol. 1, p. 519.

21 Sibṭ b. al-Jawzī 1370 h/1951, p. 691.

being dismissed from his position at the ʿAzīziyya *madrasa* and put under house arrest, the scholar was buried with much pomp in his tomb. Several medieval historians report that the cause of the conflict was al-Āmidī's occupation with the ancient sciences. But as I have shown elsewhere, this story is not only told in many contradictory versions, but is also in stark contrast to a second story which attributes the cause for the conflict to an alleged change of loyalty on al-Āmidī's side from the Ayyubids to the ruler of his native town Amida, whom al-Kamil had driven out of his principality some time before al-Āmidī was banned. Moreover, the available historical sources do not speak of other cases where either of the two Ayyubids forbade the study and teaching of the ancient sciences in Damascus as suggested by the first set of stories about al-Āmidī's fate.[22] Thus, the attribution of al-Āmidī's loss of his *madrasa* position to his involvement with the ancient sciences is less than convincing, all the more since Ibn Khallikān ascribed alĀmidī's earlier troubles in Cairo not to al-Kāmil, the Ayyubid overlord, but to a group of envious legal scholars fighting to get rid of a colleague who was a highly successful teacher and orator.[23]

The conflicts between al-Āmidī and his co-fellows as well as with the Ayyubid rulers were not exceptional. During the thirteenth and the early fourteenth centuries, conflicts among scholars and between scholars and rulers were rather the rule of the day as shown by Michael Chamberlain in his book *Knowledge and Social Practice in Medieval Damascus, 1190–1350*.[24] The reasons for such clashes and the forms of their outcome were manifold. Rulers put scholars under arrest or assigned positions at mosques, *madāris*, or other institutions to them in order to build alliances for their struggles with other rulers, to ensure the governing of the city, to guarantee the education of their sons, to win the approval of the populace, and to maintain their claim to legitimacy, or in order to suppress disturbing turbulences among the scholars themselves, to punish supporters of rivals, or to enforce the services they expected from the scholars.

The Ayyubid ruler of Damascus, al-Muʿaẓẓam (d. 624/1227), for instance, put the Ḥanafī scholar and deputy main judge of Damascus, Ismāʿīl b. Ibrāhīm al-Māridīnī, known as Ibn Fallūs, (593/1196–630/1232), under house arrest because he declined to grant the ruler permission to drink alcoholic beverages.[25] Ibn Fallūs apparently was a very pious Muslim who not only taught religious and legal subjects, but also took some of the mathematical treatises he composed to a pilgrimage to Mecca and Medina where he carried them around the holy places and performed with them the ritual duties of the *ḥājj*.[26] No wonder that he refused

22 Brentjes 1997, pp. 17–33.
23 *Ibid*., pp. 25–26.
24 Chamberlain 1992, pp. 91–107.
25 al-Nuʿaymī 1988, vol. 1, p. 540.
26 Ibn Fallūs, MS Berlin, Preussischer Kulturbesitz, Staatsbibliothek, Landberg I 99, ff. 15a-30a; in particular f. 15a.

to stretch the legal teachings of Abū Ḥanīfa (80/699–150/767) to satisfy the Ayyubid's cravings for pleasure.

Disputes over taxes, land, and water caused some of the legal scholars of Damascus to attack Ayyubid and Mamluk rulers of Syrian cities. The Shāfiʿī deputy judge Shams al-Dīn Abū l-Ḥasan ʿAlī b. M. al-Shahrazūrī (d. 675/1276) and the Ḥanafī main judge Shams al-Dīn Abū Muḥammad known as ʿAbdallāh al-Adhraʿī (595/1199–673/1274) criticized the Mamluk al-Ẓāhir Rukn al-Dīn Baybars-Bundukdārī (d. 658/1260) with harsh words. In al-Ẓāhir's presence in the court, the Shāfiʿī, facing the rich agricultural plains outside Damascus, exclaimed: "the water, the grass, and the pasture do not have an owner, but (all other things) someone owns are his. And void is a given word which contradicts this notion."[27] The Ḥanafī opposed the Mamluk with strong words in his Friday sermon: "(our) support is for the owners of property. It is not allowed for anyone to take away their possessions. And he who appropriates lawlessly what God has forbidden sins against the Lord."[28] Al-Nuʿaymī reported on the base of earlier sources: "The sultan got furious and changed his color saying: 'I sin against the Lord. Look for some other sultan.' The meeting was closed by the sultan in frosty atmosphere."[29] The power relations, however, were of a complex nature, because al-Nuʿaymī continued: "But in the evening, the sultan sent to find the judge. After he (the judge) had entered (his palace), (the sultan) came to him, honored him, presented him a robe of honor, and (told him not to perform) the obligatory salutation of honor (towards the sultan)."[30]

As a rule, no single description of a conflict can be taken at face value but has to be traced in its genesis through the entire set of historical sources available. The evaluation of conflicts given by medieval authors is usually value-laden and has to be carefully questioned. It is an open issue whether definitive conclusions can be drawn. The narratives about Rafīʿ al-Dīn al-Jīlī's (d. 641/1244) death provide an example for the difficulties concerning the stories about scholars criticizing rulers.

Al-Nuʿaymī described al-Jīlī's fate on the base of two reports by alDhahabī (d. 748/1347–8) and a third narrative told by Sibṭ b. al-Jawzī. Al-Dhahabī stated that al-Jīlī was a very good student of law, an able disputant, excellent in the rational sciences, and a *mutakallim* (dialectic theologian), but also a pseudophilosopher (*mutafalsif*), bad or arrogant with regard to religious beliefs or creeds, a slave of a sect (*dīyāna*). None of this, however, caused his being killed, which occurred late in 641/1244. Al-Dhahabī reported that al-Jīlī was thrown in a cavern in the Baqāʿ plains.[31] Sibṭ b. al-Jawzī chastized al-Jīlī in colorful terms, saying that his religious beliefs were corrupted since he was an adherent of the *Dahriyya*, the sect which al-Dhahabī did not call by name, somebody "who made fun out of the affairs

27 al-Nuʿaymī 1988, vol. 1, p. 442, 578.
28 *Ibid.*, p. 442.
29 *Ibid.*
30 *Ibid.*
31 *Ibid.*, p. 188–189.

of the *sharīʿa,* who went drunken to the Friday prayer, and whose house was like a pub."[32] Then Sibṭ b. al-Jawzī stated the reason for al-Jīlī's imprisonment. Al-Jīlī had written a letter to ʿImād al-Dīn Ismāʿīl (d. 648/1250), the Ayyubid ruler of Damascus in 1237 and 1240–1245, saying: "You carried one thousand times thousand Dinar from the property of the people into your treasury."[33] According to Ibn al-Jawzī this caused his imprisonment, the confiscation of his property, and his being killed in 642 (sic) by being thrown in a cavern of the Baqāʿ plains. The *madrasa* where al-Jīlī had taught "was given into the surveillance of Taqī al-Dīn b. al-Salāḥ, who appointed people of the religious sciences to her."[34]

When we compare these stories with al-Jīli's biography as described by Ibn Abī Uṣaybiʿa (600/1203–668/1269), we discover significant differences. In his dictionary of the classes of physicians, Ibn Abī Uṣaybiʿa reports that al-Jīlī was not a straightforward adversary of the Ayyubid ruler, but deeply involved with his vizier Amīn al-Dawla (d. 648/1250), a Muslim convert from Samaritan Judaism. Ibn Abī Uṣaybiʿa called the two friends. Nonetheless, he attributed the reason for al-Jilī's murder to his misbehavior as the main judge of Damascus. Al-Jīlī was appointed as the main judge of Damascus in 638/1240 due to Amīn al-Dawla's influence upon ʿlmād al-Dīn Ismāʿīl. In this position, al-Jīlī did injustice to many inhabitants of the city. As a consequence, doubts arose concerning his way of life. This went on until he was imprisoned and killed near Baalbek in 641, by getting his hands tied behind his back and being pushed into a huge, bottomless cavern by order of Amīn al-Dawla.[35]

A closer reading of al-Nuʿaymī's book corroborates that he too held Amīn al-Dawla to be the instigator of al-Jīlī's death.[36] It also shows that Amīn al-Dawla was accused of being the sole person responsible for ʿlmād al-Dīn Ismāʿīl's fights with his family. Since fighting for possession of towns, fortresses, villages, and influence was a rather characteristic attribute of Ayyubid family politics, this accusation is too simplistic and thus wrong. Al-Nuʿaymī seems to imply that the killing of al-Jīlī was merely another expression of Amīn al-Dawla's bad character which continuously made him give harmful advice to his Ayyubid master. He even called Amin al-Dawla "the vizier of Evil." In 643, ʿlmād al-Dīn Ismāʿīl lost Damascus to his Ayyubid overlord in Egypt, al-Ṣāliḥ Najm al-Dīn Ayyūb (d. 647/1249), and Amīn al-Dawla was taken prisoner and sent to Egypt. Al-Nuʿaymī attributed Amīn al-Dawla's downfall to the vizier's supposed lack of true faith caused by a superficiality of his conversion from Judaism to Islam which made God intervene to free the Muslim community from him.[37] Thus, his portrayal of the vizier is probably distorted. However, the continuous disagreement between ʿlmād al-Dīn Ismāʿīl

32 *Ibid.*, p. 189.
33 *Ibid.*
34 *Ibid.*
35 Ibn Abī Uṣaybiʿa 1965, p. 647.
36 al-Nuʿaymī 1988, vol. 2, p. 286.
37 *Ibid.*, vol. 2, pp. 285–286.

and al-Ṣāliḥ Najm al-Dīn Ayyūb and its accompanying battles put severe pressure upon Damascus. Hence, the murder of Rafīʿ al-Dīn al-Jīlī may indeed have been caused by his turning against Amīn al-Dawla's and ʿImād al-Dīn Ismāʿīl's bleeding out the Damascene population.

According to Ibn Abī Uṣaybiʿa, al-Jīlī was one of the most distinguished and leading scholars in the philosophical sciences, in the fundaments of religion and law, and in the natural science and in medicine. He was a legal scholar at the *madrasa* al-ʿAdhrāwiyya, at the entrance of the Naṣr gate. He held sessions on the classes of the sciences and on medicine. Ibn Abī Uṣaybīʿa studied with him some parts of the philosophical sciences. The stories of his death even if contradictory and manipulated show clearly that neither his philosophical and scientific interests nor his teaching of the rational sciences caused him any difficulties.

Other local potentates took bribes to deprive scholars of their chairs in favor of rivals as happened to the aforementioned Ibn al-Wakīl. Ibn al-Wakīl taught in Damascus at the ʿAdhrāwiyya and at the Shāmiyya *madrasas*. He lost both to two other scholars who had bribed the Mamluk governor of Damascus, Qarā Sunqur al Jūkandār al-Jarkassī al-Manṣūrī (d. 728/1328). He himself sought and got the support of Aleppo's governor Sayf al-Din Istadmūr (d. 711/1311). The positions changed their possessors several times. The conflict reached such an extent that Qarā Sunqur ordered an ambush to kill Ibn al-Wakīl. At the end, Ibn al-Wakīl lost the battle, but not before he had tried all possible means to defeat his opponents. These means included appellation to the Mamluk ruler in Egypt and the so-called blood-testimony before the Ḥanbalī main judge in Cairo. This ceremony served to attest the attitude of a person towards norms, rules, and teachings of Sunni Islam. It could end with severe punishment, including a death penalty, in case the judge found the applicant not clear and strong enough in faith. Ibn al-Wakīl thought it necessary to pass this testimony after Ḥanbalī scholars and jurists of other, unnamed affiliation in Damascus had come together to search for means to hinder his return to the two *madrasas*. They remembered old quarrels where Ibn al-Wakīl had attacked the aforementioned famous Ḥanbalī jurist Ibn Taymiyya in very harsh words. They also drew renewed attention to the dissatisfaction of Damascene people with Ibn alWakīl's Friday prayers because of his language, arguments, and behavior. One of the accusations brought up against Ibn al-Wakīl was his adherence to "fleshly lust," which seems to be somewhat surprising given the stipulation of the Prophet that each husband was obliged to take equal care of the sexual needs of his wives and that each man had the right for concubines if he could house, cloth, and feed them appropriately. However, the sexual behavior of scholars was integrated in debates of norms and codes of the profession since at least the tenth century.[38] The accusers sent their written attack to the governor of the province of Syria. The Mamluk sultan in Cairo apparently was informed about

38 I thank Jonathan Berkey for pointing this out to me.

their letter since he did not support Ibn al-Wakīl's reinstallation in Damascus, but wisely enough chose to promote him in Cairo for a while.

Al-Nuʿaymī reported variants of this case at least three times in his book. He seemed to have preferred such narratives which possessed at least a slight tone of condemnation of Ibn al-Wakīl's adversaries and their deeds. Among others, he quoted a very long extract from Ibn Kathīr's (d. 774/1372) historical book in which Ibn Kathīr claimed that Ibn al-Wakīl did not love sex overly much. On the contrary, this and other accusations were made up by people who envied Ibn al-Wakīl. He wrote:

> Among his comrades, there were those, who envied him and loved him, and those, who envied him and hated him. They told stories about him and dirtied him with incredible (things). But he himself was without constrains and had put off the robe of shame . . . Then he had to deal with the governor of (the province) al-Shām Aqūsh al-Afram (d. 719/1319) and it happened things to him which are unfit to mention and the order (to carry them out) does not lead to the right path.[39]

Al-Nuʿaymī closed Ibn Wakīl's biography with quoting Tāj al-Dīn al-Subkī's (727/1326–771/1369) *Al-Ṭabaqāt al-kubrā* on the classes of the Shāfiʿī scholars:

> When the people in Damascus learned of Ibn alWakīl's death they read for him a prayer in the mosque. And Ibn Taymiyya said (after he got this message): "May God console the Muslims (with regard to your death), oh Ṣadr al-Dīn." And several scholars (expressed) their sorrowness in poems.[40]

The message given by Ibn Kathīr's presentation of the affair supports my claim expressed in connection with the stories about al-Jīlī's death: when studying medieval biographies, in particular stories about conflicts and condemnations of scholars, we have to take into consideration internal fights among Muslim scholars as well as the application of metaphors and figures of speech which were regarded as helpful to achieve the gains the persons involved in the conflict, on the one hand, hoped to win and the author of the biography, on the other hand, intended to get by his description. Franz Rosenthal in his book on the historiography of Muslim history has shown that medieval Muslim historians were aware of the problems connected with writing biographies. They knew and admitted that biographies served to praise and to calumniate the biographee, to put the biographer's beliefs and values into the mouth of chosen participants of a conflict, and to express criticism towards their own contemporaries. As a consequence, no

39 al-Nuʿaymī 1988, vol. 1, p. 29.
40 *Ibid.*, vol.1, pp. 30–31.

stories of conflict and condemnation whatever their subject can be accepted without deconstructing and reconstructing them.

Most of the conflicts, al-Nuʿaymī described concerned social, political, and legal issues or were caused by greed or envy and covered by accusations of heresy and unfaithful behavior. Only rarely the occupation with the ancient sciences in general and the exact sciences in particular appeared to have been the cause of serious tensions in Damascus between 1100 and 1500. On the contrary, the picture al-Nuʿaymī drew testifies primarily to a relatively friendly acceptance or at worst indifferent attitude of the scholarly community towards these disciplines and their students.

References

Arberry, Arthur J. ed. *The Chester Beatty Library: A Handlist of the Arabic Manuscripts.* Vols I–VII. Dublin: Catalogue of Chester Beatty Library.

ʿAẓīmu'd-dīn Aḥmad, ʿAbdul Muqtadir, Muʿīnu'd-dīn Nadwī and ʿAbdul Ḥamīd. eds. 1937. *Catalogue of the Arabic and Persian Manuscripts in the Oriental Public Library at Bankipore*. Vol. XXII: *Science*. Patna: Catalogue Patna Library.

Brentjes, Sonja. 1997. *"Orthodoxy", Ancient Sciences, Power, and the Madrasa ("College") in Ayyubid and Early Mamluk Damascus.* Berlin: Max-Planck-Institut für Wissenschaftsgeschichte, Preprint no. 77.

Chamberlain, Michael. 1992. *Knowledge and Social Practice in Medieval Damascus, 1190–1350*. Princeton: Princeton University Press.

Endress, Gerhard. 1987. "Die wissenschaftliche Literatur." In: *Grundriß der Arabischen Philologie*. Edited by Helmut Gätje. Vol. 2: *Literaturwissenschaft.* Part 8. Wiesbaden: Dr. Ludwig Reichert Verlag.

Goldziher, Ignaz. 1915. "Die Stellung der alten islamischen Orthodoxie zu den antiken Wissenschaften." *Abhandlungen der Königlich-Preussischen Akademie der Wissenschaften. Philosophisch-Historische Klasse* 8: 357–400.

Grunebaum, Gustav E. von. 1964. "Muslim World View and Muslim Science." In: *Islam: Essays in the Nature and Growth of a Cultural Tradition.* London: Routledge & Kegan Paul Limited, pp. 111–126.

Gutas, Dimitri. 1998. *Greek Thought, Arabic Culture: The GraecoArabic Translation Movement in Baghdad and Early ʿAbbāsid Society (2nd–4th/8th–10th Centuries)*. London, New York: Routledge.

Ibn Abī Uṣaybiʿa. 1965. *Uyūn al-anbāʾ fī ṭabaqāt al-atibbaʾ*. Al-Bayrūt: Dār Maktabat al-Ḥaya.

Ibn Fallūs. *Kitāb iʿdād al-isrār fī asrār al-aʿdād.* MS Staatsbibliothek Berlin, Preussischer Kulturbesitz, Lanberg 199, ff. l 5a-30a.

Makdisi, George. 1981. *The Rise of Colleges: Institutions of Learning in Islam and the West.* Edinburgh: Edinburgh University Press.

al-Nuʿaymī, ʿAbd al-Qādir. 1988. *Al-dāris fi taʾrīkh al-madāris*. 2 vols. Edited by Ja'far al-Ḥasanī. Bayrūt: Maktabat al-Thaqāfa al-Dīniyya.

Plessner, Martin. 1931. "Die Geschichte der Wissenschaften im Islam." *Philosophie und Geschichte* (Tübingen) 31.

Ragep, F. Jamil. 1993. *Naṣīr al-Dīn al-Ṭūsī's 'Memoir an Astronomy' (al-Tadhkira fī ʿilm al-hayʾa)*. 2 vols. New York etc.: Springer-Verlag.

Rosenthal, Franz. 1952. *A History of Muslim Historiography*. Leiden: E. J. Brill.
Rosenthal, Franz. 1965. *Das Fortleben der Antike im Islam*. Zürich, München: Artemis.
Sabra, Abd al-Hamid I. 1987. "The Appropriation and Subsequent Naturalization of Greek Science in Medieval Islam: A Preliminary Statement." *History of Science* 25: 223–243.
Sabra, Abd al-Hamid I. 1996. "Situating Arabic Science: Locality versus Essence." *Isis* 87: 654–670.
Sibṭ b. al-Jawzī. 1370 h/1951. *Mirʾāt al-zamān fī taʾrīkh al-aʿyān*, Vol. VIII. Haydarābād: Dār al-Maʿārif.

3

WHAT COULD IT MEAN TO CONTEXTUALIZE THE SCIENCES IN ISLAMIC SOCIETIES OF THE PAST?

As has been long pointed out, there are many ways to contextualize since contextualizing primarily means to find the local, the specific, and the unique that together shape a piece of knowledge either as a practice or as a result. In this sense, every researcher has to find her own approach for every single research she undertakes. The unifying point to me is the view that knowledge is not an isolated island within the ocean of culture and society, but rather like an airport with many arrivals from and departures to very different shores.

One major reproach often raised against the view that knowledge is local, specific, and unique is that it implies a static form of being. Several recent projects using contextualisation as their major methodological approach, however, also study the change of place and the transfer and transformation of knowledge. The difference to earlier studies of transfer and transformation is that the task again is to trace the specific and unique way of such a process rather than focusing only on the result. A simple example for such an approach is David Reisman's attention even to such tiny details as the availability of a riding animal and a courier who transported slips of papers written by Ibn Sīnā in Isfahan within days to a student and his friends in Shiraz who were all engaged in a heated philosophical argument.[1] While you easily can say, and correctly so, that the material Reismann used for his study of philosophical debate in early-eleventh-century Iran due to its specific character lent itself much more easily to this kind of localizing and individualizing approach and that not all subjects and types of sources will be equally suited for it, there is much more material extant that can lend support and

1 David Reisman, *The Making of the Avicennan Tradition: The Transmission, Contents, and Structure of Ibn Sīnā's al-Mubāḥathāt (The Discussions)*. Islamic Philosophy, Theology, and Science, Texts and Studies, XLIX (Leiden: E.J. Brill, 2002).

DOI: 10.4324/9781003372493-4

that all too often remains outside the horizon of studies on history of science in Islamic societies.

Additionally, the acceptance or rejection of such a localizing and individualizing approach depends on basic premises and values held by each single researcher. Let us say in a very simplifying manner that the major issue at stake is whether we want to understand the history of the sciences within a concrete setting whether in a group, a community, an institution, a ritual, or in society at large or whether we aim at exploring this history within a disciplinary, intellectual, or religious framework that focuses on progress and the connection to our own knowledge. I would wish that we could find methodologically sound ways to link these two major kinds of histories of science, because contrary to the insistence on contingency as the dominant property of history by many proponents of a contextual approach, I continue to believe in a less arbitrary character of human societies, which can be described perhaps by the metaphor of stochastic process. Privileging empirical research over theoretical generalization, I will discuss six points where I find contextualization can have a tremendous impact on how we perceive the sciences and their places, roles, and fates within Islamic societies. I do not claim that this list is comprehensive. It merely reflects the new things I have learned over the last months by reading books and papers written by colleagues in the field and by colleagues of other fields such as art history, cultural studies, history of philosophy, history of theology, and history of material culture.

1 Complicating things

A first gain we can easily reap when contextualizing the objects of our study is an increase in depth, density, and complexity both of the authors and of the subject matters we investigate. When reading many years ago Joel Kraemer's books on the Renaissance of Islam in the Buyid epoch and on Abū Sulaymān al-Sijistānī, I was not so much struck by the application of terms belonging to Catholic and Protestant history of art and ideas, but by his remarks about the cultural pre-eminence of Abū l-Wafāʾ and Abū Jaʿfar al-Khāzin.[2] When Abū Sulaymān al-Sijistānī set out to travel to Sijistan one day, he made a stop with the express desire to pay homage to Abū Jaʿfar al-Khāzin "out of respect for his erudition and age."[3] When Abū Sulaymān resumed his travel, Abū Jaʿfar sent him a note about the obligations of friendship together with a gift.[4] Not paying a visit to al-Khāzin would have been a grave social blunder. Hence, al-Sijistānī applied a social etiquette, most often reserved for princes, viziers, and saints, to a mathematician and astronomer who,

2 Joel L. Kraemer, *Philosophy in the Renaissance of Islam: Abū Sulaymān Al-Sijistānī and His Circle* (Leiden: EJ Brill, 1986); Joel L. Kraemer, *Humanism in the Renaissance of Islam: The Cultural Revival during the Buyid Age* (Leiden: E.J. Brill, 2nd revised edition, 1992).

3 Kraemer, *Philosophy in the Renaissance of Islam*, p. 28.

4 Kraemer, *Philosophy in the Renaissance of Islam*, p. 29.

while well-known to us for his scholarly works, is not at all known among historians of science for his exceptional standing in polite society. How may he have gained this status? In the context of another story about the relationship between the vizier Ibn al-ʿAmīd and a number of scholars he employed, Abū Ḥayyān al-Tawḥīdī states that al-Khāzin was among them due to the esteem that the Buyid amīr Rukn al-Dawla had expressed for him.[5] Hence, al-Khāzin's social eminence may have been the result of this linkage to the first head of the Buyid dynasty. But we still do not know how he came to be favoured by the Buyid ruler and why this linkage translated into Abū Sulaymān's feeling obliged to give al-Khāzin a polite visit. The easy answer would be to say that his excellent mathematical and astronomical skills were at the heart of the matter. A third anecdote reported by Abū Ḥayyān confirms that al-Khāzin's knowledge was indeed one aspect of his reputation.[6] But another story told by Miskawayh makes clear that politics and shifting loyalties had played a much greater role in the scientist's public standing. Abū Jaʿfar al-Khāzin had served the Samanid prince Nūḥ b. Naṣr before he came to Rukn al-Dawla. When he appeared in 341/952 at the gates of Rayy, he accompanied the Samanid general Abū ʿAlī al-Ṣaghānī b. Muḥtāj who attacked, together with the Ziyarid ruler Washmgīr, the Buyid capital. When they could not conquer the town, al-Khāzin was their emissary to broker a truce.[7] Hence, al-Khāzin was already invested with a high social status while at the Samanid court. When he changed sides, he may have been invited to do so because of this status and the skills that went with it. Al-Khāzin was probably not a mere astronomer and mathematician at the Buyid court. He may have been the first mathematically skilled political envoy in Islamic history, but he was not the last, not even in the Buyid period.

Abū l-Wafāʾ commanded an equally broad and possibly even more influential status than al-Khāzin. In 970, he was the *naqīb al-majlis* (leader of the discussion circle) and *murattib al-qawm* (minister of protocol) at ʿIzz al-Dawla's court in Baghdad.[8] In 972, he headed a delegation of Baghdadi, mostly religious, dignitaries visiting ʿIzz al-Dawla in Kufa in order to negotiate how to countermand the impending threat of a Byzantine attack on Baghdad.[9] Eight years later, Abū l-Wafāʾ administered the hospital in Baghdad and was closely related to the vizier Ibn Saʿdān. In this capacity, he helped Abū Ḥayyān al-Tawhīdī, an old friend of his, to find a paid position at the hospital and then with the vizier. Kraemer suggests that Abū l-Wafāʾ made his way to the upper echelon of the Buyid court as a member of the secretarial class.[10] He does not give any evidence for this assumption except for Abū l-Wafāʾs book for secretaries. If Kraemer is right, then we should perhaps ask whether al-Khāzin too may have acquired his social eminence

5 Kraemer, *Humanism in the Renaissance of Islam*, p. 252.
6 Kraemer, *Philosophy in the Renaissance of Islam*, p. 29.
7 Kraemer, *Philosophy in the Renaissance of Islam*, p. 29.
8 Kraemer, *Humanism in the Renaissance of Islam*, p. 182.
9 Kraemer, *Humanism in the Renaissance of Islam*, pp. 100–101.
10 Kraemer, *Humanism in the Renaissance of Islam*, p. 200.

under the Samanids by climbing the ladder in the secretarial class. If we could verify that these two important mathematicians and astronomers indeed were members of the *kuttāb*, would that not change our perspective on the social and cultural meaning of the ancient sciences in the tenth century?

But Kraemer's notes on al-Khāzin and Abū l-Wafāʾ point to a second aspect, namely, that we do not possess any substantial biographies of the two. All in all, there are only very few scholars from Islamic societies such as Abū l-Rayḥān al-Bīrūnī and Naṣīr al-Dīn al-Ṭūsī that have induced historians of science to pay attention to their lives outside a specific text, instrument, or map; to their relationship with courts, religious scholars, merchants, notables, and families; and to their involvement in the major conflicts of their times. Integrating the lives of the numerous authors of scientific texts, makers of instruments, or designers of maps in the broadest sense possible will not only enrich our knowledge of their individual fates, but open our eyes to the manifold variations and profound changes that took place over time and space. In order to achieve this, we need to do at least three things. We need to read material often used only as a mine of information such as Ibn al-Nadīm's *Kitāb al-fihrist* in a different manner. I mean historians of science need to study the bibliographical collections as historical objects in their own rights and question their assumptions, narratives, and values. When we do this, we may understand better why certain sciences and scientists are present in a special bibliography, while others are not. We also may discover that the anecdotes are not merely meant as entertainment and are surely not a simple collection of objective facts. We may perhaps even gain a new understanding of the specific values that are encoded in these stories. Secondly, we need to study material often ignored by historians of science such as princely mirrors, stories, poems, inscriptions, and works of art. In such sources, we can find information about what mattered most to the social group that produced or cherished such material, what symbols they used for expressing their beliefs or preferences, and whether the ancient sciences had any linkage to these other fields of culture. Third, we need to pay attention to non-scientific aspects of scientific texts such as titles of texts and persons, structures and points of emphasis in introductions, and allusions to contemporary trends and events. Such trends and events can encompass religious, epistemological, cosmological, and other beliefs and convictions, remarks on the evolution and standing of schools of law and *kalām*, orders of Sufism, disciplines, professions, offices, and other aspects of social and cultural life. Let me point, as an example, to Muḥammad b. Mūsā al-Khwārizmī's *Algebra*.

I find it highly remarkable that to the best of my knowledge nobody ever mentioned three aspects of this important mathematical work that point to the author's effort to situate himself within major trends of his time. He presented himself first as someone who professed, at least outwardly, adherence to Muʿtazilī belief by referring in a certain manner to the throne-verse. Then, he seems to express his intellectual allegiance to al-Kindī by quoting summarily a statement of epistemological value about the intellectual relationship between scholars of the past and of the present found in al-Kindī's text about divining from a cooked shoulder blade of a sheep and described by him as a text translated from a Greek philosophical treatise. Finally, al-Khwārizmī poses as contributing to the evolution of a law school still in the making, that is, the school of Abū Ḥanīfa. There are, of course, several possible ways to interpret these three aspects. But before we can decide which interpretation is most credible, a much more detailed study of other available texts of early-ninth-century Baghdad is needed. A glance in Book XI of Euclid's *Elements* demonstrates that other translators, writers, or copyists of mathematical texts also deemed it necessary to allude to specific theological beliefs such as the belief that God was no body like a human body. From these observations we can draw the tentative conclusion that mathematicians and astronomers in early-ninth-century Baghdad apparently were not merely interested in situating the newly translated texts within the specifics of their professions, but strove to highlight linkages between the religious beliefs in law and *kalām* that prevailed at the Abbasid court in that period. If this preliminary conclusion could be substantiated in future research, we would gain a precious addition to Gutas' study of the social history of the translation movement. The flow of ideas and concepts between different intellectual camps at court is an important theme if we wish to understand where the early practitioners of the ancient sciences wished and managed to establish themselves in Abbasid courtly society.

While it is not easy to contextualize authors of the ninth century because of a lack of suitable sources, it is difficult to achieve a well-balanced portrait of authors say of the thirteenth century because the contrary is the case – we are inundated with texts by such authors and stories and information about them, at least for a number of them. Quṭb al-Dīn al-Shīrāzī is a good representative of this problem. He was a prolific writer on various subjects, including Peripatetic philosophy, *ishrāqī* philosophy, astronomy, mathematics, medicine, *ḥadīth*, and Qur'anic exegesis. So far, nobody achieved to show what linked these various writings if there was such a connection at all. John Walbridge, in his study about Quṭb al-Dīn's commentary on Suhrawardī's work *Ḥikmat al-ishrāq* (The Philosophy of Illumination), suggests that the author's various personae as a physician, Sufi, courtier, astronomer, philosopher, and wealthy Muslim often in search for patronage and students did not at all stand in conflict with each other. Walbridge argued his case mainly by two claims he brought forward. The first claim says that Suhrawardī's theory of illumination is a specific form of a substantially modified Peripatetic philosophy. This modified philosophy replaced Ibn Sīnā's basic axiom of *existence* by the basic concept of the *manifest*, that is, light known through the senses in a concrete mystical experience of an individual. It then showed that Ibn Sīnā's basic axiom was untenable because of an infinite regress inherent in its definition. The second claim says that the various forms of knowledge, belief, and lifestyle constituted a harmonized whole fully in place when Quṭb al-Dīn was born and already followed by older members of his own family with the exception of *ishrāqī* philosophy. The addition of the latter did not disturb the unity of the other components, because substantial parts of it were nothing but Peripatetic philosophy as taught by Ibn Sīnā. What Walbridge did not undertake and what needs to be clarified in future research is the respective interpenetration of the various domains of knowledge practices by Quṭb al-Dīn.[11]

The most striking point, in a sense, in Walbridge's book about Quṭb al-Dīn al-Shīrāzī was for me the two short treatises on a major aspect of *ishrāqī* philosophy given in the appendix. The point discussed in a letter written by an otherwise unknown scholar named Aḥmad and the answer given by Quṭb al-Dīn concerns the *reality of the world of images*, that is, the reality of Platonic forms. Aḥmad's *Risāla* consists of a brief introduction and three brief chapters. Among the many fascinating points of his letter, three strike me as particularly important. Aḥmad's goal is to attain happiness by achieving pure faith. He wants to do this by clarifying doubts and acquiring certain knowledge. In this letter he starts from certain statements found in the Qur'an about the time elapsing between the death of a man and the day of resurrection and the challenges it brings to what remains of man after his death:

> Many questions of this sort have disturbed me; but seeing that to list all of them would lead to prolixity, I have confined myself to those among them most important in seeking happiness . . .
>
> The True Law hath promised and threatened us with the states of the Barrier and the incarnation of deeds. – (Events experienced by the soul between death and resurrection.) Might it be the spiritual substance, now released from the attachments of the body, that conceives these forms? – (The soul, separate from the body after the death.) – But how can this determinate conception arise from something like this, which is immaterial? – (A particular can be thought only with the aid of a material organ. Immaterial substances, like the heavenly and the disembodied souls, thus can conceive only universals. If this is so, then how are the torments and pleasures of the afterlife possible?) – Is the seat of the imagination another body – a sphere or something else? . . .
>
> **Chapter (1)**
>
> It has been claimed that the human essence is a self-subsistent substance. However, why should it not be the case that composition of the human body from the elements simply necessitates the existence of a fifth thing – just as combining vitriol, gall, and ware causes blackness? This would be what causes the human states. Or this fifth thing could be a capacity occurring by the mixture of the elements. At each moment this mixture would cause a state necessitating a particular human state,

11 John Walbridge, *The Science of Mystic Lights. Quṭb al-Dīn Shīrāzī and the Illuminationist Tradition in Islamic Philosophy* (Cambridge, MA: Harvard University Press, 1992).

> in accordance with the changing states caused by the influences of the celestial bodies.[12]

He then continues to matters pertaining to animals and their likeness to human beings:

> **Chapter (3)**
>
> What is the fate of the animal spirits? Do they perish with their bodies? If so, they are not immaterial substances; and, therefore, they do not understand universals. But sometimes we see that if one of them is struck with a stick and then after a while another stick is raised against it, it will flee from it. If there is not a universal meaning remaining in its mind corresponding to every variety of that species, it would not have fled. This is not because it perceives the particular meaning, for it is impossible for the previous blow to be exactly repeated, only for something similar to recur. . . .[13]

Aḥmad investigates his doubts and possible answers to settle them by taking recourse to philosophical themes and analogies drawing on a chemical process and infant evolution:

> Thus, you see that a newborn infant is almost without perception. It does not have most faculties, despite its efforts, except for desire and anger in a weak way. The rest of its internal and external senses are not perceptive. But after some time passes, you will notice that it hears the sound when someone speaks near to it, and it sees whomever is close before it. Then, as it becomes stronger in the actions of the vegetative faculties, it will take something given to it, but it will not be disturbed by the thing's being taken away again. However, after a little while more, it is distressed when the thing is taken away. And so it is, stage upon stage, degree after degree, until the end of life.
>
> Why then might not all these be properties emanated upon it through the influences of the celestial bodies in accordance with the capacity of its constitution? At decomposition all that would cease. Or if the variation of conditions were due to an emanation reaching it from the Giver of Forms in accordance with the variation of capacities, the receptivity would vary in accordance with the variation in states of capacity. All those would be accidents perishing with decomposition.[14]

12 Walbridge, *The Science of Mystic Lights*, p. 202.
13 Walbridge, *The Science of Mystic Lights*, p. 204.
14 Walbridge, *The Science of Mystic Lights*, p. 203.

Quṭb al-Dīn's answer is much longer than the letter. It is couched in thoroughly philosophical terms and makes apparent how far the mutual permeation of religious and philosophical forms, not only of arguing and proving, but also of thinking about the universe and the afterlife, had progressed. In accordance with Suhrawardī's teachings, Quṭb al-Dīn briefly states that the universe is made up of four worlds – the world of the intellects, the world of the souls, the world of the bodies, and the world of image and imagination. The latter is called by scholars of the Qur'ān *the Barrier* and by practitioners of the rational sciences *the world of immaterial apparitions*. This latter world does not only provide believers and philosophers with themes to discuss, but also soothsayers and sorcerers can rely in exercising their professions upon its existence and its inhabitants. It is the world of divine apparitions, of miracles, of prophecy, and of the multitude of immaterial lights. Its existence is attested by the testimony of the prophets, saints, and divine sages among them Plato, Socrates, Pythagoras, Empedocles, and Ibn ʿArabī.

This general explanation of the structure of the universe precedes Quṭb al-Dīn's so-called rational proof for the existence of the world of Platonic forms quoted from his commentary on Suhrawardī's book *The Philosophy of Illumination*. This proof starts with a discussion of competing optical theories and observations of optical appearances. The theory Quṭb al-Dīn subscribes to is Suhrawardī's *ishrāq ḥuḍūrī* (presential illumination), which says:

> . . . vision occurs by the illumined thing being opposite a sound eye – nothing more. An illumination of presence thereby occurs in the soul of the lighted object, and so the soul sees it. . . . Just as the form in the mirror is not within it, so too the forms by means of which the soul perceives things are not in the retina; rather they occur when the thing and the eye are opposite each other, as we have mentioned. Thereupon a presentational illumination falls from the soul upon that lighted thing – if the thing has an external reality – and it sees it. It is a sheer apparition, it needs another locus of manifestations, such as a mirror. If the retina happens to be opposite the mirror in which the forms of the things opposite appear, a presentational illumination occurs in the soul so that it will see those things by means of the mirror of the retina and the external mirror – though only if the necessary conditions exist and hindrances are removed.[15]

Then follows a discussion of the link between the world of Platonic forms and the sleeping soul and a discussion of the conditions of human souls after their separation from the human bodies. Quṭb al-Dīn draws upon Plato, Pythagoras,

15 Walbridge, *The Science of Mystic Lights*, pp. 213–214.

the Ikhwān al-Ṣafā', the Qur'an, Muḥammad's ascension to the heavens, and Bukhārī's *Ṣaḥīḥ*.[16]

The last part of Quṭb al-Dīn's reply takes up individual questions and claims from Ahmad's letter and answers or rejects them by drawing exclusively upon Suhrawardī and his own commentary on the *The Philosophy of Illumination*.

Quṭb al-Dīn's style of arguing is exemplified in the two brief subsequent passages starting with the question of Aḥmad:"Is the seat of its imagining the body . . .?"

Quṭb al-Dīn answers: "If he means by its body its sensible body, it would, of course, decompose at death; but if he means its imaginal body, it would not, for it is not corrupted by death." Then he turns to Aḥmad's next question: "Is the seat of the imagination another body – a sphere or something else?" and writes:

> He says this because the seat of imagination of the particular for the intermediate soul is the body of the sphere. We have already explained to you that the intermediate are connected to it and not to other imaginal bodies due to their having ascended from the World of Image without reaching the World of Lights because of their imperfection. Necessarily they are connected to one of the bodies of the spheres – to a higher sphere, if they are purer, or else to a lower sphere.[17]

The reason that I was more impressed by these two small texts than by Walbridge's own writing is the profound otherness of Aḥmad's and Quṭb al-Dīn's worldview in comparison to what I was taught in school and at university about the universe, physics, and biology. This may sound all too trivial. But what I mean to say here is that I doubt we can ever hope to understand less lofty texts by scholars of past Islamic societies when we do not take seriously their own general views of the world and try to figure out how they relate to those more mundane matters. It is in this sense that I fully agree with Walbridge's point made in the conclusions to his book about Quṭb al-Dīn. He wrote:

> As I pursued this study, it became clearer to me just how difficult it is to evaluate an Islamic philosopher in isolation and how necessary it is to take seriously his methodology and its expression in his works as a whole. To make any sense at all of Quṭb al-Dīn's works, I found I had to understand the central ideas of Suhrawardī's *The Philosophy of Illumination*. Most modern studies, however, I found to be almost exclusively concerned with mystical aspects. For my purposes, I found I had to work out from the beginning the logical and conceptual structure of the work in the light of the canons of scientific method as expounded in the Islamic

16 Walbridge, *The Science of Mystic Lights*, pp. 216–217, 219, 222.
17 Walbridge, *The Science of Mystic Lights*, p. 224.

tradition of logic. Having done this, I was rewarded by the discovery that the work has a clear and logical structure, that Suhrawardī had something very precise and reasonable in mind when he said that his book was based on mystical intuition, and the principles of *The Philosophy of Illumination* have a clear and specific relation to other Islamic philosophy. This understanding supports the interpretation of Suhrawardī held by later Islamic philosophers against that of Henry Corbin and his school.[18]

2 Shifting places and roles

The second gain we can achieve when contextualizing the sciences of past Islamic societies is a fundamental shift in understanding their places and roles within these societies. This gain already surfaced in the stories told by Abū Ḥayyān and Miskawayh about Abū Sulaymān al-Sijistānī, Abū Ja'far al-Khāzin, and Abū l-Wafā'. It also shines through the quotations I have just given from Aḥmad's letter and Quṭb al-Dīn's reply. Rather than marginalized and banned, philosophy, mathematics, astronomy, and other sciences were integrated parts of the educational landscapes of Islamic societies, of the intellectual controversies waged between their scholars, of the social fights over offices and incomes, and of the cultural rituals of honorific name-giving, burial ceremonies, and formal letter writing to officeholders. In the times of the Ayyubid dynasty, a process was in full speed that secured spaces for the ancient sciences in the *madrasas* and *khānaqāhs*, transformed some of them and certain of the so-called modern sciences into the so-called rational sciences, gave teachers of these rational sciences full rights to office-holdings at *madrasas* and to participation in choosing other officeholders, and led to the donating of official professorships at a number of *madrasas* and mosques mainly for physicians, but also for *muwaqqit*s and occasionally for adepts of a few other disciplines by rulers of various dynasties including the Abbasid caliphs in Baghdad.

This process of social transformation of the ancient and modern sciences into a new body of disciplinary structure and affiliation followed a process of intellectual reform or better several processes of intellectual controversy and rethinking. Historians of theology and philosophy like Wilferd Madelung, Joseph van Ess, Dimitri Gutas, Sabine Schmidtke, or Hossein Zia have edited, translated, and interpreted major theological and philosophical texts written in Arabic and Persian between the tenth and the fourteenth centuries that all have one thing in common – the trespassing of what was previously thought of as rigid boundaries between logic and philosophy on the one hand and *kalām* mostly in its Ash'arī versions on the other hand. Sunni and Shi'i *mutakallimūn* alike have studied, criticized, borrowed from, and acknowledged as rival, but proper knowledge what Ibn Sīnā and his followers had written. Certain important parts of Ibn Sīnā's philosophy were incorporated into major *kalām* texts such as his epistemology, the theory of

18 Walbridge, *The Science of Mystic Lights*, p. 163.

syllogism, his concept of the necessary existent, and the concept of emanation. This does not mean that the *mutakallimūn* took over these parts without changing and adapting them to their own needs. Neither does it mean that philosophy stopped with this kind of integration into *kalām*. Dimitri Gutas claimed in a recent paper that philosophy reached its "golden age" in the thirteenth century. While I find such epithets potentially misleading, biographical dictionaries of scholars under the Ayyubids and Mamluks confirm that logic and certain works of Ibn Sīnā were, indeed, studied at the schools in Cairo, Damascus, Aleppo, and other towns of their states. The innovative character of the madrasa systems in the Arabic Middle East and in Iran that served as the institutional basis for the transformation of the ancient and modern sciences into the rational and traditional disciplines is not exhausted by these processes of transformation. The madrasa and cognate institutes spread learning and teaching to provincial towns, even occasionally to a village and a remote spot. A survey showing where the rational sciences were taught with which canonical texts would not only show that philosophy was taught in fourteenth-century Mecca, but would uncover the discrepancies between regions and times, the connections between various centres and between centres and peripheries, and the ways by which scholarly networks were created that formed the institutional basis for the migration of scholars, their ideas, and their works from Central Asia, Khurasan, and Azerbaijan to Iraq, Syria, and Egypt, and from Azerbaijan to Khurasan and Central Asia, to name only two examples important for the transfer of philosophy as well as planetary theory.

Contextualizing the mathematical sciences yields, however, not only for the centuries after the Buyids, valuable new insights in the place and role of some or all mathematical sciences in a specific Islamic society. Reading well-known texts of the ninth and tenth centuries also can teach us things we did not yet grasp fully. Abū Ḥayyān al-Tawḥīdī, for instance, did not only write the descriptions of nightly sessions with the vizier Ibn Saʿdān. He also wrote an epistle on the fields of knowledge he considered relevant at his day and place. From his view, only astrology and arithmetic need to be mentioned. The two other and higher-ranking ancient sciences he discussed in his treatise are logic and medicine. Astrology shares with medicine its composition of two branches – a theoretical and a practical. But Abū Ḥayyān's description of both of these branches is not exactly what one would expect. The theory is described in a way that gives pre-eminence to theoretical astronomy. Only then typical astrological terms and aspects are enumerated. Practical astrology surprises by its being called very difficult. The difficulties range from *tamāzuj ṣuwar al-kawākib* (the intermingling of the images of the stars) to *ikhtilāf ashkāl al-falak* (the differences regarding the figures of the orbs), followed by *iʿtiyāṣ asrār al-faḍāʾ* (the difficulty [to discover] the secrets of the empty space), and end with *buʿd marām al-qadar wa-l-jabar al-mawjūd fī l-ʿālam* (the great desire for power and force that exists in the world). The poor follower of this practical branch has to work very hard, but achieves very little. His errors surpass his successes and his successes cause more harm than his ignorance. Abū Ḥayyān al-Tawhīdī's remarks on the *ḥāsib* are of an even greater interest to contextualization, since he points out first that the *ḥāsib* was the amanuensis of the astrologer, although he had to do all the work of calculating single-handedly.[19] This situation apparently had changed quite considerably in Abū Ḥayyān's own time because he closes with the remark: "fa-ḥinā'dhin lā yastaḥ saraf al-ʿulamā' li-annahu yakūnu fī darajat al-ṣunnāʿ ka-l-kātib wa-l-masīḥ."[20]

3 Belief and knowledge

A closely related point to the one discussed previously is a modified understanding of the relationship between belief and knowledge. Books and papers about texts written by various Islamic scholars, in particular al-Bayḍāwī, Sayf al-Dīn al-Āmidī, Ṣadr al-Dīn al-Qunawī, al-ʿAllāma al-Ḥillī, and Ibn Abī Jumhūr, as well as thematic monographs such as Frank Griffel's book on the evolution of the related concepts of calumny, apostasy, and heresy from the Qur'ān to *ḥadīth* to al-Ghazālī have made clear that the relationship between belief and knowledge is one major point that lies at the heart of many fierce discussions among Islamic scholars. In some sense, one could say that the relationship between the two categories was inverse to what we are accustomed to since the Enlightenment. Revealed knowledge, hence belief, is certain knowledge, whereas rationally acquired knowledge, while for many Islamic scholars a necessary good to strive for, often was seen as uncertain knowledge and a matter of conviction, hence belief. Revealed knowledge because of its absolute certainty and its direct transmission from God to the possessor of the perfect rational soul, that is a prophet or the perfect man, ranked higher than rationally acquired knowledge, the domain of the scholars. Most mainstream philosophers and *mutakallimūn* subscribed to these two positions.

But things were of course not that simple. The encounters between belief and knowledge did not stop with classifying souls and intelligibles. There were certain domains of rational knowledge that delivered certain knowledge, too, but did not bear upon belief or revealed knowledge. The most famous kind of such knowledge is geometry. Thābit b. Qurra and al-Ghazālī if they ever could have met clearly would have found at least this conviction that geometry was certain knowledge as common ground. They also would have agreed that not all geometry was certain knowledge, but only that following the style of Euclid's *Elements*.[21] On the other hand, al-Ghazālī apparently felt ill at ease with this kind of certainty,

19 *Min Rasā'il Abī Ḥayyān al-Tawḥīdī*, Ikhtiyār wa-dirāsat al-duktūr ʿIzzat al-Sayyid Aḥmad, Manshūrāt wizārat al-thaqāfa, Dimashq, 2001, pp. 235–236.

20 *Min Rasā'il Abī Hayyān al-Tawḥīdi*, p. 236: And at that time, he did not crush the immoderateness of the *ʿulamā'*, because he was in the degree of the artisans like the secretaries and the surveyors.

21 Sabit ibn Korra, *Matematičeskie traktaty*, Naučnoe nasledstvo, tom 8 (Moskva: Izdatel'stvo Nauka, 1984), pp. 54–59; Ulrich Rebstock, *Rechnen im islamischen Orient. Die literarischen Spuren der praktischen Rechenkunst* (Darmstadt: Wissenschaftliche Buchgesellschaft, 1992).

since it contained the potential to distract the believer.[22] Ṭāshköprüzāde made a somewhat different point. He saw geometry at par with philosophy, since both were at distance from the sciences of the world to come and, hence, only those who prefer this life to the next occupy themselves with them.[23] On the other hand, Ṭāshköprüzāde maintained that logic was an indispensable tool for proving the existence of the *wājib al-wujūd*, (i.e., God).[24]

In the ninth century, either al-Ḥajjāj b. Yūsuf b. Maṭar or a commentator of Euclid's *Elements* thought it necessary to add an explanation to the Euclidean definition of a mathematical solid: "al-shakl al-mujassam huwa lladhī lahu ṭūl wa-ʿarḍ wa-sumk; wa-kull mā kānat lahu juththa."[25]

It is obvious that the addition of *juththa* does not make the concept of a mathematical solid any clearer nor would it seem to us appropriate to describe such a notion of the mind as endowed with a corpse and hence capable of living, bleeding, and dying. There are two possibilities that come to mind immediately for explaining the unhelpful addition. One is that the interpolation comes from an earlier set of texts using a different kind of mathematical terminology. The other is that it referred at the time of its incorporation into the *Elements* to another type of discussion about bodies and their properties to which the interpolator wished to point and whose relationship to the mathematical solid he wished to characterize. In our case, both possibilities are available. Ibn al-Muqaffaʿ's introduction into logic written in the early years of the Abbasid dynasty works with a number of highly specific terms uncommon in later texts of this genre, among them *juththa*. It is very clear that *juththa* in Ibn al-Muqaffaʿ's epistle has nothing to do with a corpse. It is rather and surprisingly so one of the five representatives of the subclass of *al-manẓūm* (the ordered) within the category *ʿadad* (number). He defines it together with the line and the plane as follows:

> wa-amma al-khaṭṭ wa-l-basīṭ wa-l-juththa, fa-inna kull shayʾ min al-ashyāʾ yadhraʿu innamā huwa ʿalā aḥad thalāthat wujūh: ammā ṭūl la-ʿarḍ maʿahu, wa-innamā yusammā al-khaṭṭ, wa-ammā ṭūl wa-ʿarḍ, fa-yusammā al-basīṭ, wa-ammā ṭūl wa-ʿarḍ wa-ghilaẓ, fa-yusammā al-juththa.[26]

22 Rebstock, *Rechnen im islamischen Orient*, pp. 20–1.

23 Ms Berlin, Preussischer Kulturbesitz, Staatsbibliothek, Sprenger 1823, 6a,15–6.

24 Ms Berlin, Preussischer Kulturbesitz, Staatsbibliothek, Sprenger 1823, 5b,12–14.

25 Ms Madrid, Escorial, Ar. 907, f 141b,3: "The solid figure is that, which has length and breadth and depth, and it is all that has a corpse."

26 *Al-Manṭiq li-Ibn al-Muqaffaʿ. Ḥudūd al-manṭiq li-Ibn Bihrīz*, M.T. Dānishpazhūh (ed.) (Teheran: Anjumān-i shāhanshāhī wa-falsafa-i Īrān, 1978), p. 12: What concerns the line, the surface and the corpse, they are each of the things that extends, however, it is of three kinds: either it is length without breadth together with it and so it is truly called the line; or it is length and breadth, thus it is called the surface; or its length, breadth, and thickness, thus it is called the corpse.

But the difference between Ibn al-Muqaffaʿ's text and the interpolation in the *Elements* while subtle is nonetheless important. For Ibn al-Muqaffaʿ, *juththa* is a mathematical solid, while for the interpolator of the *Elements* everything that has *juththa* is a mathematical solid. Hence, it was most likely not Ibn al-Muqaffaʿ's introduction to logic that the interpolator wished to allude to with his remark.

The other field of intellectual debate where *juththa* was used as a major concept was religion. Here, scholars of different religious disciplines engaged in fierce battles over the right interpretation of the Divine with regard to His word in His book, His properties, and His relationship to His creation. The so-called anthropomorphists such as the early Hanbalis believed in the literalness of His word and hence in His human-like properties. For them, God indeed spoke, heard, and sat on a material throne and hence had a real body made from flesh and blood. This body they called *juththa*. The Muʿtazilis, in contrast, rejected this literalism in favour of a rational interpretation of God's message. To them, God's speaking, hearing, and sitting on a throne were metaphors for his fundamentally abstract properties such as Knowledge, Power, and Might. When they talked of His body, they preferred the use of *jins*, since it did not allude to flesh and blood. Hence, there can be no doubt that the addition to Definition XI,1 of the *Elements* is situated within this debate about correct belief. But did its author wish to express his support for Aḥmad b. Ḥanbal or rather make a point against what he may have considered a mere philological separation between *juththa* and *jins*? Or did he mean to say that the corpse and the solid were both bodies of this world, while God and His body were of a completely different quality than the two, belonging, as they did, to the other world? Only a serious study of texts on *kalām* and *fiqh* and their terminology for the body may help us to understand the interpolation of the Euclidean text in its proper meaning. The same applies to other apparently innocent terms like *given* or *being*, which occur in different renditions in Arabic versions of the *Elements* and the *Data*. Could it be that the variants were chosen because the translators or editors had a preference either for philosophy or for *kalām* or for none of the two?[27] The difference in the two variants of definition VII,1 edited by De Young indicate very clearly that neither Isḥāq nor Thābit were very much interested in philosophical or religious terminology. All that Isḥāq b. Ḥunayn obviously was interested in was to translate the Greek text(s) he worked with as literally as possible. This is confirmed by my studies of the first four books of the Arabic manuscripts of the *Elements*. The change from *kull wāḥid min al-mawjūdāt*, which renders indeed the Greek *hen hekaston ton onton* as literally as one could wish, into *kull al-mawjūd* then reflects what Thābit was doing when

27 Al-Kindī: *mutʿa*/*al-ays*; al-Ḥajjāj or derivative: *maʿlūm*/*al-shayʾ*; Isḥāq b. Hunayn: *mafrūḍ*/*kull wāḥid min al-mawjūdāt*/Thābit b. Qurra: *mafrūḍ*/*al-mawjūd*. See for the religious significance of *al-shayʾ* and *maʿlūm* and its impact on philosophy, for instance, Robert Wisnowsky, Notes on Avicenna's Concept of Thingness (*ayʾiyya*), *Arabic Science and Philosophy* 10 (2000), 181–221.

he corrected Isḥāq's translation – he turned Isḥāq's Greekish Arabic into a simpler, less cumbersome Arabic.[28]

Another example that indicates the fascinating variability that the relationship between belief and knowledge assumed in different Islamic societies concerns the activities undertaken when a lunar or solar eclipse came up or other events occurred on the heavens. Surprisingly many historical chronicles report with a remarkable calm and in a straightforward manner that such and such an event as an eclipse or an earthquake occurred in this or that year. Only a few among them ever make an explicit link to astrological beliefs and describe such appearances as being an omen for the demise of an unjust ruler or the outbreak of a war or an epidemic in a well-defined geographical region. Given this overwhelming limitation of such statements in Arabic histories to the mere fact that the event happened, one could surmise that astrological interpretations of the kind just mentioned and well-documented, for instance, in a nineteenth-century notebook of a Mandaean priest from southern Iraq were not very fashionable among Muslim elites in the Arabic Middle East from the earliest times on.[29] This is, however, somewhat difficult to believe since astrological geography, for instance, was a kind of thinking about the relationship between the stars and the regions on earth that was widespread among the very same elite groups. What is needed is an analysis of the rules that governed the genre of historical chronicles at a given time and place.

The survey about eclipse reports in translated Arabic and Syriac historical chronicles prepared for astronomical purposes by Said, Stephenson, and Rada uncovers a second fascinating aspect connected with such events.[30] Rather than discussing the astrological implications of an eclipse, Muslims would gather in a mosque for a special prayer, also called the eclipse-prayer. This prayer was led by a particularly chosen, honourable scholar in town. It was directed towards countermanding any possible bad effect of the event. It was considered particularly effective if held in connection with the Friday prayer. The eclipse prayer, according to the sources, could be held shortly before an eclipse occurred or after it had passed. It rarely, if ever, took place during an eclipse. The time and character of the eclipse also contributed to the decision whether the prayer was to be performed. The one question not discussed in the sources used by Said, Stephenson, and Rada nor in additional historical chronicles checked by myself is: if the eclipse prayer took place before the eclipse took place who informed the notables in town of its

28 Gregg De Young, *The Arithmetic Books of Euclid's Elements in the Arabic Tradition: An Edition, Translation, and Commentary*, Ph. D. Dissertation, Harvard University, Cambridge, MA, 1981, Volume I, Part I, p. 1, Volume I, Part II, p. 318; Evclides II, Elementa V-IX. Post I.L. Heiberg Edidit E.S. Stamatis (Leipzig: BSB B.C. Teubner Verlagsgesellschaft, 1970), p. 103.

29 *The Book of the Zodiac (Sfar Malwašia)*, Translated by E.S. Drower (London: The Royal Asiatic Society, 1949).

30 Said F. Said, F. Richard Stephenson, Wafiq Rada, Records of Solar Eclipses in Arabic Chronicles, *Bulletin of the School of Oriental and African Studies* 52.1 (1989), 38–64; F. Richard Stephenson, Said S. Said, Records of Lunar Eclipses in Medieval Arabic Chronicles, *Bulletin of the School of Oriental and African Studies* 60.1 (1997), 1–34.

impending occurrence and who decided on which grounds whether the prayer was necessary or not. Hence, I am unable to say whether *muwaqqit*s were directly, that is, in person, or indirectly, that is, through their tables, instruments, or classes, part of this peculiar religious ritual that may have evolved in Syria and Egypt sometime around the twelfth century (i.e., slightly before the professional timekeeper emerged in the same region). How then should we read the character of the society at that time – as first turning to belief and then to knowledge or as creating domains where the two spheres became finally separated from each other? Or do we need to come up with some other kind of analytic concept for understanding what happened in Ayyubid and Mamluk Egypt and Syria?

4 Neglected fields of relevance

Fields that historians of science in Islamic societies often neglect are the arts, literature, popular knowledge, ritual, and the occult disciplines. Since several decades art historians have discussed whether translations of Greek illuminated manuscripts stood at the beginning of the arts in Islamic societies, in particular for painting and the art of the book. While opinions clash mostly over how much credence should be given to a Byzantine and Greco-Egyptian ancestry of wall-painting and miniature painting in Islamic societies, there can be no doubt that a substantial number of extant early illuminated manuscripts written in Arabic belong to the field of the ancient sciences such as pharmacology, botany, zoology, medicine, and astronomy. Until the thirteenth century, a substantial effort of illustrating manuscripts went into scientific products. More than 50% of all extant illustrated Arabic manuscripts produced before the end of the thirteenth century are scientific manuscripts. Although several of these manuscripts carry no information about the place of their origin, it is clear that the Abbasid, Buyid, Hamdanid, Artuqid, and Fatimid courts patronized the artful illustration of scientific texts. The Artuqid courts of Northern Iraq even seem to have turned the art of the book into a successful business.

During the Ilkhanid period, the influence of Chinese art forms also reached scientific, in particular medical and agricultural, manuscripts. Produced under the patronage of the vizier Rashīd al-Dīn and through his personal cooperation with the Mongol warrior and ambassador of the Yüan court, Bolad Ch'eng-Hsiang, Chinese medical and agricultural treatises were translated into Persian and illustrated accordingly.[31] But Rashīd al-Dīn did not only pay attention to cultural exchange with the new rulers. In the team, which produced under his supervision the *Jāmiʿ al-tavārikh*, worked at least one Italian monk who translated, as part of the *Jāmiʿ*, a Latin history of Europe into Persian. The Italian monk was part of a larger group of Dominican friars who worked as missionaries among the Turkic

31 Thomas T. Allsen, Biography of a Cultural Broker: Bolad Ch'eng-Hsiang in China and Iran, in: *The Court of the Il-Khans 1290–1340*. Edited by Julian Raby and Teresa Fitzherbert (Oxford, New York, Toronto: Oxford University Press, 1994), pp. 7–22.

and Mongol tribes north of the Black Sea and along the shores of the Caspian Sea. They brought home, among other things, the so-called Cuman dictionary preserved today in the Biblioteca Marciana in Venice. Here we have a direct link of knowledge transfer from the Ilkhanid court westwards, which may possibly have included information about, texts on, or drawings of the works of the astronomers at the Ilkhanid court. A closer study of the relationship between the Mongols in Iran and Venice, Genova, or Pisa may, perhaps, yield new insights in the westward move of Ilkhanid *ʿilm al-hayʾa*.

The courtly patronage of the Ilkhanids for the art of the book inspired their successors like the Jalayirids and Muzaffarids to direct funding to this cultural domain too. Scientific manuscripts continued to be among the texts illuminated at court. The major scholarly domains included this new symbiosis between knowledge and art were astrology, magic, pharmacology, and natural history. The first three of these fields had immediate practical meaning. The last one served general educational purposes, entertainment, and edification. As studies by art historians have shown, the texts as well as the illustrations did not follow imitative patterns, but reflected local, specific interests and hence changed in substantial ways. Courtly arts evolved as a major carrier of scientific information and contributed to its spread among and acceptance by diverse groups at court and the urban wealthy. Timurid illuminated manuscripts indicate that locality remained a major point in this respect. While historians of science usually consider Ulugh Beg's patronage for astronomy and mathematics as the only scientific achievement of the Timurid dynasty worth mentioning, Shāhrukh and Iskandar Sulṭān also funded copies of scientific texts. Shāhrukh sponsored a fine copy of Ḥamdallāh Mustawfī's *Nuzhat al-qulūb*. He also made Ḥāfiẓ-i Abrū write a new geographical overview of the inhabited world. During the five years (1409–1414) of his rule in Shiraz, Iskandar Sulṭān spent quite a fortune for including geographical, medical, mathematical, astronomical, astrological, and alchemical treatises into the anthologies of literary, historical, religious, and scientific works produced at his court. His court astrologer was one Maḥmūd al-Kāshī, who compiled his splendid horoscope in 1411. Keshavarz, who edited the horoscope, identified him as Ghiyāth al-Dīn al-Kāshī's grandfather.[32]

Ignoring such activities impoverishes the history of the sciences in Islamic societies and hinders us to understand the variety of forces that contributed to it. The comparison of the activities of diverse Timurid princes and their viziers not only highlights the local character of courtly patronage for the sciences. It also documents that courts could sponsor educational institutions such as *madrasas* and private tutors either included in their funding for science-related activities, as is the case of Ulugh Beg, or could treat these domains as separate areas of sponsorship, as is the case of Shāhrukh. Furthermore, the illuminated scientific texts of

32 Fateme Keshavarz, The Horoscope of Iskandar Sultan, *Journal of the Royal Asiatic Society of Great Britain & Ireland* 2 (1984), 197–208, p. 198.

the Ilkhanid, Jalayirid, Muzaffarid, and Timurid courts prove that novelty was not excluded from this domain. It did not happen only in the arts. It also occurred with regard to previously not illustrated scientific texts, which could be taken up as recipients of images as was the case of Shāhmardān b. a. l-Khayr's *Nuzha-yi nāma-yi ʿĀlāʾī* and Quṭb al-Dīn al-Shīrāzī's *al-Tuḥfat al-Shāhiyya*. Or new texts, mainly in the fields of pharmacology, natural history, astrology, astronomy, geography, cartography, and technology, were written, which were adorned with various kinds of illustrations.[33] Finally, the formal presentation of high-quality manuscripts, whether in religion, literature, or science, began to include precious ornamentation of the cover page, the title, and the margins. As a result, a manuscript of Ibn Sīnā's *al-Qānūn fī l-ṭibb* could now be mistaken for a Qurʾān at first glance. The artistic taste of the wealthy erased the visible difference between religion and science. The symbiosis between art and science turns out to be a delightful indicator for the profound merger between different areas of knowledge and belief in a number of later Islamic societies.

5 Cultural relevance of specific scientific methods and theories

Specific scientific methods and theories assumed an impressive range of cultural meaning within different Islamic societies, as David King and Jamil Ragep have shown for *ʿilm al-mīqāt* and *ʿilm al-hayʾa* or Ulrich Rebstock for *ʿilm al-farāʾiḍ, ḥisāb*, and *misāḥa*.[34] But my impression is that despite all our best efforts, we are still guided by our own perceptions of what science is and what good science should look like. Hence, we have a penchant for the theoretical and the result. We tend to privilege intellectual exchange between different domains of knowledge. We overlook texts on surveying in well-known manuscripts from major collections in Europe, even when such texts seem to come from the early Abbasid period. We rarely study the collection of texts bound into one manuscript as a coherent historical document in order to analyse the interests and orientations of the collector or author. We believe that a study of texts on *ʿilm al-farāʾiḍ* has, if not nothing, then very little to contribute to history of mathematics and the exact sciences. In contrast, I tend to believe that these preferences lead us unto wrong tracks. My impression from reading a few texts on arithmetic written by legal scholars is that these scholars tried very seriously and in interesting ways to come to terms with

33 See, for instance, Nasrollah Pourjavady (editor general), C. Parham (volume editor), *The Splendour of Iran*, 3 vols. (London: Booth-Clibborn Editions, 2001), vol. III: *Islamic Period*.

34 David A. King, *In Synchrony with the Heavens: Studies in Astronomical Timekeeping and Instrumentation in Islamic Civilization*, 2 vols., vol. I: *The Call of the Muezzin. Studies I-IX*, Study VIIc (Leiden: E.J. Brill, 2004); Jamil Ragep, Freeing Astronomy from Philosophy: An Aspect of Islamic Influence on Science, *Osiris* 16 (2001), 49–71; Ulrich Rebstock, *Rechnen im islamischen Orient. Die literarischen Spuren der praktischen Rechenkunst* (Darmstadt: Wissenschaftliche Buchgesellschaft, 1992); Ulrich Rebstock, *The Kitāb al-Kāfī fī mukhtaṣar (al-ḥisāb) al-hindī* of al-Sardafī, *Zeitschrift für Geschichte der Arabisch-islamischen Wissenschaften* 13 (1999–2000), 189–204.

the various arithmetical systems available since the early ninth century. In particular, their efforts to incorporate and to adapt the Indian decimal system seem to me worthwhile of much more solid investigation than available today. Equally, the repeated references to and quotations from theoretical texts of ancient geometry and number theory in texts on *ʿilm al-ḥisāb*, *ʿilm al-misāḥa* and even *ʿilm al-farāʾiḍ* are not only an interesting object for future study per se. One novelty introduced, for instance, by al-Sardafī, a Yemeni legal scholar who died in the same year as al-Ghazālī, was arranging the numerous individual inheritance cases according to arithmetical fractions.[35] Another element we can learn from Rebstock's studies is that al-Sardafī lived and worked east of Janad, a town known to readers of King's and Schmidl's works on Yemeni treatises on the determination of the *qibla*, prayer times, shadow schemes, and other astronomical items. But we could go much further. When analyzing why these scholars, often attached to a *madrasa* or another teaching institute, found it necessary to base their particular methods for naming and handling fractions and integers on a mixture of Nicomachean and Euclidean number theory, we could gain a kind of new access to the question as to which relationship emerged over time and at different places between the ancient heritage and the contemporary needs of an Islamic teaching environment.

6 Larger cultural and social patterns and their changes

Focusing on the results of study and research, whether in forms of manuscripts or material objects, as we often do, limits us to a specific part of past Islamic societies – their intellectual sphere. While I do not wish to deny that it is reasonable to investigate this sphere seriously, I do think that excluding the practical sides of the intellectual sphere as well as everyday life allows us to reach indeed a reasonable understanding of what happened in the intellectual sphere itself. Reading general histories or literature by authors from various Islamic societies of the past shows very quickly that for most of these authors at best two sciences were worth mentioning – medicine and astrology. A few authors have a word or two to say about alchemy, geomancy, arithmetic, geography, astronomy, and geometry. But these statements very often refer back to the intellectual world, to achievements by cultural heroes and sages, or to clashes between scholars. In contrast, certain aspects of technology are very prominent during the first five or so centuries. Niẓām al-Mulk, the second Saljuq vizier, for instance, points in his princely mirror *Siyāsatnāma* to the very practical duties of the ruler in an Islamic society such as creating channels; cleaning major river beds; building bridges, fortresses, and edifices; founding new cities; and caring for the safety and comfort of caravans by establishing hostels alongside the major trading routes.[36] Other

35 Rebstock, *The Kitāb al-Kāfī fī mukhtaṣar (al-ḥisāb) al-hindī* of al-Sardafī, p. 190.

36 Niẓāmulmulk, *Das Buch der Staatskunst Siyasatnama. Gedanken und Geschichten.* Aus dem Persischen übersetzt und eingeleitet von Karl Emil Schabinger, Freiherr von Schowingen. Manesse Verlag, Zürich, s.d., p. 162.

sources too point to such official duties seen as the major aspects constituting the reputation of the just ruler and indeed taken up from the earliest time of Islamic expansion as can be seen in al-Ṭabarī's history. All of these tasks provided outlets for experts in the mathematical sciences and technology such as *muhandisūn*, *munajjimūn*, and *miʿmarūn*. One source known for many years that contains a few short texts on the works of such experts in the region of Ahwaz and southern Iraq is P 169 in the BnF, a collection of short treatises and drawings made in the sixteenth century for a man from Isfahan, possibly a merchant. Until now, it was used only for its anonymous Persian text on ornaments, the map showing the *qibla*, and the incomplete variant of Book II of Euclid's *Elements*.

A second princely mirror, Aḥmad b. ʿUmar Niẓāmī Arużī's *Cahār Maqāla*, written in the middle of the twelfth century, devoted an entire book to astrology adding to it information about which authors should be read in the mathematical sciences. The source of this information is believed to have been al-Bīrūnī's *Tafhīm al-awāʾil fī ṣināʿat al-tanjīm*. But Aružī's list also may indicate that in his time, that is, only a few decades after Niẓām al-Mulk's generous patronage for Shāfiʿī *madrasas* throughout the Saljuq empire, there existed an accepted set of mathematical, astronomical, and astrological texts that was recommended for study not only to the potential professional, but also to the educated public in general. Something, which remains a desideratum of some urgency, is to map the scientific translocation across centuries. I am strongly convinced that a number of surprises and new possibilities will show up once we will be the happy owners of such maps.

4

THE MATHEMATICAL SCIENCES IN SAFAVID IRAN

Questions and perspectives

Introduction

The issue, which I wish to address in this paper, can be formulated in a variety of ways. It results from the lack of interest in this period among historians of science in Islamic societies and their conviction that research should focus on earlier periods since it was then when truly interesting work took place in the mathematical sciences. As a result, very little is known about the activities in the mathematical sciences in Iran between 1500 and 1700. The little that is known has been dismissed as elementary, repetitive and boring. Hence, one way to pose the problem is to ask whether there should be a history of the mediocre and if so how to study and write about it.

Recently a *mutakallim* of the early Safavid period has been found whose contributions to planetary theory George Saliba praised as outstanding – Šams al-Dīn Ḫafrī (d. ca. 957/1550), one of the favored scholars of the first Safavid Šāh.[1] David A. King appreciated the Mecca-centered world map on three compound astrolabes from the late Safavid period as of great value since it was made with a projection that keeps angles and distances invariant. King argued that Ḥabaš al-Ḥāsib (third/ninth century) might have invented this projection.[2] Hogendijk suggested on the basis of newly found manuscript evidence from the tenth century that the arcs seen on the instruments might in fact not be arcs of circles as standard on medieval instruments, but arcs of ellipses.[3] These discoveries challenge our perceptions of and attitudes towards the mathematical sciences in Safavid Iran. Hence, another way to phrase the issue I am going to address in this paper is to enquire which mathematical ideas, theories or methods created by scholars in earlier periods were taken up and perhaps even rediscovered by writers and instrument makers of the Safavid Empire.

1 Saliba 2004, 55–66.
2 King 1999, 345–358.
3 King 2004, 842–845.

 DOI: 10.4324/9781003372493-5

The simplest form of my question is to ask who worked where in the Safavid Empire with which mathematical sources. A more difficult question is to investigate on which problems these students of mathematical literature worked. Its higher difficulty results from the fact that most extant mathematical texts seem to be teaching texts and hence allow no direct access to the problems that were of interest. The most complicated version of this question is to ask for the cultural contexts of the mathematical sciences between 1500 and 1700. This question can be made more accessible by dividing it into sub-questions, such as in which institutions did the authors of mathematical texts work in this period, who gave the means for their work and which place did their mathematical work occupy in their entire scholarly biographies. Questions resulting from the particular intellectual configurations under the Safavids are whether the turn to philosophers of the classical period and their works had an impact on the study of the mathematical sciences and whether changes and conflicts in the religious sphere influenced these sciences.

Since there is little to no systematic research available, I will begin with a discussion of why we know so little about the mathematical sciences in the Safavid period, (i.e., I will briefly summarize the main methodological approaches applied today in history of science at large and the issues linked to the approach that is preferred in history of science in Islamic societies). The main body of this paper presents the preliminary results of my search for mathematical texts written in the Safavid period and addresses the question how to evaluate these texts. I focus on texts since I do not work about scientific instruments such as astrolabes.

The prevalent approach to the history of the mathematical sciences in Islamic societies operates with a vertical perspective. It looks at history as a sequence of events embodying either progress or decline across time. It investigates and writes this history as a process that aims to find objective truth about the past, which is independent of our interests, theories, methods and values. It focuses on intellectual achievements and their inner-scientific relevance. It concentrates its attention on works by major scholars, a qualifier determined primarily in relationship to ancient Greek contributions to the mathematical sciences and occasionally in relationship to results achieved by scholars in early modern and modern Catholic and Protestant societies in Europe. As a consequence, it downplays or overlooks the thousands of manuscripts of an elementary nature, considers many contextual issues as either superfluous or beyond the boundaries of the discipline and ignores many aspects of representation.

In the last forty years, the vertical approach to history of science in Islamic societies has been concerned primarily with restoring the sciences – in a modern understanding of the word – to history and public knowledge, with chasing evidence for innovation in these sciences and with pushing the time limit for progress and the beginning of decline through the centuries closer towards the present.[4] Two opposite trends characterize the works done in this period, one that argues

4 For a substantial, but not exhaustive bibliography see: http://facstaff.uindy.edu/~oaks/Biblio/IslamicMathBiblio.htm.

for understanding the sciences in Islamic societies as universally acceptable and accepted bodies of culturally unspecific knowledge; the other emphasizing the impact of scientific problems derived from and to be solved for religiously prescribed behaviour. The success of scholars working within this vertical approach in regard to locating new scientific theories, methods and results in manuscripts and instruments produced in Islamic societies, above all between the third/ninth and the seventh/thirteenth centuries, but in specific areas of knowledge also long after the thirteenth century, is remarkable. The body of knowledge about the sciences in past Islamic societies and the achievements of their practitioners has considerably deepened and broadened. Three issues inherited from early twentieth-century historiography, however, continue to plague the field to which Muslim scholars added a fourth issue since the 1960s:

- the belief in a Golden Age or a renaissance;
- the acceptance that the sciences in Islamic societies should be situated primarily in relationship to ancient Greek sciences and medieval Latin sciences;
- the identification of the sciences in terms of language, ethnicity or modern nationhood;
- the identification of the sciences in terms of religion.

As a result, very little work has been done on the mathematical sciences in Islamic societies in Iran after the Mongols with the exception of Uluġ Beg's activities in Samarkand. A few texts such as Bahā' al-Dīn 'Āmilī's (954–1030/1547–1622) *Ḫulāṣat al-ḥisāb* were edited and commented on. But we know next to nothing about the why, how and what of the mathematical practices under the various post-Ilkhanid dynasties that ruled in various parts of Iran; the relationship of those who studied, taught and used mathematical knowledge to instrument makers, scholars of other disciplines, administrators, artists and the military elites; or the exchange of texts, instruments and knowledge with the neighbours of the Safavids and visitors to the country.

A different approach to history of science characterizes today the study of the sciences in Western societies that operates with a horizontal perspective. It looks at history as contingent, discontinuous and local. It understands investigating and writing history as socially constructed and resulting in competing narratives told from different perspectives. This approach is prevalent in the study of the natural and the life sciences, but has gained entrance into the study of the mathematical sciences too. It focuses on practices of knowing, doing and speaking in the various scientific fields and thus studies science in and as culture. Its practitioners are fascinated with many contextual aspects, with power relations, language and diverse practitioners of scientific knowledge independent of their visibility, status and locus in regard to institutionalized science.

Applying such a horizontal approach to the study of the mathematical sciences in Safavid Iran or any other Islamic society can considerably enrich the knowledge gained from applying the vertical approach. Such an approach necessitates surveying, identifying and analyzing the scientific activities and products in a specific period and locality. It asks for contextualizing these activities and products (i.e., to understand them in the terms of their specific time and locality). It allows discovering and appreciating changes in visual and discursive representation, function, usage, ownership and reputation of the mathematical sciences at large or some of its disciplines in addition to the focus on content privileged by the vertical approach.

Part I

1 How to situate the mathematical sciences of the Safavid period?

From the perspective of a horizontal approach, there are two basic possibilities for situating the mathematical sciences of the Safavid period, which are not mutually exclusive, but should be applied concomitantly. One approach compares the mathematical sciences in the Safavid period with those of the dynasty's immediate predecessors in Iran and adjacent territories, in particular the Timurids and Aq Koyunlus, one of which the Safavids emulated and with the other of which they were affiliated by marriage. The other approach compares the mathematical sciences under the Safavids with those under their contemporary neighbours, rivals and allies, in particular the Ottomans, the Mughals and the Quṭb Šāh. The purpose of this survey is not to do justice to the activities in the mathematical sciences under these predecessors and contemporaries of the Safavid dynasty. Such a goal is not only beyond the scope of a paper, but it is also impossible to achieve because of lack of reliable research results for all of them except the Ottomans and one Timurid prince, that is, Uluġ Beg (r. 851–853/1447–1449). All that I am setting out to do here is to outline some of the trends that I see in regard to these sciences based on the partial results available in history of science, art history and manuscript catalogues. Given the current research situation, I will focus primarily on the Timurids and the Ottomans. In addition, I will sketch what I know about the Aq Koyunlu and Mughals. I will have to leave aside the mathematical sciences under the Quṭb Šāh of Golconda since I am not aware of any research done on them, except for a few brief remarks based on information I found while searching for data about the Safavid period.[5] Responding to beliefs held among historians about the exclusion of the non-religious sciences from the endowed teaching institutes and the disappearance of courtly patronage for these sciences after either the Ilkhanids or Uluġ Beg, I will outline trends for two main social loci of the mathematical sciences – the courts and the endowed teaching institutes.

5 S. R. Sarma, an expert in in medieval and early modern astronomy and astrology in India, confirmed that there is no research done on the mathematical sciences in the reign of the Quṭb Šāh. I thank S. R. Sarma for his help concerning the Mughal court. G. de Young's paper on the impact of Bahā' al-Dīn 'Āmilī's *Ḫulāṣat al-ḥisāb* focuses primarily on the eighteenth century, de Young 1986, 1–15.

2 *The mathematical sciences under the Timurids*

The Timurids had strong interests in astrology and its theoretical basis defined as *'ilm al-hay'a*, that is, planetary theory. These two interests brought with them attention for geometry, trigonometry and arithmetic as the tools needed for calculating horoscopes and developing planetary models. Numerous Timurid rulers and princes, beginning with Timur, patronized scholars at their courts who cast horoscopes for male and female members of the family, determined the beginnings of military campaigns and battles or found the auspicious day for a marriage and other festivities. Horoscopes of several members of the Timurid dynasty have survived, among them those for Iskandar Sulṭān (r. 811–818/1409–1414), his half-brother Rustam ibn 'Umar Šayḫ (r. 818–828 or 829/1414–1423 or 25) and Uluġ Beg's son 'Abd al-'Azīz (d. 854/1449). All these activities included a solid amount of theoretical knowledge of planetary theory and practical skills in geometrical constructions, calculations and the usage of tables and instruments.[6] Scholars skilled in astrology, astronomy, geometry and arithmetic wrote introductions or surveys on planetary theory, astronomical instruments and astrology for individual Timurid princes and members of their courts. Ġiyāṯ al-Dīn Kāšī (d. 833/1429) composed between ca. 814/1411 and 818/1415 for Iskandar Sulṭān at least two such introductory texts on *'ilm al-hay'a*, a text on instruments for astronomical observations and possibly a brief extract of Euclid's *Elements*.[7] Qāżīzāda Rūmī (d. 815/1412) wrote in 815/1412 a commentary on the introduction into mathematical cosmography, *al-Mulaḫḫaṣ fī 'ilm al-hay'a*, by Maḥmūd al-Čaġmīnī (fl. ca. 620/1223), for Uluġ Beg. Rukn al-Dīn ibn Šaraf al-Dīn Ḥusaynī Āmulī (ninth/fifteenth century) wrote in 860/1456 a treatise on the astrolabe in fifty chapters and dedicated it to Abū l-Qāsim Bābur Ḫān Bahādur (r. 851–862/1447–1457). A commentary on Naṣīr al-Dīn Ṭūsī's *Bīst bāb dar ma'rifat al-asṭurlāb* was dedicated to the Timurid vizier 'Alīšīr Navā'ī (845–907/1441–1501). In addition to writing introductions and elementary surveys for the purpose of education, scholars at Timurid courts in Shiraz, Samarkand and Herat composed astronomical handbooks that served as database for ephemerides, horoscopes etc. Ġiyāṯ al-Dīn Kāšī wrote at Iskandar Sulṭān's court in Shiraz parts of his *Zīj-i Ḫāqānī* with the aim to improve the Ilḫānī Zīj. Uluġ Beg's cooperation with Qāżīzāda Rūmī, Ġiyāṯ al-Dīn Kāšī, Mu'īn al-Dīn Kāšī (ninth/fifteenth century) and 'Alī Qušjī (d. 879/1474) for producing a new astronomical handbook based on new observations and its outcome, the *Zīj-i jadīd-i sulṭānī*, are well known. 'Alī Qušjī

6 See for instance van Dalen 2023, "Commentary on the Horoscope of Iskandar Sultan: An Introduction to Islamic Mathematical Astronomy," *ZGAIW*.

7 Kāšī's *Muḫtaṣar fī 'ilm al-hay'a* and the extract of the *Elements* are parts of the anthology MS London, British Library, Add. 27261, ff. 340b–342b and 343a–344a. His *Lubāb-i Iskandarī* is extant in MS Qom, Kitābḫāna-yi 'umūmī-yi Ḥażrat-i Āyatollāh al-'uẓmā Mar'ašī Najafī, 1015, ff. 272b-278b. His *Ālāt-i raṣad*, written in 818/1415 on order of Iskandar Sulṭān, is extant in MS Hyderabad, Andhra Pradesh Government oriental Manuscript Library and Research Institute, Riyāḍī 129.

wrote one of the most important commentaries on it.[8] A set of simplified planetary tables derived from this *Zīj* was prepared by ʻImād al-Dīn ibn Jamāl al-Dīn Buḫārī (ninth/fifteenth century) who started his work possibly on order of Uluġ Beg.[9] He apparently finished it only after Uluġ Beg's death, since a copy preserved in Qom is dedicated to Abū Saʻīd Gürägān (r. in Samarqand 855–863/1451–1459 and Herat 864–874/1459–1469).[10] Two other Zījes were compiled in Shiraz and Herat by Rukn al-Dīn Āmulī who dedicated one of them to Abū Saʻīd Gürägān.[11] Only one *taqwīm* is known that was dedicated to a Timurid prince and contained his horoscope, the *Zā'iča* compiled by Muʻīn al-Dīn Kāšī's son ʻAbd al-Razzāq (ninth/fifteenth century) for Uluġ Beg's younger son Jalāl al-Dawla ʻAbd al-ʻAzīz.[12] That the interests of these scholars and Uluġ Beg did not stop here, but included planetary theory beyond the elementary level, has been discovered in the last years.[13] In Mashhad is a unique copy of a work on planetary theory by Ġiyāṯ al-Dīn Kāšī, which so far nobody has analyzed.[14] ʻAlī Qušjī wrote at least two, if not three treatises that tackle issues of planetary theory. He refers explicitly to Quṭb al-Dīn Šīrāzī's *al-Tuḥfa al-Šāhiyya* in *al-Risāla fī aṣl al-ḫārij yumkinu fī al-sufliyyayn.*[15] It is possible that the commentary on this work by Šīrāzī that is attributed in some copies to ʻAlī Qušjī and in other copies to an unidentified name was indeed written by Qušjī.[16] Qušjī dedicated one of these texts dealing with Mercury to Uluġ Beg (i.e., the *Risāla fī ḥall iškāl muʻaddil li-l-masīr).*[17] Iskandar Sulṭān, ʻAbd al-Laṭīf ibn Uluġ Beg and ʻAlīšīr Navā'ī may also have shared this kind of theoretical interest, since all of them received formal education in astronomy. ʻAlīšīr Navā'ī's teacher, for instance, was Fasīḥ al-Dīn Muḥammad ibn ʻAbd al-Karīm Niẓāmī Kūhistānī (d. 937/1530) who wrote a commentary on Čaġmīnī's introduction into astronomy, three treatises on instruments, a supercommentary on Qāżīzāda Rūmī's commentary on Šams al-Dīn Samarqandī's *Aškāl al-ta'sīs* and a book on astrology.[18] Furthermore, Ṭūsī's *Takmila* and *Zubda* and Quṭb al-Dīn Šīrāzī's *Tuḥfa* were available and copied in Shiraz, Herat and Samarkand during the ninth/fifteenth century.[19]

In addition to astronomical and astrological works, a few geometrical and arithmetical texts were also part of Timurid courtly patronage of the mathematical

8 I rely in this evaluation on Benno van Dalen's as of yet unpublished new survey of Arabic, Persian and other *Zīj* works.
9 I thank Benno van Dalen for this information.
10 Marʻašī – Ḥusaynī 1373, vol. 23, 169, no 9014.
11 I thank Benno van Dalen for this information.
12 Marʻašī – Ḥusaynī 1374, vol. 24, 178–179, no 9402.
13 See, for instance, Saliba 1993 and Ragep 2005.
14 Matvievskaya – Rozenfel'd 1985, 486.
15 *Osmanlı Astronomi Literatürü Tarihi* 1997, 35.
16 *Osmanlı Matematik Literatürü Tarihi* 1999, 36; King 1986, 153, dH7, G88.
17 *Osmanlı Astronomi Literatürü Tarihi* 1997, 35; Saliba 1993.
18 Matvievskaya – Rozenfel'd 1985, 533–534.
19 Dāniš-Pažūh – Anvarī 1976, 365, nr 578.

sciences. One of Iskandar Sulṭān's anthologies contains a very brief extract from Euclid's *Elements* taken from Muḥammad ibn Maḥmūd Āmulī's (d. ca. 753/1352) encyclopedia *Nafā'is al-funūn fī 'arā'is al-'uyūn* dedicated to the last Inju ruler Abū Isḥāq (d. 754/1353). Ġiyāṯ al-Dīn Kāšī dedicated his major mathematical work, the *Miftāḥ al-ḥisāb*, in 830/1427 to Uluġ Beg. He also dedicated his commentary on Šams al-Dīn Samarqandī's *Aškāl al-ta'sīs*, a revision of Books I and II of Euclid's *Elements*, to this ruler.[20] But in general, although works on geometry, arithmetic, number theory or algebra served as tools in astronomy and astrology, they did not occupy a major position in courtly cultural politics.

In contrast, texts on astronomy and astrology together with instruments were a visible part of this politics. They were integrated into courtly patronage of the arts and princely gift giving. Expression of this integration of astronomy and astrology in Timurid cultural politics was also the copying of several astronomical and astrological texts of scholars from earlier periods, in particular those who had worked at the Ilkhanid courts, for Timurid princes because this served to buttress Timurid claims of their descent from the Ilkhanids and the continuation of their politics. The *Zīj-i Īlḫānī* by Naṣīr al-Dīn Ṭūsī (597–674/1201–1274) and his collaborators were, for instance, copied on order of Šāhruḫ (r. 808–851/1405–1447) and Iskandar Sulṭān. Several Timurid copies of 'Abd al-Raḥmān Ṣūfī's (290–376/903–986) *Kitāb ṣuwar al-kawākib al-ṯābita* are extant, one of them produced for Uluġ Beg's library allegedly based on a copy made by or for Naṣīr al-Dīn Ṭūsī.[21] Uluġ Beg's art workshop also produced fine copies of his *Zīj*.[22] Abū Sa'īd Gürägān followed this example and sent a beautiful version of the *Zīj-i jadīd-i sulṭānī* as a gift to Abū l-Qāsim Bābur.[23] Beside the instruments built for the observatory in Samarkand an astrolabe and brass rings were commissioned by or dedicated to Uluġ Beg.[24]

In contrast to this comparatively rich evidence for Timurid courtly patronage of the mathematical sciences, there is little information known so far that illustrates the study of these disciplines and their works at madrasas and other endowed teaching institutes in the Timurid realm. One exception is the well-known case of the madrasa sponsored by Uluġ Beg in Samarkand where Qāżīzāda Rūmī and Ġiyāṯ al-Dīn Kāšī taught and 'Alī Qušjī first studied and then taught. In ninth/fifteenth-century Herat several astronomical texts, among them Naṣīr al-Dīn Ṭūsī's *Zubda fī l-hay'a* and *Bīst bāb dar asṭurlāb* and Quṭb al-Dīn Šīrāzī's *al-Tuḥfa al-Šāhiyya*, were copied.[25] In two cases, the copyist was affiliated with a madrasa, the Jalāliyya Madrasa and the Madrasa-yi marḥūm-i Sulṭān Āqā.[26] In ninth/fifteenth-century Yazd, the more practical astronomical texts of Ṭūsī like the *Bīst*

20 *Osmanlı Matematik Literatürü Tarihi* 1999, 7.
21 Richard 1997, 78.
22 Lentz – Lowry 1989, 374.
23 Lentz – Lowry 1989, 367.
24 Lentz – Lowry 1989, 374; Charette 2006.
25 Mar'ašī – Ḥusaynī 1372, vol. 21, 33–34, nr 8021; Dāniš-Pažūh – Anvarī 1976, 365, no 578.
26 'Abbāsi 2001, pp. 151, 158; Dāniš-Pažūh – Anvarī 1976, 365, no 578.

bāb and the *Sī faṣl* were copied in the Ḥāfiẓiyya Madrasa together with an anonymous *Zawā'id al-taqwīm*.[27]

3 The mathematical sciences under the Aq Koyunlu

The Aq Koyunlu appear to have had little interest in the mathematical sciences, although this claim needs to be taken with strong caution since next to no research has been done so far on this subject. Despite their strong interest in the arts and their non-negligible support for philosophical works, no copies of astronomical, astrological and mathematical texts produced in a courtly art workshop seem to be extant. Only one commenting edition of Naṣīr al-Dīn Ṭūsī's *Taḥrīr* of Euclid's *Elements* is known that was made for an Aq Koyunlu ruler, the work by Ilḥāq ibn Abī Isḥāq dedicated in 886–7/1481–2 to Sulṭān Ya'qūb Bahādur Ḫān (r. 884–896/1479–1490).[28] Twelve years before Ibn Abī Isḥāq wrote his notes on the *Elements*, 'Alī Qušjī spent time at Ūzūn Ḥasan's (r. 857–883/1453–1478) court as an honoured guest scholar. He brought with him a copy of the *Zīj-i jadīd-i sulṭānī*, which he then took with him when he went from Tabriz to the Ottoman court of Mehmed Fātih (r. 848–850/1444–1446, 855–886/1451–1481). This transport of the mathematical sciences from the Timurid to the Ottoman realm and the travel of the *Zīj* produced in Samarkand may explain an otherwise impossible claim in a copy of this work preserved today in Cairo according to which Uluġ Beg sent a copy of this *Zīj* to the Ottoman Sultan Bayezid II (r. 886–918/1481–1512).[29]

There is almost nothing known about the presence of the mathematical sciences at madrasas and other endowed teaching institutes in the sphere of power of the Aq Koyunlus. The vivacious intellectual climate in fourteenth-century Tabriz, for instance, included the study of texts from all the mathematical disciplines as shown by the material comprised in the *Safīne-yi Tabrīzī*. It is likely that this was also the case in the following century, but I am not aware of manuscripts that would support this assumption.

4 The mathematical sciences in the Ottoman Empire

Our knowledge on the mathematical sciences under the Ottomans is much richer than that for the two previous dynasties. We owe this better knowledge to the work mainly of Turkish colleagues and David A. King on whose books I rely in the following reflections. The mathematical sciences in the Ottoman Empire in the period I am discussing here (i.e., the second half of the fifteenth century to the early eighteenth century) were shaped to a considerable degree by four traditions from outside the Empire. The mathematical sciences as exercised in the Mamluk

27 Mar'ašī – Ḥusaynī 1372, vol. 21, 33–34, no 8021

28 MS London, British Library, or. 1514. There are two other copies of this text extant in Istanbul in the Archaeological Museum and the Süleymaniye Library, which I have not seen though.

29 King 1986, 119, no D139.

realm constituted one of these traditions. A second major influence came from Timurid scholars. The third component was formed by scholars from Iran who migrated in the early years of the emergence of the Safavid dynasty. The fourth influence stems from a new group of migrants in the seventeenth century who introduced works of Safavid scholars to Ottoman circles.

A major heritage of the Mamluk mathematical sciences was the new profession of the *muwaqqit*. A person called *muwaqqit* held a position at a mosque, a madrasa or a Sufi *ḫānaqāh* and was responsible as an expert in spherical astronomy, trigonometry and the design of astronomical and mathematical instruments for determining prayer times, prayer directions and the new month. The term evolved in the late seventh/thirteenth and early eighth/fourteenth century in Mamluk Egypt and Syria. In the second half of the ninth/fifteenth century, the first *muwaqqit*s appeared at the Ottoman court, apparently without being attached to a mosque or another religious institute. After the conquest of the Mamluk realm, *muwaqqit*s can be found at many Ottoman mosques and madrasas, in particular in Istanbul, Edirne, Bursa, Mecca and in continuation of Mamluk practice in Syria and Egypt. *'Ilm al-mīqāt* became a regular part of teaching in the Ottoman Empire and emerged as one of the disciplines in which treatises were increasingly written in Turkish.

The Timurid component came above all with 'Alī Qušjī to the Ottomans. After he had arrived in Istanbul and received a professorship for teaching the mathematical sciences in one of the eight madrasas founded by Mehmed Fātih, he translated two of his works written in Samarkand and dedicated to Uluġ Beg, his introduction into arithmetic and geometry and his elementary survey on mathematical cosmography, from Persian into Arabic for Mehmed Fātih.[30] In addition to this straightforward rewriting of formerly successful texts for the new environment, 'Alī Qušjī also took up courtly discussions about specific mathematical points. His *Risāla fī l-zāwiya al-ḥādda* reflects one kind of interest in mathematical knowledge that reigned supreme in the courtly sphere – entertainment provided by problems that sounded strange, surprising or inaccessible to common sense. The problem that 'Alī Qušjī had discussed in a courtly session with Mehmed Fātih was "the strange problem . . . that when one side of an acute angle was moved towards the side of obtuseness an obtuse angle resulted" without an intermediary, that is, right, angle coming between the two.[31] It relates to theorem III,15 in Euclid's *Elements* (i.e., the angle formed between an arc of a circle and a tangent). It was an issue that had drawn the attention of the *mutakallimūn* since the ninth century who used it in their debates about atomism and continuity. When Mehmed Fātih had heard Qušjī's explanations he was amazed and demanded him to write it up for all the scholars and virtuous people in his realm.[32] Sinān Paşa (d.

30 *Osmanlı Astronomi Literatürü Tarihi* 1997, pp. 21, 25; *Osmanlı Matematik Literatürü Tarihi* 1999, 29, 33.

31 *Osmanlı Matematik Literatürü Tarihi* 1999, 26–27.

32 *Osmanlı Matematik Literatürü Tarihi* 1999, 26.

891/1486) who had studied astronomy and other mathematical sciences possibly in Sivrihisar, Bursa and Istanbul participated in this discussion between the sultan and the scholar and wrote a commentary on Qušjī's treatise.[33] Almost one hundred years later, Muṣṭafā ibn Maḥmūd al-Tosyavī (d. 1004/1596) was puzzled by comments on this problem by Sa'd al-Dīn al-Taftazānī (722–793/1322–1390) in his commentary on Najm al-Dīn al-Kātibī's (seventh/thirteenth century) *Šamsiyya* on logic. He discussed it with students in Bursa and found access to it after reading Abū l-Rayḥān Bīrūnī's (364-after 444/973-after 1053) *Kitāb al-tafhīm* on astrology.[34] This implies a limited understanding of elementary mathematics and a lack of familiarity with basic source works on geometry such as Euclid's *Elements*.

Translations of Persian texts into Arabic and of Persian or Arabic texts into Ottoman Turkish formed an important field of Ottoman courtly patronage for the mathematical sciences. One of the first mathematical texts in Ottoman Turkish was dedicated to Bayezid II by Muḥammad ibn Hacı Atmaca al-Kātib (fl. 899/1494). Called *Majma' al-Qavā'id*, it addressed primarily administrators and other practitioners giving a survey on Indian arithmetic and administrative practices of calculation and *siyāq* based on Arabic and Persian texts.[35] The same sultan was also the addressee of Persian and Arabic texts on arithmetic and planetary theory, of ephemerides in Persian, commentaries on Uluġ Beg's *Zīj* and of texts on *'ilm al-mīqāt* and instruments.[36] The texts on planetary theory ranged from commentaries on the introductory *al-Mulaḫḫaṣ fī l-hay'a* by Čaġmīnī to a commentary on the advanced models discussed by Ṣadr al-Šarī'a (d. 748/1347) in his *Ta'dīl al-'ulūm*.[37] Two presenters of texts on instruments and *'ilm al-mīqāt* were *muwaqqit*s at the court of Bayezid II (*al-muwaqqit bi-bāb al-Sulṭān*) and his successors Murad and Süleyman – Muḥammad ibn Yūsuf (fl. 916/1510), possibly the first known Ottoman *muwaqqit*, and Muḥammad ibn Kātib Sinān (d. 930/1523–4).[38]

Similar types of texts were dedicated to Ottoman sultans in the sixteenth and seventeenth centuries, although – given the dedications mentioned in the surveys of Ottoman astronomical and mathematical literature on which most of my claims are based – the number of dedicated texts decreased significantly in these two centuries.[39] This steady decrease – if this is indeed what happened – may reflect the moving away of the *muwaqqit* from the courts into the mosques and madrasas in the course of the sixteenth century and the extension of the net of endowed teaching institutes. In both cases, regular salaries may have decreased the need for the scholars to ask or pay for courtly patronage with dedicated books. Such

33 *Osmanlı Matematik Literatürü Tarihi* 1999, 104.

34 *Osmanlı Matematik Literatürü Tarihi* 1999, 27–28.

35 *Osmanlı Matematik Literatürü Tarihi* 1999, 29–30.

36 *Osmanlı Matematik Literatürü Tarihi* 1999, 35, 47; *Osmanlı Astronomi Literatürü Tarihi*1997, 40, 42, 47, 65, 91.

37 *Osmanlı Astronomi Literatürü Tarihi*1997, 95.

38 *Osmanlı Astronomi Literatürü Tarihi* 1997, 70–71, 85–87, 94, 95–97, 98–100.

39 *Osmanlı Matematik Literatürü Tarihi* 1999, 55, 96–97, 99, 153–154; *Osmanlı Astronomi Literatürü Tarihi*, 85, 87, 89, 97, 110, 195, 212, 214, 241, 251–252, 282, 292.

an interpretation may find support in the fact that the few listed works of the seventeenth century that were dedicated to a sultan were ephemerides produced by the *müneccimbašı* who lived in the palace. The *müneccimbašı* was the head astrologer, an office installed at court in the early Ottoman period. He was responsible for horoscopes, the production of *taqwīm*s, the determination of fortunate and unfortunate days and the supervision of the lower-ranking astrologers and *muwaqqit*s. Horoscopes of several sultans are extant today, among them that of Murad II (r. 824–855/1421–1451) in Paris.[40] Occasionally, *muwaqqit*s such as Muṣṭafā ibn ʻAlī (d. 945/1538) also wrote and dedicated works on mathematical geography to a sultan.[41] In addition to the sultans, Ottoman princes such as Ahmad and Korkut ibn Bayezid II; viziers and other courtiers such as Ayas Paša (d. 946/1539), Sokullu Mehmed Paša (911–987/1505–1579), Köprülü Fazıl Ahmed Paša (1045–1087/1635–1676) and Kara Mustafa Paša (d. 1095/1683); and religious dignitaries such as the Šeyḫülislam or the Sadr also received dedicated astronomical or mathematical works.[42] In the late ninth/fifteenth and early tenth/sixteenth centuries, the court was not the only social locus for the mathematical sciences, including discussions about planetary theory, in the Ottoman Empire. Scholars working at madrasas also studied planetary models. The main basis for this interest was formed by Ṭūsī's and Šīrāzī's texts. The intermediaries through which, for instance, the knowledge of the *Taḏkira* and Šīrāzī's *Nihāyat al-idrāk* was channelled were the commentaries by al-Sayyid al-Šarīf Jurjānī (741–816/1340–1413) and Niẓām al-Dīn Nīšābūrī (thirteenth/fourteenth century) and texts by Ġiyāṯ al-Dīn Kāšī and ʻAlī Qušjī.[43] Since Niẓām al-Dīn Birjandī (d. 935/1528) was in 911/1505 in Trebizond where he wrote a treatise on the city's *qibla* for Selim I, he probably composed his commentary on Ṭūsī's *Taḏkira* written two years later also in the Ottoman Empire.[44] It was a further text on planetary theory that was studied repeatedly by later Ottoman scholars.

Other texts studied, used and, in some cases, translated into Turkish or Arabic include Qusṭā ibn Lūqā's treatise on the astrolabe; texts by *muwaqqit*s from Mamluk Syria and Egypt such as Šams al-Dīn al-Ḫalīlī (720?–782?/1320?–1380?), Ibn al-Majdī (767–850/1365–1447), Ibn al-Šāṭir (694–777/1304–1375) and Sibṭ al-Māridānī (d. 912/1506); texts on arithmetic and algebra by Ibn al-Hā'im (d. 815/1412) from Mamluk Jerusalem; Uluġ Beg's Zīj; Ġiyāṯ al-Dīn Kāšī's treatise on sin 1°, his *Miftāḥ al-ḥisāb* and one of his texts on *'ilm al-hay'a* written for Iskandar Sulṭān; Naṣīr al-Dīn Ṭūsī's *Sī faṣl* and *Risāla fī manāzil al-qamar*; Nẓām

40 MS Paris, BnF, Supplément persan 367.

41 *Osmanlı Astronomi Literatürü Tarihi* 1997, 162.

42 *Osmanlı Matematik Literatürü Tarihi* 1999, 74, 100, 155; *Osmanlı Astronomi Literatürü Tarihi*1997, 70, 73, 161, 266, 269, 292, 336, 352.

43 *Osmanlı Astronomi Literatürü Tarihi* 1997, 44–45, 48.

44 *Osmanlı Astronomi Literatürü Tarihi* 1997, 110. According to Ziriklī, he was in Istanbul in 1525, but Ziriklī gives no source for this claim, Ziriklī 1979, 30. I thank Lutfullah Gari, Yanbu', for pointing this out to me. Other authors believe he lived in Herat or at the court of Ismā'īl. See for instance Matvievskaya – Rozenfel'd 1985, 541.

al-Dīn Nīšābūrī's *Šamsiyya* on arithmetic; and 'Alī Qušjī's Muḥammadiyya on arithmetic and geometry.[45] Classical texts like Euclid's *Elements* and the *Kutub al-mutawassiṭāt* are conspicuously rare or absent. Over time, the number of texts written on *'ilm al-mīqāt* topics and treatises based on Uluġ Beg's *Zīj* grew considerably, while works on planetary theory apparently became less relevant except for Čaġmīnī's introductory text and its commentaries, in particular that by Qāżīzāda Rūmī.

In the early eleventh/seventeenth century, migrants from Safavid Iran such as the physician Muḥammad Bāqir ibn 'Imād al-Dīn Maḥmūd (eleventh/seventeenth century) brought copies of Bahā' al-Dīn 'Āmilī's treatises to the Ottoman Empire and wrote commentaries on them.[46] The texts that became the most prominent were *Ḫulāṣat al-ḥisāb* and *Tašrīḥ al-aflāk*.[47] Among the newly written texts by authors who worked in the Ottoman Empire, the well-known work of Taqī al-Dīn (d. 993/1585) is outstanding, but other texts like the *Miṣbāḥ al-ṭālib wa-munīr al-muḥibb al-kāsib* of the ophtalmologist Mūsā ibn Ibrāhīm al-Yaldāvī (d. ca. 926/1520) also deserve attention. Al-Yaldāvī not only surveyed the mathematical rules and methods needed for designing astronomical instruments, but he also gave their proofs and discussed philosophical problems related to the question whether a line was composed of points and to the fourth postulate about the equality of right angles.[48]

A certain preference for ancient classical names of individual mathematical sciences seems to surface in the eleventh/seventeenth century in manuscripts as well as in biographical dictionaries. This may have been an aspect of a renewed interest in Ibn Sīnā's works among some Ottoman scholars. An example for this link is the treatise on number theory written by the head astrologer Ahmed Dede (d. 1113/1702). He called the discipline *'ilm al-arithmāṭīqī* and relied in its discussion on Ibn Sīnā's *Kitāb al-Šifā'* and 'Ubaydallāh Jūzjānī's mathematical chapter in the *Kitāb al-Najāt*.[49] He was also one of the very few Ottoman scholars in this period who wrote a commentary on Euclid's *Elements*.[50]

The arts were a further domain where the mathematical sciences received Ottoman patronage, albeit in a limited manner. A few astrological poems and prose texts were beautifully illuminated and dedicated to the court. A well-known example is 'Alā' al-Dīn Manṣūr Šīrāzī's poem *Šāhinšāh-nāma*, which he dedicated to Murad III (r. 982–10003/1574–1595), discussing the founding and shutting down of the observatory.[51] Another example is Mehmed Su'udi's *Maṭālī' al-Sa'ādāt*,

45 *Osmanlı Matematik Literatürü Tarihi* 1999, 55–56, 90–93, 155; *Osmanlı Astronomi Literatürü Tarihi* 1997, pp. 65, 74, 89, 213, 222, 225, 239, 249, 293, 309, 345, 352, 403.

46 *Osmanlı Astronomi Literatürü Tarihi* 1997, 336.

47 *Osmanlı Matematik Literatürü Tarihi* 1999, 138, 140, 145, 151, 155, 159; *Osmanlı Astronomi Literatürü Tarihi* 1997, 336.

48 *Osmanlı Astronomi Literatürü Tarihi* 1997, 76–83.

49 *Osmanlı Matematik Literatürü Tarihi*1999, 162.

50 *Osmanlı Matematik Literatürü Tarihi*1999, 163.

51 *Osmanlı Astronomi Literatürü Tarihi* 1997, 191.

produced in Istanbul in 990/1582 and illuminated by the painter Osman with 122 wonderful miniatures, among them the zodiacal signs and a portrait of Murad III in his study with a European clock and an astrological or astronomical manuscript in front of him.[52] However, ʿAbd al-Raḥmān Ṣūfī's *Kitāb ṣuwar al-kawākib al-ṯābita* was only rarely copied in the Ottoman period. A manuscript produced in Egypt in 922/1516 is preserved in Paris.[53] Other, less expensive artistic decorations of mathematical texts include colourful *'unwans* and some calligraphy. Such adorned texts were destined for personal libraries in the palaces as well as in the houses of the urban elites. Collecting and displaying finely illustrated scientific manuscripts was part of Ottoman elite cultures in the tenth/sixteenth and eleventh/seventeenth centuries. A new element in Ottoman scientific culture emerged in the tenth/sixteenth century, when the exchange of gifts and knowledge, including scientific instruments, in particular telescopes, and books, with scholars and diplomats from Catholic and Protestant countries in Europe began to flourish. Ottoman princes ordered world maps in Venice. Italian scholars, printers, officers, newly established rulers and ambassadors sent manuscripts of their works, maps and translations to the Ottoman court in hope for patronage, recognition and alliances.[54] The vivacious mutual exchange of knowledge and its products as part of commerce, embassies and other kinds of travel was particularly effective in cartography where both sides profited from each other. In the eleventh/seventeenth century, elements of Western astrology were incorporated by Ottoman practitioners of the field into their writings, among them head astrologers of the court. One Ottoman scholar is credited for having used a telescope in his observations, although the used term – *dürbünü* – does not need to signify necessarily this instrument, since it was also used for the simple, long known sighting tube mounted occasionally on an astrolabe or quadrant.[55] In the second half of the eleventh/seventeenth century, French astronomical handbooks were translated into Arabic and Turkish.[56] Visitors from various Catholic and Protestant countries in Europe observed eclipses, measured solar and stellar altitudes and calculated latitudes in Rhodes, Cairo, Istanbul, Trebizond and other Ottoman cities. Other travellers questioned scholars in Istanbul about their interests in mathematics and planetary theory. They reported about various new scientific developments in their own countries and the Republic of Letters at large such as Galileo Galilei's (1564–1642) new mechanical works and Tycho Brahe's (1546–1601) new cosmological

52 MS Paris, BnF, Supplément turc 242, ff. 7v, 8v, 10v, 12v, 14v, 16v, 18v, 20v, 22v, 24v, 26v, 28v, 30v.

53 MS Paris, BnF, Arabe 2490.

54 An example is the fascinating history of the multiple dedications of Francesco Berlinghieri's *Geographia* (1482) to Mehmed Fātih, Federico da Montefeltro, Bayezid II, Sultan Cem and Lorenzo de' Medici.

55 *Osmanlı Astronomi Literatürü Tarihi* 1997, 280.

56 *Osmanlı Astronomi Literatürü Tarihi* 1997, 304–305, 327, 340–345; *Osmanlı Matematik Literatürü Tarihi* 1999, 153.

system. Merchants sold telescopes, microscopes and clocks across the Empire.[57] The main loci of this exchange were the houses of Christian ambassadors and their secretaries in Istanbul, houses of consuls and missionaries in the provinces and the bookshops and other market places in major Ottoman cities.

5 The mathematical sciences in Mughal India

The Mughals brought Muslim and Hindu astrologers to their courts and regarded horoscopes as relevant to matters of state. As in other areas of courtly politics, Mughal support for the mathematical sciences consisted of three main components – the continuation of Timurid precedence, the integration of mathematical works by scholars from Iran and the enrichment of the mathematical sciences through cross-cultural exchange primarily with their Hindu subjects, but also with some of their visitors from Europe. The court-sponsored translation of mathematical and astronomical works from Hindu scholars into Persian began in the reign of Akbar (963–1014/1556–1605) and was continued under his grandson Šāh Jahān (1037–1068/1628–1658), but was of very limited scale. Its impact on Muslim scholars has not been well studied. Winter and Mirza have pointed to the existence of several copies of the Persian translation of Bhāskara's *Lilāvatī* for Akbar by the poet Fayżī and his partners in 995/1587 in manuscript collections in London, Manchester and Berlin, most of which were produced in the late eighteenth century.[58] The Hindu scholar Dhārma Narāyāṇ wrote in 1074/1663 a Persian commentary on the translated text, which he dedicated to Awrangzīb (r. 1658–1707).[59] But it is not known whether Muslim scholars studied either of these two texts or the Persian translation of Bhāskara's *Bījagaṇita* made for Šāh Jahān in 1044/1634–5 by 'Aṭā Allāh Rašīd ibn Aḥmad Nādir (eleventh/seventeenth century), the eldest son of the architect of the Tāj Maḥal.[60] Akbar's opening of the endowed teaching institutes to Hindu students and the prescription of a broad range of disciplines to be taught there, including some of the mathematical sciences, was unique in the early modern Islamic world. Nothing comparable with regard to the inclusion of non-Muslim students into the endowed teaching institutes took place in Safavid Iran, although the inclusion of various philosophical and mathematical disciplines or texts into the teaching at madrasas in the Mughal capital resulted from the recommendations of Akbar's great vizier Faṭḥ Allāh Šīrāzī (d. ca. 998/1588).[61] Faṭḥ Allāh Šīrāzī had studied in Shiraz with a student of Jamāl al-Dīn Davvānī (828–907/1423–1501) and with Ġiyāṯ al-Dīn Manṣūr Daštakī Šīrāzī (d. 948/1542), among other things, the mathematical sciences. He taught in his native town where one of his students

57 Paul Lucas (1664–1737), for instance, sold during his travels at the beginning of the eighteenth centuries such items together with jewelries and other goods.

58 Winter – Mirza 1952, 2–3.

59 Ethé 1903, c. 1233.

60 Maulavi 1925, cc. 1112–1113; Rahman 1982, 391.

61 de Young 1986, 7.

was Farīd al-Dīn Mas'ūd Dihlavī (d. 1039/1629), who later worked at the Mughal court in Lahore.[62] Farīd al-Dīn compiled for Šāh Jahān the new *Zīj-i Šāh-Jahānī*.[63] Thus Akbar's commands for changing the teaching content at the madrasas in his realm reflect experiences and standards of Safavid and indirectly also Timurid Shiraz. It is not very likely though that the inclusion of Hindu students into the madrasa training had a greater impact on how the mathematical sciences were practiced by Muslim scholars in Mughal India than the Persian translations of Sanskrit mathematical and astronomical texts.

Although all Mughal rulers of the tenth/sixteenth and eleventh/seventeenth centuries patronized the endowed teaching institutes and a number of disciplines, there are only a few works in the mathematical sciences known that were dedicated to them by scholars who worked at the court. Examples are a poem summarizing Euclid's *Elements* written by 'Aṭā Allāh Qārī for Dārā Šikōh (1025–1069/1615–1659), the *Zīj* written for Šāh Jahān mentioned above and a beautifully written commentary on Bahā' al-Dīn 'Āmilī's *Tašrīḥ al-aflāk* by Muḥammad Rašīd al-Dīn for Rāḍī al-Dīn 'Alī, a grandson of Jahāndār Šāh (r. 1124–1125/1712–1713).[64]

Encyclopedias were a literary genre that Mughal scholars considered suitable for dedications to rulers. Muḥammad Fāżil ibn 'Alī ibn Muḥammad Miskīnī Qāżī Samarqandī eulogized Humāyūn (r. 937–963/1530–1556) in the preface of his encyclopedia *Jawāhir al-'ulūm-i Humāyūnī,* which he dedicated to the ruler shortly after his return to power in 962/1554.[65] He modeled his encyclopedia after the work of major predecessors, in particular the *Nafā'is al-funūn* by Muḥammad ibn Maḥmūd Āmulī and the *Ḥadā'iq al-anwār* by Faḫr al-Dīn Rāzī (d. 606/1209). In its third book, he devoted thirty-three chapters to the mathematical sciences including planetary theory, astronomical instruments, Ṣūfī's star constellations, astrology, mathematical geography, arithmetic, surveying, Euclidean geometry, the *Kutub al-mutawassiṭāt* (Middle Books), mechanics, optics, music and magic squares. He also integrated a series of occult sciences such as geomancy, dream interpretation, alchemy or a Hindu type of occult science called *dam o daham* into the class of mathematical sciences.[66] Another multi-volume encyclopedia dedicated to a Mughal ruler is the *Šāhid-i Ṣādiq* by Muḥammad Ṣādiq ibn Muḥammad Ṣāliḥ Iṣfahānī Āzādānī (1018–1059/1609–1651). It covers a vast array of topics from religion over philosophy, ethics and cosmography to astronomy, astrology, the occult sciences, geography and history. It is well known today for its numerous maps, although they are not much appreciated due to their lack in precision and accuracy. In Book III, which is named *On Reason, Knowledge, Efficiency and Deficiency*, seven out of eighty sections deal with planetary theory, astrology, the astrolabe, arithmetic, surveying and administrative practices of numeration and

62 Matvievskaya – Rozenfel'd 1985, 572.

63 Matvievskaya – Rozenfel'd 1985, 594.

64 Rahman 1982, 424.

65 Maulavi 1925, cc. 144–150.

66 Maulavi 1925, c. 149.

calculation. Book V on the universe, time, life, death and other topics dedicates thirty out of ninety-six sections to astronomy and geography. Thus, while some basic elementary knowledge on arithmetic and geometry is included in this encyclopedia, the primary interest of the author was directed towards mathematical cosmography.[67]

Despite the lack of dedications – which may reflect rather the limitations of cataloguing styles than Mughal practice – Mughal court astrologers produced a number of texts on mathematical cosmography, astronomical handbooks, astrology and ephemerides. Examples are the *Risāla dar hay'a* by 'Abd al-Majīd ibn Muḥammad Quṭb al-Dīn, an astrologer of Akbar; Mullā Chand's astronomical handbook *Tashīlāt* written as court astrologer of Akbar; Ṭayyib ibn Ibrāhīm Dihlavī's *Risāla dar taqvīm*, a brother of Farīd al-Dīn Dihlavī and like him affiliated to Akbar's court; or Mullā Tarzī's astronomical handbook *Ma'ādin al-jawāhir* compiled for Jahāngīr (1014–37/1605–27).[68] An important family of scholars of the mathematical sciences was founded by Jahāngīr's architect Aḥmad Lāhūrī (d. 1059/1649). His eldest son was 'Aṭā Allāh mentioned above. His second son Luṭf Allāh Muhandis (eleventh/seventeenth century) studied with his father and brother. He translated Ṣūfī's *Ṣuwar al-kawākib* into Persian, wrote about number theory and commented on Bahā' al-Dīn 'Āmilī's *Ḫulāṣat al-ḥisāb*.[69] Luṭf Allāh's sons commented on Bahā' al-Dīn 'Āmilī's *Tašrīḥ al-aflāk* and Birjandī's commentary on Ṭūsī's *Bīst bāb*, translated Ṭūsī's *Taḥrīr* of Euclid's *Elements* and of Ptolemy's *Almagest* into Persian and wrote a treatise on the astrolabe.[70]

Mughal princes and courtiers on their side did not always wait for dedications, but ordered translations of well-known mathematical texts from Arabic into Persian, asked for copies and commanded introductions into arithmetic, planetary theory and other topics. In 1034/1624, a certain Muḥammad Rustam ibn Muḥammad Ḫalīfa asked in Agra for a copy of Niẓām al-Dīn Nīsābūrī's *Tawḍīḥ al-Taḏkira*.[71] In 1092/1681, Luṭf Allāh Muhandis wrote an abridged Persian version of Bahā' al-Dīn 'Āmilī's *Ḫulāṣa* for some Mīr Sayyid Muḥammad Sa'īd Mīr Muḥammad Yaḥyā.[72] In 1107/1696, a Mughal governor called 'Abd al-Waḥḥāb-Ḫān Bahādur ordered a copy of a Persian translation of Bahā' al-Dīn's work as well as a copy of a commentary on the *Ḫulāṣa* written by Luṭf Allāh Muhandis in 1092/1681.[73]

Similar to the Ottoman Empire, the mathematical sciences in Mughal India profited from the migration of Iranian scholars. In the later sixteenth century, Muṣliḥ al-Dīn Lārī (d. 979/1571), another student of Ġiyāṯ al-Dīn Daštakī Šīrāzī, moved to Humāyūn's court and dedicated to him his commentary on

67 Maulavi 1925, cc. 159–169.
68 Rahman 1982, 276, 335, 368.
69 Rahman 1982, 404–405.
70 Matvievskaya – Rozenfel'd 1985, 614–615; Maulavi 1927, 62–63.
71 Maulavi 1937, 38–39.
72 Ethé 1903, c. 1229.
73 Ethé 1903, cc. 1227, 1229.

'Alī Qušjī's introduction into mathematical cosmography.[74] Humāyūn himself is also credited with a mathematical treatise which he wrote for his son Jalāl al-Dīn, the later Akbar. The text explains the concept of great circles on a sphere.[75] In the early eleventh/seventeenth century, Nūr Allāh Šuštarī (d. 1019/1610) moved from Iran to the Mughal court. In addition to his qualification in Shi'i law and *kalām*, he also was the author of a treatise on the astrolabe.[76] With him came a manuscript to Mughal India that contained a text on arithmetic and algebra much appreciated by Safavid scholars, *al-Fawā'id al-Bahā'iyya* by 'Abd Allāh ibn Muḥammad al-Ḫuddāmī Baġdādī (eighth/fourteenth century). The manuscript included also a commentary on the text by a student of the author as well as a commentary by Kamāl al-Dīn Fārisī (d. 720/1320), a student of Quṭb al-Dīn Šīrāzī. The special value of the manuscript is marked by the fact that it was written by Niẓām al-Dīn Birjandī.[77] In the second half of the seventeenth century, works by Bahā' al-Dīn 'Āmilī became influential among Mughal scholars, primarily in the form of commentaries by 'Āmilī's students like Muḥammad Javād ibn Sa'd al-Kāẓimī (eleventh/seventeenth century) and Šams al-Dīn Muḥammad ibn 'Alī Ḫalḫālī (eleventh/seventeenth century) or Muḥammad Ašraf ibn Ḥabīb Allāh al-Ḥasanī al-Ḥusaynī Ṭabāṭabā'ī (eleventh/seventeenth century), a student of a student of 'Āmilī.[78] 'Iṣmat (or: 'Āṣim) Allāh ibn 'Aẓīm ibn 'Abd al-Rasūl (eleventh-twelfth/seventeenth-eighteenth centuries) for instance stood in this tradition with his commentaries on the *Ḫulāṣa* (1086/1675) and the *Tašrīḥ* (1087/1676). He is seen as one of the leading scholars of the mathematical sciences in Mughal India in the second half of the eleventh/seventeenth century.[79]

In the eleventh/seventeenth century, Jesuit missionaries resided in Agra. Some of them brought mathematical and astronomical printed books and instruments to the Mughal Empire and taught some of their knowledge to converts and other interested students. One outcome of these efforts was the translation of Christopher Clavius' (1538–1612) *Gnomonices* and of some theorems from his new edition of Euclid's *Elements* into Persian by Rustam Beg Ḥārithī Badaḫšānī (eleventh/seventeenth century).[80]

Mughal patronage of the arts included several lavishly illuminated works on astrology.[81] Astrologers and geomancers are repeatedly depicted in Mughal

74 Subḥānī – Āq Sū 1374, 176.
75 Matvievskaya – Rozenfel'd 1985, 553.
76 Rahman 1982, 342.
77 Maulavi 1937, 11–13.
78 Maulavi 1937, 18, 20, 49.
79 Maulavi 1937, 18–19, 49.
80 MSS London, British Library, or 975 and Add 14332.
81 An example is a copy of *Kitāb-i Sā'āt* made near Patna on 21 Šawwal 991/7 November 1583 for Akbar's foster brother Mīrzā 'Azīz Koka (d. 1035/1624). Sothebys, Wednesday 23 April 1997, 26–29.

miniatures.[82] A copy of Rašīd al-Dīn's (645–618/1247–1318) *Jāmi' al-tavāriḫ* made for Akbar in 967/1559–60 shows Ṭūsī and his colleagues in a rich setting observing the sky.[83] 'Abd al-Raḥmān Ṣūfī's *Ṣuwar al-kawākib* was copied and illuminated by Mughal painters.[84] But I have not come across illuminated Mughal manuscripts of other astronomical and mathematical texts in my studies in Indian manuscript libraries or in books on Mughal art history. Mughal rulers and their administrators were, however, keenly interested in such texts, even if only as objects for their libraries. Several mathematical and astronomical manuscripts carry ownership marks from Mughal rulers and courtiers. An example is al-Šarīf Jurjānī's commentary on Naṣīr al-Dīn Ṭūsī's *Taḏkira* written in 811/1409 in Shiraz that is extant today in the Khuda Bakhsh Library in Patna. It carries a seal of Maḥābat Ḫān, a noble at the court of Šāh 'Ālam Bahādur Šāh (r. 1119–1124/1707–1712).[85] Some of these ownership marks indicate that substantial efforts had been undertaken for obtaining such manuscripts, in particular when they were several centuries old. Mathematical and astronomical manuscripts seem to have served as cultural objects linking the Mughals to previous Islamic dynasties, in particular in India, as well as to their Timurid relatives. While research on Mughal texts of the mathematical sciences has not gone very far yet, our knowledge of Mughal scientific instruments is much more solid, primarily thanks to the work of Ś. R. Sarma. He has surveyed extant specimens and information found on and about them regarding their makers and their relationship to the Mughal dynasty. He showed that Humāyūn had not only a substantial interest in these instruments, but possessed a solid knowledge of their composition and functions and patronized the emergence of a rich and long tradition of astrolabe making in Lahore.[86]

The court of the Quṭb Šāh (1518–1687) at Golconda patronized apparently, in the tenth/sixteenth and eleventh/seventeenth century, several scholars interested in the mathematical sciences. Abū Isḥāq ibn 'Abd Allāh composed in 964/1555 a commentary on Niẓām al-Dīn Nīšābūrī's *Šamsiyya*, which he dedicated to Amīr 'Abd al-Karīm.[87] Texts of prominence in the mathematical sciences in Safavid Iran attracted also the attention of scholars in Golconda. Beside the *Šamsiyya*, texts copied, studied or commented on in Golconda were Quṭb al-Dīn Šīrāzī's encyclopedia *Durrat al-Tāj*, including its mathematical chapters (in 1017/1607) and Bahā' al-Dīn 'Āmilī's *Tuḥfa-yi Ḥātimiyya*, *al-Ṣafīḥa* and *al-Kur* and Ṭūsī's *Bīst bāb* (in 1073/1662–3, all in one manuscript).[88] But the migration of texts and scholars did not go exclusively in one direction, that is, from Iran to India.

82 See for instance Hajek 1960, plate 18, Náprstek Museum, Prague, 82.
83 This copy is preserved in the Golestan Museum, Tehran.
84 Schmitz – Desai 2006, I.2 for a Mughal copy produced between 1000/1590 and 1019/1610.
85 Maulavi 1937, 40.
86 Sarma 2003, 7–9, 18–19, 34–52.
87 Rahman 1982, 384.
88 Maulavi 1927, vol. 11, 139–142.

The manuscript compiled in Golconda in 1073/1662–3, for instance, was owned four decades later, in 1119/1707–8, by Muḥammad Taqī Ḥusaynī ibn Muḥammad Bāqir in Mashhad.[89]

Part II

1 The mathematical sciences during the Safavid period

The mathematical sciences in Safavid Iran show several similarities, but also differences in comparison to what took place under the predecessors and contemporaries of the dynasty. There were no *muwaqqit*s in Safavid Iran and hence mathematical treatises devoted specifically to the determination of prayer times and related issues, except for determining prayer directions, are relatively rare although some knowledge of writings by Mamluk *muwaqqit*s can be found in Safavid Iran. These topics were often included in texts about astrolabes and later in the eleventh/seventeenth century in works that discuss prayer times or auspicious and inauspicious days from the perspectives of *ḥadīṯ*, *fiqh*, astrology and astronomy. Among the instruments, Safavid scholars apparently preferred the astrolabe and thus texts about the numerous instruments developed by Mamluk *muwaqqit*s are also not widespread. But like the Ottomans and Mughals, the Safavid dynasty patronized astrology directly at their courts. Several astrologers counselled the Šāhs about auspicious days, cast horoscopes and dedicated astrological as well as astronomical writings to them. The position of the head astrologer was institutionally recognized by its integration into the courtly protocol and the list of officially prescribed titles.

As under the Timurids and Ottomans, a good number of astronomical, astrological, geometrical and arithmetical treatises were dedicated to Safavid šāhs and occasionally to a vizier, a governor or a eunuch. In contrast to what is known so far for the Timurids and Ottomans, the primary focus of texts dedicated to the Safavid court is on descriptions on how to work with an astrolabe; on astrological compendia; on copying, translating and newly illustrating ʿAbd al-Raḥmān Ṣūfī's *Ṣuwar al-kawākib*; and on ephemerides. Unique for the Safavid period is the integration of images of star constellations in the style of Rīżā ʿAbbāsī in two astronomical texts, Quṭb al-Dīn Šīrāzī's *al-Tuḥfa al-šāhiyya* and an anonymous text on the astrolabe. Occasionally, an edition of Euclid's *Elements* was also dedicated to a member of a Safavid court. From 1070/1660 on, several beautifully illustrated and written copies of a collection of extracts of writings on a broad array of disciplines, among them geometry, number theory, algebra, arithmetic, astrology, mathematical cosmography, mathematical geography and images of constellations, were produced in Tabriz for and with the participation of the vizier

89 Monzavī 1341, vol. 2, 294.

of Azerbaijan, Mīrzā Ẓahīr al-Dīn Muḥammad Ibrāhīm (d. 1102/1691), in Isfahan for Šāh Sulaymān (r. 1077–1105/1666–1694), as well as in other cities.[90]

Predominantly of very elementary quality and often incomplete, these texts cannot have served to educate the readers. The spatial arrangement of the manuscript is of a kind that reading them as continuous texts is not an easy task. Reading one page after the other was not the purpose of the collection since most single pages are covered by three sets of text in different sizes – one text with big letters and one text with small letters written between the lines of the first text are placed in the center of the page. One or more texts are written in the margins, running from top to bottom. The two center texts move in different speed (i.e., begin and end at different folios) and hence cannot be read concurrently. If one wished to read the marginal texts, the manuscript needed to be turned in different directions. The artistic quality of the illustrations and the fascinatingly complex spatial arrangements derive from Timurid and Safavid precedents, which were less complex in their arrangements, but in the case of some of the Timurid anthologies more suited for education. Examples for less complex, but related spatial divisions can be found in one of Iskandar Sulṭān's anthologies extant in the library of Istanbul University and in a copy of Sa'dī's *Gulistān* in the British Library.[91] The impact that this kind of artistic presentation of intellectual matters and its inscribed practices of reading may have had on the mathematical sciences has not even been raised as a question, let alone analysed.

Further dedicated texts combine subjects of the mathematical sciences with *ḥadīṯ* and literature. One such text is the *Angušt-i šumārī* written for Šāh Ṣafī (r. 1038–1052/1629–1642) by a student of Bahā' al-Dīn 'Āmilī. It comments on a couplet by Firdawsī (325–410/935–1020) taking it as a starting point for explanations of arithmetical rules.[92] Other texts bring together what David King has called folk astronomy, i.e., rules based on pre-Islamic traditions, *ḥadīṯ* and legal interpretations for determining prayer times, the directions for prayer and the beginning of the new month, with standard themes of mathematical astronomy and geography. An example for this kind of text is Muḥammad Bāqir ibn Muḥammad Taqī Majlisī's (1037–1110/1628–1699) *Iḫtiyārāt-i ayām*. It consists of forty chapters in which recommendations for electing days are combined with quotes from *ḥadīṯ* about solar and lunar eclipses and the rules for Muḥarram and a number of scientific themes among them astronomical rules for eclipses and explanations of their causes, the movement of the sun and the position of the stars, the determination of the *qibla* and other themes of mathematical geography such as distances on earth and the size of its diameter, a survey of the seven climates and a description

90 See Reza Pourjavady and Ahmad-Reza Rahimi-Riseh, "The Late Safavid Riddle Codices: Portrayals of Ẓahīr al-Dīn Mīrzā Ibrāhīm's Worldview," in *Safavid multi-text manuscripts: The Nihāyat al-aqdām fī ṭawr al-kalām by Ẓahīr al-Dīn Mīrzā Muḥammad* Ibrāhīm, edited by Sonja Brentjes, *manuscript cultures* 20 (forthcoming in 2023).

91 MSS Istanbul, Istanbul University Library, IÜF 1418; London, British Library, or 24944.

92 Subḥānī-Āq Sū 1374, p. 137.

of the seas, rivers and lakes.[93] While no new knowledge is expected to be found there – the text has not been studied yet by a historian of the mathematical sciences – the integration of this kind and amount of scientific knowledge into a work about astrological practices of daily life in the Safavid period and its justification on religious grounds is worthy of attention.

In addition to the study of dedicated texts and instruments, a study of dated copies of texts brings several new information about the state of the mathematical sciences in the Safavid period. It reveals a focus on works by Naṣīr al-Dīn Ṭūsī, Maḥmūd ibn Muḥammad Čaġmīnī, Qāżīzāda Rūmī, Abū Rayḥān Bīrūnī, Ibn Abū Šukr al-Maġribī, Ġiyāṯ al-Dīn Kāšī, ʻAlī Qušjī, Niẓām al-Dīn Birjandī, Šams al-Dīn Ḫafrī and Bahā' al-Dīn ʻĀmilī.

Texts by a few other early scholars of the mathematical sciences such as Qusṭā ibn Lūqā, Ṯābit ibn Qurra, Abū Maʻšar, ʻAlī ibn ʻĪsā al-Asṭurlābī, Abū Sahl Kūhī, Aḥmad ibn Muḥammad ibn ʻAbd al-Jalīl Sijzī and Kūšyār ibn Labbān were also copied and occasionally commented on in the Safavid period, albeit less often than those by the previously named writers.

A study of the colophons, ownership marks and waqf certificates indicates that students of Bahā' al-Dīn ʻĀmilī and Mullā Ṣadrā (980–1050/1571–1640) were among those who copied or collected such mathematical texts during the eleventh/seventeenth century. Students of the former with mathematical interests like Muḥammad Bāqir ibn Zayn al-ʻĀbidīn (d. ca. 1047/1637) wrote themselves treatises on mathematical themes and worked as astrologers. Students of the latter like Muḥsin ibn Murtaḍā Fayż Kāšānī (d. 1091/1680) worked at madrasas and wrote primarily about religious topics. The study of colophons also reveals a rich geography of copying mathematical, astronomical and astrological texts across Safavid Iran. The most significant observation among all these results may however be the discovery of a continued interest in planetary theory. Copies of Ḫafrī's commentary on Ṭūsī's *Taḏkira* were not only made until the later eleventh/seventeenth century, but Safavid astrologers and madrasa teachers, among them Muẓaffar ibn Muḥammad Qāsim Gunābādī (d. 1024/1615), Bahā' al-Dīn ʻĀmilī, Muḥammad Bāqir ibn Zayn al-ʻĀbidīn, Muḥammad Ṣanīʻ and ʻAlī Rīżāvi, studied Ḫafrī's work together with the commentaries by commentators from earlier periods and wrote themselves glosses and supercommentaries.

2 *Mathematical texts with miniatures and other artistic illustrations*

Safavid courtiers and other wealthy circles had a remarkably strong interest in depictions of the star constellations in the tradition of ʻAbd al-Raḥmān Ṣūfī's *Kitāb ṣuwar al-kawākib al-ṯābita*. Several Safavid copies of the Arabic version of the work were produced in the tenth/sixteenth and eleventh/seventeenth centuries.[94] Around 1040/1630, a sumptuously illustrated Arabic copy of the *Ṣuwar*

93 Monzavī 1341, vol. 1, p. 11.

94 Examples are MSS St Petersburg, Public Library, Ar. 119, Paris, BnF, Arabe 4670 and Geneva, Sadruddin Aga Khan collection, MS 9. For the last manuscript see Welch, 1972, vol. 2, 69.

al-kawākib was produced. It is today preserved in Majlis Šūrā' Library of Tehran.[95] As I have argued elsewhere, the colour palette and the background decoration link its paintings rather with Indian, possibly Deccani, miniature painting of the late tenth/sixteenth and early eleventh/seventeenth centuries, although the human figures and their dress point to a Safavid model.[96] Safavid manuscripts that are clearly related to this model contain Persian translations of Ṣūfī's text. One of these translations was made in 1043/1633 by Ḥasan ibn Sā'd Qā'inī for Manūchihr Ḫān (d. 1046/1636), the governor of Mashhad. Two copies of this fresh translation by Manūchihr's court astrologer are known to exist, one in New York and the other in Cairo.[97] A third manuscript that has been linked earlier to these two copies is MS Tehran, Malik Library, 6037. Dated 1008/1598 the manuscript is more than thirty years older than either of the two other manuscripts. Hence, it contains another Persian version. Whether it is the same version as in MS Paris, BnF, Supplément Persan 1551, which is said to be an anonymous abridged version, needs to be investigated. A copy of a possibly further anonymous translation is preserved in New York.[98] Miniatures of a distinct colouring and figurative style from another Safavid version in Persian of Ṣūfī's book are held by the Royal Ontario Museum.[99]

In addition to these complete copies and translations, fragments and summaries of Ṣūfī's work were integrated into collections and other astronomical texts or exist as single leaf. Examples are the copies of the collection of Mīrzā Muḥammad Ibrāhīm extant in the Arthur M. Sackler Museum at Harvard University, the British Library in London, the Süleymaniye Library in Istanbul or the Malik Museum in Tehran; the copies of Quṭb al-Dīn Šīrāzī's *Tuḥfa* in the Riḍā 'Abbāsī Museum in Tehran and in the Bayerische Staatsbibliothek in Munich; the copy of an anonymous treatise on the astrolabe at the British Library; and the single leaf of Boötes in the former Riefstahl collection.[100] The integration of constellations from Ṣūfī's book in a style very similar to that of the *Tuḥfa* and the leaf of the former Riefstahl collection into the anonymous treatise on the astrolabe has come to my attention only recently (2009) when I checked the manuscript for its chapters on timekeeping and astrological matters. The paper on which the images were painted as well as the changes incorporated in the original text make clear that the illustrations were added at a later stage into a formerly unadorned manuscript. The new owner wished to make it appear that the images were a genuine part of the manuscript. This suggests that he considered the images as a fitting addition to a straightforward, elementary description of the astrolabe and its areas of application. Such an evaluation of image and educational text may have

95 Vesel 2001, 296.

96 Brentjes 2014.

97 MSS New York, New York Public Library, Spencer 6; Cairo, Dār al-Kutub, MMF9; see King 1986, pl. III; Edwards 1982, 13; Schmitz 1992, 123.

98 MS New York, New York Public Library, Spencer 25.

99 MS Toronto, Royal Ontario Museum, Inv. no. 971.292.13. I thank the curator of Islamic Decorative Arts, Karin Rührdanz (now retired), for this information.

100 I completed the data known to me in 2010 by new information acquired since 2018.

been the result of a new attitude towards the mathematical sciences and the arts in the eleventh/seventeenth century. So far only these two astronomical works with high quality paintings of images from another text are known to exist. The only text whose author already incorporated several constellations from Ṣūfī's book is Zakariyyā' Qazvīnī's *'Ajā'ib al-maḫlūqāt*. As a rule, the images of these constellation are small-scale and often of a different iconographic style than those of the Ṣūfī traditions. There are, however, two folios of a ninth/fifteenth-century Iranian manuscript of an Arabic version of Qazvīnī's work that show a constellation taken indeed from the Ṣūfī cycle.[101]

Although this is not an exhaustive survey of all possible Safavid copies of Ṣūfī's *Ṣuwar al-kawākib* that exist today in manuscript libraries across the world, these few bits of information available from published literature and manuscript catalogues indicate that there was a strong interest in the text and its illustrations. The strength of this twofold interest in astronomical information and artistic presentation seems to be unique in the period, although there are also several copies of the work made in different Indian regions such as Gujarat, the Deccan, Delhi and possibly the Panjab. As under the Timurids and Ottomans, Safavid copies of Euclid's *Elements* or treatises on planetary theory could be adorned with beautiful *'unwan*s and set in gilded frames. An example for this integration of mathematical and astronomical texts in the production of luxury goods is the copy of Ḫafrī's *Muntahā al-idrāk fī mudrāk al-aflāk*, a commentary on Šīrāzī's *Tuḥfa* preserved in Qom.[102]

3 Activities in the mathematical sciences in the Safavid period

In order to gain an overview on the various kinds of activities undertaken in the mathematical sciences in Safavid Iran, I extracted information on dated manuscripts from catalogues of manuscript libraries in Iran, in particular Qom, Mashhad, Tehran, Qazvin and Yazd that were available to me. I supplemented the study of these catalogues by information about manuscripts from collections in London, Paris, Berlin and Cairo. This survey is not exhaustive since catalogues of major Iranian libraries were not available to me. It is also not meant to be exhaustive since my aim here is rather modest. I want to offer some preliminary views on possible focal points, trends and affiliations in the mathematical sciences during the Safavid period. In addition to the practical problems, my approach suffers under several intrinsic limitations caused by the different styles in which catalogues were produced in the last two centuries. In a first round, I chose manuscripts that are dated, affiliated to a place and identified by a scribe, addressee, owner or donor in the Safavid period. Since not all catalogues provide however this range of information, I included in a second cycle manuscripts that lacked one or two of these elements. Manuscripts that lack an affiliation to a place are, of course, difficult to situate. Scribes, owners or some of the addressees with a nisba linking

101 MS Cambridge, Mass., Arthur M. Sackler Museum, Harvard University, 1919.131a-b.
102 Mar'ašī – Ḥusaynī 1358, vol. 4, 65, no 1260.

them to localities in Iran may not have lived in an Iranian city or village. I have tried to limit the uncertainties caused by this lack of information about localities by including in such cases only manuscripts from Iranian libraries. This choice does not completely remove the problem since libraries like the Public Library of Āyatullāh Mar'ašī in Qom bought manuscripts from abroad. The disciplines and topics which I included in this survey are geometry, surveying, arithmetic, algebra, number theory, magic squares, mathematical cosmography, astronomical handbooks, instruments, ephemerides, prayer times, *qibla*, new moon and astrology. The results of this survey are found in Table 4.1 in the Appendix of this chapter. This table reflects interesting similarities with the mathematical sciences under the predecessors and contemporaries of the Safavids as summarized above. It also highlights some major differences.

The Timurids, Ottomans and Safavids shared major authorities in the mathematical sciences, in particular Naṣīr al-Dīn Ṭūsī, Uluġ Beg, Qāżīzāda Rūmī, Ġiyāṯ al-Dīn Kāšī and 'Alī Qušjī. The authors and copyists of mathematical texts in the Ottoman and the Safavid Empires shared a major interest in texts on the astrolabe and ephemerides. In difference to the situation under the Ottomans, the impact of Mamluk mathematical sciences in Safavid Iran is almost negligible. The lack of the institution of the *muwaqqit* is reflected by the almost complete lack of specialized texts on topics from *'ilm al-mīqāt*, which are treated rather in texts on the astrolabe and in texts on astrological themes. The scholars in Safavid Iran, like their Timurid predecessors, paid continuous attention to texts on planetary theory; dedicating texts on the mathematical sciences to rulers and a few of their relatives and courtiers is a further feature common to scholarly practices under the three dynasties. Like during Ottoman times, dedication patterns vary significantly over time and tend to focus in the later decades more and more on astrological themes relevant for matters of daily life and on ephemerides. The interest of Safavid scholars in classical texts, mostly in their edition by Naṣīr al-Dīn Ṭūsī, seems to have been stronger than among their peers under the Timurids, Ottomans and Mughals. In difference to Mamluk patterns where interest in theoretical geometry focused primarily, albeit not exclusively on Šams al-Dīn Samarqandī's *Aškāl al-ta'sīs* and parts of Euclid's *Elements*, Safavid scholars and scribes in particular in the eleventh/seventeenth century had a broader outlook that included several texts of the so-called *Kutub al-mutawassiṭāt*. The differences between the texts contained in the various collections of these *Kutub al-mutawassiṭāt* deserve further study and reflection.[103]

In difference to the efforts of some patrons, dignitaries and scholars in the Timurid, Ottoman and Mughal Empires to produce new astronomical handbooks and carry out new astronomical observations, the inclination among Safavid rulers for engaging in these kind of projects was apparently feeble and even scholars were rarely interested in compiling their own *Zīj* works. A few previous *Zīj* works though were occasionally copied by Safavid scholars or scribes. Migration

103 Compare also Kheirandish's report on copies of the *Kutub al-mutawassiṭ*āt in Iranian Libraries, Kheirandish 2000.

patterns of texts and scholars into Safavid Iran also seem to have differed from those moving out of it. There was apparently only one scholar, who had left Iran very early in the Safavid era for the Ottoman Empire, whose texts were of major relevance to the work of Safavid scholars – Niẓām al-Dīn Birjandī. No information was registered in the catalogues that would describe the people and ways through which Birjandī's texts returned to Iran. The table indicates merely that it happened very soon after Birjandī had composed his works.

The material compiled in the table also suggests that the major locus for theoretical texts on the various mathematical disciplines in Safavid Iran was the madrasa, not the court as apparently was the case in Timurid Iran and Central Asia. Furthermore, it indicates that the mathematical sciences were well established in a few centres, but spread – at least on the more elementary levels – beyond them. This geography of the mathematical sciences in Safavid Iran deserves further investigation.

When going beyond the criteria of date and location including more broadly texts that are affiliated with one or more names of patrons, scholars or scribes from the Safavid period, a few more interesting aspects for a future history of the mathematical sciences in Safavid Iran can be discovered. In the eleventh/seventeenth century, several scholars who did not work as astrologers or are not known (yet?) as teachers of any of the mathematical sciences acquired a basic education in these sciences; collected a set of texts; occasionally wrote their own treatises on the one or the other topic from arithmetic, algebra, geometry, astrology or astronomy; and donated them as *waqf*. Among these scholars were students of Mullā Ṣadrā who became leading representatives of the rational sciences in their own rights such as Muḥsin Fayḍ Kāšānī and sons, grandsons and great grandsons of well-known scholars such as Qiwām al-Dīn Ḥusayn, a son of Šams al-Dīn Ḫafrī; Muḥammad Hādī, a son of Bahā' al-Dīn 'Āmilī; Kalb 'Alī, a son of Javād Kāẓimī; Muḥammad, a son of Muḥsin Fayż Kāšānī or Muḥammad ibn Ḥusayn; and Muḥammad Bāqir ibn Muḥammad Ḥusayn, a grandson and great grandson of Muḥammad Bāqir ibn Zayn al-'Ābidīn. Other names appear repeatedly as scribes or owners of texts on the mathematical sciences, which suggests that these texts were copied probably for teaching or studying purposes. Examples are Šayḫ Asad Allāh Muḥammad ibn Ḫātūn (*waqf* 1067/1656), Muḥammad 'Alī Māzandarānī, Qurbān 'Alī ibn Ramaḍān Šams al-Dīn Ṭabasī (fl. between 1080 and 1090/1670 and 1680), Sayyid Muḥammad Masīḥ Ḥusaynī Šīrāzī (fl. between 1070 and 1090/1660 and 1680) and Āqā Zayn al-'Ābidīn Ḫādim Šarīf Iṣfahānī (*waqf* 1166/1753).[104] A closer study of such names, the persons they represent and the mathematical, astronomical and astrological texts they owned, studied or copied is an important task for the future. A further group of buyers of treatises on topics from the mathematical sciences can be identified through ownership marks – physicians. While this is not surprising in the light of earlier centuries, the fact that so few physicians appear

104 Ibn Ḫātūn went from Iran to India where he became vizier in Haydarabad, Deccan. He donated 400 manuscripts to the Shrine of Imām Riżā in Mashhad; Šākirī 1367, 92–93. For Āqā Zayn al-'Ābidīn see Velā'ī 1380, 37.

as owners of mathematical, astronomical and in particular astrological texts raises interesting questions about their access to the needed information, their qualification to interpret such information and their relationship to the astrologers.

A topic that I have not pursued due to lack of access to material and time is the engagement of a few princes, governors and viziers with the mathematical sciences not merely on the level of education and entertainment, but as authors in their own rights.

4 Studies of planetary theory and astronomical handbooks

As already indicated briefly in my introductory summary on the mathematical sciences in Safavid Iran, I consider the observation that texts on planetary theory by Ṭūsī, Šīrāzī, Niẓām al-Dīn Nīšābūrī, Birjandī and Ḫafrī were copied, studied and commented on throughout the tenth/sixteenth and eleventh/seventeenth centuries as one of the major results of my exploration of catalogues and their information about colophons, ownership marks and glosses. The texts that drew the attention of Safavid scholars and copyists were Ṭūsī's *Taḏkira, Zubda, Mu'īniyya* and *Ḥall-i muškilat-i Mu'īniyya*; Šīrāzī's *Tuḥfa* and *Nihāya*; Nīšābūrī's *Tawḍīḥ al-Taḏkira*; Birjandī's *Šarḥ al-Taḏkira;* and Ḫafrī's *al-Takmila fī l-Taḏkira*. They were studied in Isfahan, Shiraz and Mashhad, mainly in madrasas. A particular interest in planetary theory seems to have existed in Shiraz. If the information given in a manuscript extant today in Patna is reliable, this interest started with Šīrāzī himself when he wrote in this city in 682/1283 his *Nihāyat al-idrāk* for the vizier Šams al-Dīn Muḥammad Juvaynī (d. 683/1284). A copy made from the autograph in 687/1289 remained for some 300 years in Shiraz. In the ninth/fifteenth century it was in the hands of Jalāl al-Dīn Davvānī, who was, however, not the only scholar in Shiraz at that time who studied Šīrāzī's astronomical works.[105] Šāh-Mīr Šīrāzī (d. 898/1492–3) studied in Shiraz and wrote a commentary on 'Alī Qušjī's *Hay'a* called *Tanqīḥ-i maqāla va-tavżīḥ-i risāla*. It consists of a *maqṣid* on the supralunar world, followed by a *maqṣid* on the sublunar world and a survey of the knowledge needed from geometry and natural philosophy.[106] Some of his knowledge Šāh-Mīr derived from Ṭūsī's *Taḏkira* and Šīrāzī's *Tuḥfa* and *Nihāya*.[107] It may be however that he wrote this text in Gujarat.[108] In the first half of the tenth/sixteenth century, Ġiyāṯ al-Dīn Manṣūr Daštakī Šīrāzī acquired Šīrāzī's manuscript of *Nihāyat al-idrak* copied in 687/1289. During the rule of Akbar, the manuscript was brought to India. For almost one hundred years, nothing is known about its whereabouts. Then, in 1104/1692, a member of Awrangzīb's court bought it. After his death, it was sold to a scholar from Lucknow.[109] Whether the manuscript was studied at the Mughal court is unknown. But in Shiraz the interest in planetary theory did not stop with the

105 Maulavi 1937, 42–43.
106 Mar'ašī – Ḥusaynī 1364, vol. 11, 131–132.
107 Mar'ašī – Ḥusaynī 1357, vol. 6, 141.
108 See Rahman 1982, 313.
109 Maulavi 1937, 41–43.

move of this manuscript to Mughal India. Šams al-Dīn Ḫafrī is said to have studied the mathematical sciences with Ṣadr or Ġiyāṯ al-Dīn Manṣūr Daštakī Šīrāzī, including planetary theory. Since Ḫafrī wrote a commentary on Šīrāzī's *Tuḥfa*, it may well be that he also studied Šīrāzī's *Nihāya*. While no later information about the study of Šīrāzī's texts in Shiraz was available to me, Ṭūsī's texts continued to be of interest to teachers and students until the later eleventh/seventeenth century. In 1074/1663, for instance, a certain Muḥammad Rīżā copied Ṭūsī's *Taḏkira* in the Ismā'īliyya Madrasa of Shiraz for Mīrzā Ibrāhīm, his teacher and head of the madrasa.[110]

In Isfahan, Šīrāzī's *Tuḥfa* was copied in 990/1582.[111] In the same city, Ṭūsī's *Zubda, Mu'īniyya* and *Ḥall-i muškilat-i Mu'īniyya* were copied in 1068/1658 in combination with two parts of Ibn Sīnā's *Kitāb al-Najāt*, the parts on natural philosophy and metaphysics.[112] While this alone does not justify a claim about a possible impact that interest in medieval philosophy in eleventh/seventeenth-century Isfahan had upon the study of the mathematical sciences, other manuscripts also bring together these two fields. Thus, the possibility of such a link should at least be formulated as a question for further research.

Ḫafrī's commentary on Ṭūsī's *Taḏkira* proved more attractive to Safavid scholars than the master's work itself. A copy of Ḫafrī's *al-Takmila fī Šarḥ al-Taḏkira* written in Ṣafar 932/November-December 1525, only a few weeks after the autograph (Muḥarram 932/October-November 1525), was owned by Bahā' al-Dīn 'Āmilī, who annotated it.[113] A further copy was produced in 960/1552–3. It contains glosses by unspecified authors.[114] A third copy was made in Isfahan in 1074/1664. It contains glosses and extracts from other commentaries on the *Taḏkira* by Niẓām al-Dīn Nīšābūrī, Sayyid Šarīf Jurjānī, Birjandī, Muẓaffar Gunābādī, Muḥammad Bāqir Yazdī, Muḥammad Ṣanī' and 'Alī Rīżāvi.[115] In addition to manuscripts with glosses and marginalia, other copies of Ḫafrī's *al-Takmila fī l-Taḏkira* were produced in Isfahan, Shiraz or elsewhere in Iran until the end of the Safavid period, for instance in 1001/1591, 1059/1649, 1063/1653, 1066/1656, 1074/1664, 1112/1700 and 1120/1708.[116]

Niẓām al-Dīn Nīšābūrī's and Niẓām al-Dīn Birjandī's commentaries on Ṭūsī's *Taḏkira* were also copied by Safavid scholars. 'Alī ibn Mas'ūd Ḥusaynī Tafrīšī, who seems to have lived in Qom, produced a copy of the former in 1041/1631, and Muḥammad Sa'īd ibn Muḥammad Amīn Kāšānī made a copy of the latter in 1061/1651.[117] Some scholars of the Safavid period also wrote their own texts on *'ilm al-hay'a* among them Ġiyāṯ al-Dīn Manṣūr Daštakī Šīrāzī, Muṣliḥ al-Dīn Lārī, Abū

110 Maulavi 1937, 41.
111 University of Pennsylvania, Schoenberg Center, Smith Collection.
112 Dāniš-Pažūh – Anvarī 1976, 14.
113 Dāniš-Pažūh 1351, *Kitābḫāne-yi Gawharšād,* no 186.
114 al-Naqšbandī – Abbās 1982, 46, no 933,2.
115 Dāniš-Pažūh – Anvarī 1976, 268, no 464; see also Mar'ašī – Ḥusaynī 1372, vol. 21, 167–168; 1376, vol. 27, 129.
116 Mar'ašī – Ḥusaynī 1368, vol. 17, 95; Afšār – Dāniš-Pažūh 1352, 164–165.
117 Mar'ašī – Ḥusaynī 1367, vol. 15, 31; 1370, vol. 20, 159.

al-Ḫayr Taqī al-Dīn Muḥammad al-Fārisī (tenth/sixteenth century), Bahā' al-Dīn 'Āmilī and Mīr Dāmād (d. 1041/1630).[118] Bahā' al-Dīn 'Āmilī also is ascribed a work on the *Almagest*. But almost none of these texts have been analyzed yet.

In contrast to the continued interest in planetary theory, new astronomical handbooks were only rarely produced in the Safavid period. One such *Zīj* was made by an anonymous scholar in Tabriz around 957/1550.[119] Two other new astronomical handbooks were compiled by Maẓhar al-Dīn Muḥammad ibn Bahā' al-Dīn 'Alī Qārī in Shiraz in 960/1553.[120] A fourth new *Zīj* was produced more than one hundred years later in 1078/1668 in Mashhad for Šāh Sulaymān by Zamān ibn Šaraf al-Dīn Ḥusayn Mašhadī.[121] However, the study of major Ilkhanid and Timurid *Zīj* works was paid more attention to among scholars of the Safavid period. Commentaries on Uluġ Beg's *Zīj* were written by Ġiyāṯ al-Dīn Manṣūr Daštakī Šīrāzī under the title M*afātīḥ al-munajjimīn* in the first half of the tenth/sixteenth century and Maẓhar al-Dīn Qārī in Shiraz. A commentary on Ṭūsī's *Zīj-i Īlkhānī* was composed by Muḥammad Ašraf ibn Muḥammad Ja'far Iṣfahānī in 996/1586 in Yazd.[122] Kāšī's *Zīj-i Ḫāqānī* and Uluġ Beg's *Zīj* were copied once each in the eleventh/seventeenth century in Isfahan. The person who copied Kāšī's tables was Zayn al-'Ābidīn.[123] Uluġ Beg's *Zīj* was copied in Isfahan's Masjid-i Jāmi'-i Jadīd-i 'Abbāsī.[124]

5 Dedicated works on the mathematical sciences

In the first decades of Safavid rule, no mathematical, astronomical or astrological treatise is known at the moment that was dedicated to a šāh except for one text on *siyāq* dedicated to Ismā'īl.[125] Patronage of these disciplines may not have yet focused on the court. A few treatises on planetary theory, on the one end of the spectrum, and on surveying, on the other, were dedicated to local dignitaries.[126] In the late tenth/sixteenth and early eleventh/seventeenth century dedicating texts of the mathematical sciences to a Šāh became more widespread. But not every Šāh was considered equally worthy of the honour. As in the case of the Ottoman sultans, certain Safavid Šāhs received more dedications than others. In difference to the picture projected by Ottoman manuscripts, there is no clear trajectory that could link this fact to other social and cultural developments. It rather seems that in the Safavid case the number of dedications reflected the support of certain of the mathematical sciences given or hoped for by a particular ruler. Most of the

118 Matvievskaya – Rozenfel'd 1985, 552, 553, 559, 582, 594.
119 I owe this information to Benno van Dalen.
120 Monzavī 1341, vol. 1, 58; Matvievskaya – Rozenfel'd 1985, 568.
121 King 1986, no G76.
122 Dāniš-Pažūh 1351, *Kitābḫāne-yi Gawharšād*, no 338; Matvievskaya – Rozenfel'd 1985, 552, 568.
123 Mar'ašī – Ḥusaynī 1372, vol. 21, 136.
124 Mar'ašī – Ḥusaynī 1357, vol. 7, 195.
125 See Appendix, Table 1.
126 See Appendix, Table 2.

mathematical, astronomical and astrological texts were apparently dedicated to two Šāhs – ʿAbbās I and Ḥusayn. Some other male members of the family and occasionally a great vizier or governor also received dedicated texts. The dedications mostly recognized received patronage. Occasionally they were part of a relationship that provided education. Court astrologers such as Mullā Muẓaffar ibn Muḥammad Qāsim Gunābādī wrote texts on the astrolabe (1005/1595; a commentary on Ṭūsī's *Bīst bāb*), astrology (1014/1604) and the *qibla* (1019/1609) for Šāh ʿAbbās I and his great vizier Ḫwāja Naṣīr al-Dīn Ḥātim Beg.[127] Gunābādī's treatise on the *qibla* outlines four elementary methods that were known since a long time. Ḥātim Beg was the recipient of another dedicated astronomical treatise written for him by Bahā' al-Dīn ʿĀmilī. It explains in seventy chapters the components of the astrolabe and their functions. It emulates Naṣīr al-Dīn Ṭūsī's *Bīst bāb* and is seen as part of the astronomical teaching Bahā' al-Dīn gave to the interested great vizier.[128]

While Safavid courtly patronage for the mathematical sciences in the eleventh/seventeenth century appears to have been fairly stable, if also without major highlights, in the tenth/sixteenth century dedications of one and the same text could address different members of the dynasty. This fluidity of dedications may reflect instable patronage conditions that occurred in particular when a patron died. Bahā' al-Dīn ʿĀmilī's famous *Ḫulāṣat al-ḥisāb*, for instance, carries in some manuscripts a dedication to Šāh Ṭahmāsb and in others a dedication to Mīrzā Ḥamza ibn Muḥammad Ḫudābanda, a grandson of Ṭahmāsb.[129] Bahā' al-Dīn's *al-Kur* on *fiqh* and arithmetic was originally dedicated to Ṭahmāsb. In some manuscripts this dedication has been changed to ʿAbbās I.[130] Bahā' al-Dīn ʿĀmilī was the most successful writer of mathematical treatises in the Safavid period in terms of patronage relationships he accrued and in terms of posterior influence and recognition of his texts. In addition to the works he dedicated to Ṭahmāsb, Ḥamza and Ḥātim Beg, he wrote two astronomical treatises for ʿAbbās I, *Tašrīḥ al-aflāk* and *al-Ṣafīḥa*. Both explain the astrolabe, its construction and functions.[131]

Šāh Ḥusayn was the addressee of dedicated texts on the solar and lunar months (*Miftāḥ al-šuhūr*, 1109/1698) by Muḥammad Bāqir ibn Muḥammad Taqī Majlisī, several ephemerides by court astrologers and a Persian translation of Muḥammad Bāqir ibn Zayn al-ʿĀbidīn's opus on arithmetic, number theory and algebra, *ʿUyūn al-ḥisāb*.[132] In addition to dedicated texts, a few astronomical works were asked for or commanded by a šāh, a vizier or a governor. This applies, in particular, to the lavishly illustrated copies of the Persian translations of ʿAbd al-Raḥmān Ṣūfī's *Ṣuwar al-kawākib* and the collection of texts by Mīrzā Muḥammad Ibrāhīm discussed previously.

127 Marʿašī – Ḥusaynī 1355, vol. 5, 213; 1368, vol. 17, 75; 1368, vol. 18, 117.

128 Marʿašī – Ḥusaynī 1357, vol. 7, 78.

129 Maulavi 1937, 14; Ethé 1903, nr 758.

130 Monzavī 1341, vol. 2, 203; Maulavi 1937, 16; a manuscript with a dedication to Ḥamza is for instance MS London, British Library, I.O. Islamic 758.

131 Marʿašī – Ḥusaynī 1354, vol. 1, 169.

132 Marʿašī – Ḥusaynī 1354, vol. 1, 187; 1364, vol. 11, 145; Monzavī 1341, vol. 1, 61–62.

6 Copies of mathematical texts by ancient scholars and scholars from pre-Safavid Islamic societies

The texts by ancient authors copied in the Safavid period comprise of Euclid's *Elements*, most often, but not exclusively, in the edition by Ṭūsī, and a good number of texts from the *Kutub al-mutawassiṭāt* (i.e., treatises by Theodosios, Aristarchos, Archimedes, Autolykos, Menelaos, Hypsikles and several scholars from Islamic societies). An anonymous paraphrase of the *Elements* was dedicated in 1003/1593 to a eunuch of Šāh 'Abbās' I's court in Qazvin. This is the only information available at the moment about an interest in the *Elements* by a member of the Safavid court.[133]

Much greater was the interest in the *Elements* in Safavid cities and their madrasas, for instance, in Yazd and Isfahan and presumably also in Shiraz. Šams al-Dīn Ḫafrī commented on Ṭūsī's edition.[134] MS Add. 2357 of the British Library contains a copy of the Persian translation of Samarqandī's *Aškāl al-ta'sīs* and copies of several texts from the *Kutub al-mutawassiṭāt*. While the copy of the Persian translation of the commentary on the *Elements* is undated, the editions of texts by Autolykos and Theodosios were copied in Yazd by Abū al-Qāsim Yaḥyā Astarābādī in 1014/1605 for his own use.[135] The astrologer Muḥammad Bāqir ibn Zayn al-'Ābidīn Yazdī commented in Isfahan on the *Elements*. He may have done this in different formats, since three manuscripts with different titles are extant in Tehran – one called a commentary on the *Taḥrīr* of the *Elements* (i.e., most likely Ṭūsī's edition, another one on Books I to X and the third on Book X alone).[136] He also commented on and copied other classical texts such as Menelaos' *Spherics*, Theodosios' *Spherics*, Archimedes' *On Sphere and Cylinder* and Ṭūsī's edition of Apollonios' *Conics*.[137] He obviously acquired a very solid training in the basic textbooks of ancient mathematics. In this, he may have been a lonely bird among Safavid astrologers, but we also may simply know too little about the education of these professionals yet. That the latter may perhaps be the case is indicated by an ownership mark by Muẓaffar Gunābādī in MS Mashhad, Āstān-i Quds Riżavī 3157. This manuscript contains a copy of Ṭūsī's edition of Menelaos' *Spherics* made in the tenth/sixteenth century.[138] Ḏu al-Ḥijja 1068/ August 1658, twenty years after Yazdī's death, Ṭūsī's edition of the *Elements* was copied in Isfahan in the Šayḫ al-Islām Madrasa by al-'Abd al-Ġarīq for his professor al-'Ālim ibn 'Abd al-Ġanī 'Abd al-Karīm Raštī Gīlānī.[139] MS India office

133 This may, however, change when more catalogues and manuscripts will be checked. A reason for this assumption is the fact that Nādir Šāh donated in 1154/1741 several mathematical texts to the Āstān-i Quds Riżavī library in Mashhad which include two short extracts from Euclid's *Elements*, two texts from the *Kutub al-mutawassiṭāt* (Hypsikles and Abū Sahl Kūhī), Ṯābit ibn Qurra's treatise on two mean proportionals and an anonymous text on the *qibla*.

134 Sezgin 1975, 113.

135 MS London, British Library, Add 2357, ff. 31–133.

136 Matvievskaya – Rozenfel'd 1985, 590.

137 Matvievskaya – Rozenfel'd 1985, 590.

138 Afšār – Dāniš-Pažūh 1352, 113.

139 MS London, British Library, Add. 21952, ff. 149a, 9–15.

Islamic 923 contains another series of texts from the *Kutub al-mutawassiṭāt* (i.e., Autolykos' *On the Moving Sphere*, Theodosios' *On Habitations* and *On Days and Nights*, Aristarchos' *On the Sizes and Distances of the Sun and the Moon* and Euclid's *On Heaviness and Lightness)*. It also includes a Persian version of Book I of Euclid's *Elements*. Although the manuscript is undated and has no colophons, its calligraphy and illustration speak for a possible Safavid origin, maybe even a courtly environment or at least the wealthy environment of urban bibliophiles. The Persian version of Book I in this manuscript is a modified copy of Šīrāzī's Persian text of the *Elements*. The modifications are mostly philological in nature, replacing certain of Šīrāzī's Arabic terms by Persian expressions.[140]

There are several other Safavid copies of Ṭūsī's *Taḥrīr* of the *Elements*, Šīrāzī's Persian translating edition of this *Taḥrīr*, Ṭūsī's edition of the *Kutub al-mutawassiṭat*, Ptolemy's *Almagest* and Apollonios' *Conics*, both too mostly in Ṭūsī's editions, available in libraries in Tehran, Mashhad, Qom and other libraries in Iran. Table 4.3 in the appendix gives an impression of when, where and occasionally also for whom such texts were copied.

Beside works of ancient authors, Safavid students of the mathematical sciences also read books by scholars who had lived between the third/ninth and the eighth/fourteenth centuries in Iraq, Egypt, Iran, central Asia and northern India. They privileged scholars from the seventh/thirteenth to the early tenth/sixteenth centuries, in particular Naṣīr al-Dīn Ṭūsī, Ibn Abū Šukr al-Maġribī, Quṭb al-Dīn Šīrāzī, Qāżīzāda Rūmī, Ġiyāṯ al-Dīn Kāšī, ʻAlī Qušjī and Niẓām al-Dīn Birjandī. In addition to Ṭūsī's *Taḥrīr*s of works by ancient and medieval authors as well as his and Šīrāzī's works on planetary theory, Ṭūsī's texts on the astrolabe and the ephemerides were often copied or commented on. This attention was seconded by copying commentaries on Ṭūsī's *Bīst bāb dar asṭurlāb*, in particular that by Birjandī. Ibn Abī Šukr's astrological writings were most often copied, but his text on the astrolabe also attracted some attention. *Miftāḥ al-ḥisāb* was the text of Kāšī that was most often copied by Safavid scribes, students, and scholars. His *Muḥīṭ al-dā'ira* was also studied by at least one Safavid scholar who copied it from an autograph – Bahā' al-Dīn ʻĀmilī.[141] Since ʻĀmilī also owned a very early copy of Ḫafrī's autograph of his *Takmila,* it will be useful to pay more attention in future to the type of manuscripts owned by Safavid scholars. It may well be possible that an antiquarian interest stimulated their collecting and copying of various works on the mathematical sciences. Qāżīzāda Rūmī's commentary on Čaġmīnī's elementary introduction into planetary theory seconded by Birjandī's supercommentary and ʻAlī Qušjī's elementary introductions into arithmetic and planetary theory were the most often copied and commented upon works of these two colleagues of Kāšī.

The only author who lived before the seventh/thirteenth century and one of whose treatises on the astrolabe was repeatedly copied in the Safavid period is Abū l-Rayḥān Bīrūnī. Texts by a few other early scholars of the mathematical

140 MS London, British Library, India Office, Islamic 923, ff. 71b – 95a.

141 Hogendijk 2007, 190, 199.

sciences were also copied, albeit less often than those by the previously named writers. Some of them appear in collections of the *Kutub al-mutawassiṭāt* such as a text by Ya'qūb ibn Isḥāq al-Kindī (d. ca. 260/873) about the errors that happen in optics, the text of the Banū Mūsā (third/ninth century) about plane and solid figures, Ṯābit ibn Qurra's constructions of two mean proportionals or Abū Sahl Kūhī's treatise on the size of what can be seen of the heaven and the sea. Others such as Qusṭā ibn Lūqā's treatise on the astrolabe, Ṯābit ibn Qurra's text on parabolas, Abū al-Wafā' Būzjānī's book on geometrical constructions for craftsmen respectively commentaries on or abbreviations of it or Sijzī's treatise on the crab-like astrolabe are transmitted as independent works. These copies deserve to be investigated for glosses and other signs that could elucidate whether the copying reflected also an interest in these works that went beyond the preservation of old manuscripts.

7 On the geography of the mathematical sciences in Safavid Iran

The data compiled for this paper show that the mathematical sciences were present through copying, teaching and studying in major cities of the Safavid realm. Centers of previous dynasties continued to offer opportunities for the mathematical sciences under Safavid rule as did the capitals of the Safavid šāhs. Religious centers, in particular Mashhad, also attracted scholars skilled in these sciences. Several mathematical, astronomical and astrological texts were copied, taught, commented on or dedicated by scribes and scholars in Tabriz, Qazvin and Isfahan. Scholars with interest in the mathematical sciences also lived in Shiraz, Mashhad, Yazd, Kashan, Shirvan, Herat, Bistam, Tiflis, Ganja, Shamakhi and Barvan.[142] While there is little surprise that this was the case in cities with famous scholars and madrasas like Shiraz, Mashhad and Herat, there is no a priori reason to assume that all madrasas across Iran had a teacher competent in these disciplines and students interested in studying them. There were, nonetheless, a good number of places and madrasas where the mathematical sciences were taught and studied. Ṭūsī's *Bīst bāb* was copied in 996/1598–9 in Bistam.[143] Between 1020/1611 and 1024/1615, Abū l-Qāsim Istajlū studied two poems on astrology by or ascribed to Naṣīr al-Dīn Ṭūsī, a *taqvīm*, a treatise on music and letters by contemporary Janid and Safavid rulers in the capital of Daylam.[144] Mathematical, astrological and astronomical texts were even copied in Šāh 'Abbās' I summer residence Farahabad and in villages near Qa'in, Bistam and Lahijan. 'Alī Qušjī's *Risāla fī l-ḥisāb*, for instance, was copied in a village near Qa'in.[145] In 1039/1629 Muḥammad Taqī Mašhadī Ḥusaynī copied Qāżīdzāda's commentary on Čaġmīnī's *Mulaḫḫaṣ* in a

142 See Appendix, Table 1.
143 Monzavī 1341, vol. 1, 38.
144 Monzavī 1341, vol. 1, 12, 226–227, 235, 252, 256, 293–294.
145 Mar'ašī – Ḥusaynī 1374, vol. 24, 276–277.

fortress called Aḥmad Bakr.[146] On 28 Ṣafar 1031/18 January 1622 a scribe finished in Farahabad the copy of a text on astrology by Muḥammad ibn Ayyūb Ḥāsib Ṭabarī (fourth or fifth/tenth or eleventh century), one of the earliest authors who wrote mathematical and astrological texts in Persian.[147] Almost four months later, he copied the Persian translation of Kūšyār ibn Labbān's work *Mujmal al-uṣūl* in the same town.[148] In Rajab 1089/August 1678 Ṣādiq ibn Maḥmūd Kiyādehī copied 'Alī Qušjī's *Hay'a* in a madrasa in Khayrabad near Lahijan.[149] A further extension of this kind of search will surely bring forth other locations where texts on the mathematical sciences were copied, read, commented on or bought.

8 Contacts with aspects of the early modern mathematical sciences from Catholic and Protestant Europe

A major difference to the activities undertaken by some scholars and courtiers in the capitals of the Ottoman and Mughal states is the lack of impact that the encounters between Safavid scholars and Catholic and Protestant visitors from various European countries left on the mathematical sciences in eleventh/seventeenth-century Iran. Although scientific instruments and occasionally a mathematical book were part of the gifts brought by embassies to the Safavid court and despite the fact that several travellers and numerous missionaries undertook observations of astronomical events such as the repeated appearances of comets and eclipses, studied Persian astrological texts and collected information about the mathematical literature and authors studied by Safavid scholars, discussed occasionally matters of planetary theory with local astrologers and claimed to have found their Safavid hosts often very keen in learning about the mathematical sciences in Europe, there are no substantial traces found in Safavid writings on such subjects. Neither the court astrologers with whom some of the missionaries claimed to have conversed nor the scholars from Lar who in 1032/1622 took care of the sick Pietro della Valle (1586–1652) reflect these encounters in their writings. The two texts, for instance, that Quṭb al-Dīn 'Abd al-Ḥayy ibn 'Izz al-Dīn Ḥusaynī Lārī (fl. in the 1030s/1620s) wrote between 1017/1607 and 1037/1627 stand within the tradition of discussing astronomical and astrological events based on Uluġ Beg's *Zīj*.[150] The more comprehensive treatise of the two, *Ḥall va-'aqd*, was very successful as the many copies indicate that are extant today in libraries in particular in Iran and India, but also in Europe and North America. It consists of two chapters and a postface talking on the creation of an ephemeris, the casting of a horoscope and the ascent of the horoscope.[151] It remains to be seen if this picture will change when more Safavid mathematical and astronomical texts will be investigated.

146 Ṭāyār 1378, vol. 1, 154, no 119.
147 Monzavī 1341, vol. 1, 16.
148 Monzavī 1341, vol. 1, 304.
149 Mar'ašī – Ḥusaynī 1374, vol. 25, 188.
150 Mar'ašī – Ḥusaynī 1355, vol. 5, 210–211.
151 Mar'ašī – Ḥusaynī 1355, vol. 5, 210.

Perspectives

I hope I could illustrate the rewards that a horizontal approach brings to a history of the mathematical sciences during the Safavid period. The information that I compiled for this paper, while incomplete and in need of methodological refinement, shows that the activities in the mathematical sciences undertaken in Safavid Iran were in various ways surprisingly rich. Their comparison with analogous activities under some of the immediate predecessors as well as contemporary neighbours of the dynasty opens new ways for looking at the mathematical sciences in Safavid Iran and, thus, frees them from the focus on mathematical creativity and progress as the only options for evaluation. The comparison also indicates that there is no unified history of the mathematical sciences in the Islamic world at large, not even in its core territories. There were rather histories specific to times and localities. Once the similarities and differences of these histories are better known the difficult question of whether these variant histories were shaped by the politics of the particular dynasties which I took here as my very preliminary criterium of classification or whether the impact of these policies was rather negligible in comparison to much more local forces such as families of scholars and notables in a particular city can be addressed.

The comparison that I carried out for the sake of this paper shows that the activities in the mathematical sciences in Safavid Iran shared features with those undertaken under some of the dynasty's main predecessors and contemporaries. The Safavid šāhs and scholars took up Timurid patterns, in particular in regard to the fields, texts and authorities studied, preferred and preserved. Several Timurid works that are known today in unique copies were, for instance, preserved thanks to the interest that Safavid patrons and practitioners of the mathematical sciences took in them. But Safavid šāhs and scholars did not follow Timurid precedence in all respects. The little interest in *Zīj* literature and astronomical observations is one of the major differences. If the Timurid madrasas indeed failed to include the mathematical sciences on a larger scale as the current scanty data about teaching, studying or copying mathematical, astronomical or astrological texts in such environments may imply, this would be another feature that sets the mathematical cultures under the two dynasties apart from each other. The differences between the mathematical sciences under the Safavids and the Ottomans concern the professionalization of parts of astronomy at mosques, the study of theory and the kind of activities that were undertaken at courts and those that were pursued in madrasas. The spreading of mathematical literature through the cities of the realm, in particular to the northwest, perhaps following the expansion into the Caucasus Mountains, and their presence even in some of the villages may be another particularity of the mathematical sciences in the Safavid period. Peculiar features could be seen in an antiquarian interest in old manuscripts of privileged texts and in an interest in apocryphal astrological treatises that surfaced without predecessors or successors in 1084/1674, but may, perhaps, be contextualized within a survey of apocryphal texts on the occult sciences. A more detailed and comprehensive tracking of these and other specific features of the mathematical sciences in the two

Safavid centuries is a much-needed work for the future. Only then the suggestions and preliminary interpretations that I have offered in this paper can be substantiated, rejected or accepted.

Major issues that need to be clarified relate to the loci of the activities, the persons involved in these activities, the content of the activities and their dependence on their specific contexts, the place of the mathematical sciences in the general intellectual landscape of the Safavid period and the connection of their study with political, military and religious events and changes. It remains to be seen whether the current preliminary impression of a relatively late interest of the Safavid dynasty in the mathematical sciences and the limitation of Safavid courtly patronage for these sciences to some of the šāhs, viziers and governors will be sustainable. The role of specific families and their support for madrasas and other endowed teaching institutes in regard to the preservation of texts and instruments and the continuation of opportunities for teaching and studying the mathematical sciences is a topic that needs to be investigated for gaining a better understanding of the local particularities of the mathematical sciences in the Safavid period and for elucidating the ways in which scientific activities from earlier times were continued, altered or recast. Another important question that needs to be investigated is to what extent the relatively great number of copies of texts by ancient authors as well as by scholars who lived before the eighth/fourteenth century led to new mathematical studies and results among Safavid scholars or whether this care for old texts reflects other cultural trends such as an antiquarian interest in manuscripts. The documentation of the geographical spread of the mathematical sciences across Safavid Iran and the clarification of the functions of the mathematical sciences in different localities is a precondition before other issues of political, military, economic or religious contexts outside the main centres of Safavid rule can be addressed. The continuous production and reproduction of mathematical texts and their occasional co-existence with texts on natural philosophy, metaphysics and logic as documented in the catalogues of Qom, Mashhad and Tehran challenge the widespread belief among historians of Safavid intellectual life that the mathematical sciences and natural philosophy were of little relevance to Safavid scholars. The relationship between these disciplines and the attitudes that different of their representatives held over time is a topic that deserves attention in the future. The migration of illumination from one type of astronomical literature to other kinds of astronomical texts as a peculiar phenomenon of the relationship between the arts and the mathematical sciences during the Safavid period is an issue that will profit from a more systematic exploration of this relationship in general, a domain where some research was done in the last years. The patterns of migration of scholars and texts from and to Iran and their impact on the activities in the mathematical sciences within Iran are another domain which only recently began to attract some attention. So far, the research focused primarily on the moves of scholars from Iran to India and the Ottoman Empire. Itineraries of texts and ideas coming from the neighbours of the Safavid dynasty and their overseas visitors are a field of study that has remained much less explored but deserves serious attention. And finally, the situation of the mathematical sciences during the last years of the Safavid dynasty is a topic that has not been dealt with yet.

Bibliography

A Manuscripts

Clavius, Christopherus, *Elementa*. Arabic translation, extract, MS London, British Library, Add 14332.

———, *Gnomonices*. Arabic translation, acephalous, MS London, British Library, or 975.

Horoscope of Murad II. MS Paris, BnF, Supplément persan 367.

Ilḥāq ibn Abī Isḥāq, *Šarḥ Taḥrīr Uqlīdis li-l-Ṭūsī*. MS London, British Library, or. 1514.

Jung. MS Cambridge, MA, Arthur M. Sackler Museum, Harvard University, 1984.463.

Kāšī, Ġiyāṯ al-Dīn, *Ālāt-i raṣad*. MS Hyderabad, Andhra Pradesh Government Oriental Manuscript Library and Research Institute, Riyāḍī 129.

———, *Lubāb-i Iskandarī*. MS Qom, Kitābkhāna-yi ʻumūmī-yi Ḥażrat-i Āyatollāh al-ʻuẓmā Marʻašī Najafī, 1015, ff. 272b–278b.

Majmūʻa. MS Istanbul, Istanbul University Library, IÜF 1418.

Majmūʻa. MS London, British Library, Add. 27261.

Majmūʻa. MS London, British Library, India Office, Islamic 923.

Menelaos, *Spherics*. MS Mašhad, Āstān-i Quds Rižavī, 3157.

Qazvīnī, Zakarīyā' ibn Muḥammad, *ʻAjā'ib al-maḫlūqāt fī ġarā'ib al-mawjūdāt*. MS Cambridge, MA, Arthur M. Sackler Museum, Harvard University, 1919.131 and 1919.131v.

Saʻdī, *Gulistān*. MS London, British Library, or. 24944.

Samarqandī, Šams al-Dīn, *Aškāl al-taʼsīs*. MS London, British Library, Add 2357.

Ṣūfī, ʻAbd al-Raḥmān, *Kitāb ṣuwar al-kawākib al-ṯābita*. MS Cairo, Dār al-Kutub, MMF9.

———, *Kitāb ṣuwar al-kawākib al-ṯābita*. MS Geneva, Sadruddin Aga Khan Collection 9.

———, *Kitāb ṣuwar al-kawākib al-ṯābita*. MS New York, New York Public Library, Spencer 6.

———, *Kitāb ṣuwar al-kawākib al-ṯābita*. MS New York, New York Public Library, Spencer 25.

———, *Kitāb ṣuwar al-kawākib al-ṯābita*. MS Paris, BnF, Arabe 2490.

———, *Kitāb ṣuwar al-kawākib al-ṯābita*. MS Paris, BnF, Arabe 4670.

———, *Kitāb ṣuwar al-kawākib al-ṯābita*. MS Paris, BnF, Supplément Persan 1551.

———, *Kitāb ṣuwar al-kawākib al-ṯābita*. MS St Petersburg, Public Library, Ar. 119.

———, *Kitāb ṣuwar al-kawākib al-ṯābita*. MS Toronto, Royal Ontario Museum, Inv. no. 971.292.13.

Suʻudi, Mehmed, *Maṭālīʻ al-Saʻādāt*. MS Paris, BnF, Supplément turc 242.

al-Ṭūsī, Naṣīr al-Dīn, *Taḥrīr Kitāb al-uṣūl fī l-handasa wa-l-ḥisāb*. MS London, British Library, Add. 21952.

B Other sources

ʻAbbāsī, Jalāl ʻAbbās, 2001: *Catalogue of Manuscripts in the Niẓāmi Collection at the Maulana Azad Library*. Delhi, Idārah-i Adabīyat-i Delhi.

Afšār, Īrāj; Dāniš-Pažūh, Muḥammad Taqī (eds.), 1352š: *Fihrist-i Kitābhā-yi Ḫaṭṭī-yi Kitābḫāna-yi Millī-yi Malik dābeste beh Āstān-i Quds Rižavī*, vol. 1. Kitābhā-yi ʻArabī va-Turkī, Tehrān, Čāpḫāna-yi Dānišgāh.

——— (eds.), 1354š: *Fihrist-i Kitābhā-yi Ḫaṭṭī-yi Kitbāḫāna-yi Millī-yi Malik dābeste beh Āstān-i Quds-i Rižavī*, vol. 2. Kitābhā-yi Fārsī, Tehrān, Čāpḫāna-yi Dānišgāh.
Brentjes, Sonja, 1999: "The Interests of the Republic of Letters in the Middle East, 1550–1700," *Science in Context*, 12, 435–468.
———, 2001: "On the Relation between the Ottoman Empire and the West European Republic of Letters (17th–18th centuries)," in: A. Çaksu, (ed.), *International Congress on Learning & Education in the Ottoman World*, Istanbul, April 12–15, 1999. Istanbul, IRCICA, 121–148.
———, 2003: "'Renegades' and Missionaries as Minorities in the Transfer of Knowledge," in: E. Ihsanoglu; K. Chatzis; E. Nicolaidis (eds.), *Multicultural Science in the Ottoman Empire*. Brepols, Turnhout, 63–70.
../../../../Geek Squad Data Backup 7.27.2020/Users/Paige/Desktop/15031s/15031-5677 Brentjes/03 from CE/15031-5677-Ref Mismatch Report.docx - LStERROR_89———, 2013: "Giacomo Gastaldi's Maps of Anatolia: The Evolution of a Shared Venetian-Ottoman Cultural Space?," in: A. Contadini; C. Norton (eds.), *The Renaissance and the Ottoman World*. Farnham (Surrey), Ashgate, 123–141.
———, 2014: "Safavid Art, Science, and Courtly Education in the Seventeenth Century," in: N. Sidoli; G. Van Brummelen (eds.), *From Alexandria, Through Baghdad*. Springer, Berlin, Heidelberg, 487–502.
Broton, Jerry, 1998: *Trading Territories*. Ithaca, Cornell University Press.
Charette, François, 2006: "The Locales of Islamic Astronomical Instrumentation," *History of Science*, 44, 123–138.
Dāniš-Pažūh, Muḥammad Taqī (ed.), 1351š.: *Fihrist-i Nusḫahā-yi Ḫaṭṭī dar Čahar Kitābḫāna-yi Mašhad*. Tehrān, Čāpḫāna-yi Bahmān.
Dāniš-Pažūh, Muḥammad Taqī; Anvarī, Bahā' al-Dīn 'Ilmī (eds.), 1976: *Fihrist-i Kitābhā-yi Ḫaṭṭī-yi Kitābḫāna-yi Majlis-i Ṣafā'*, vol. 1. Tehrān: Čāpḫāna-yi Majlis-i Shūrā-yi Millī.
de Young, Gregg, 1986: "The Ḫulāṣat al-ḥisāb of Bahā' al-Dīn 'Āmilī and the Dār-i Nizāmī in India," *Gaṇita Bhārati*, 8, 1–15.
Edwards, Holly, 1982: *Patterns and Precision: The Arts and Sciences of Islam*. Washington, DC, National Committee to Honor the Fourteenth centennial of Islam.
Ethé, Hermann, 1903: *Catalogue of Persian Manuscripts in the Library of the India Office*, 2 vols., Oxford, Oxford University Press.
Fabris, Antonio, 1991: "Artisanat et culture: recherches sur la production vénitienne et le marché ottoman au XVIe siècle," *Arab Historical Review for Ottoman Studies*, 3–4, 51–60.
Hajek, Lubor, 1960: *Indische Miniaturen vom Hof der Mogulkaiser*. Prague, Artia.
Hogendijk, Jan P., 2007: "Similar Mathematics in Different Cultures: Jamshīd Al-Kāšī and Ludolph van Ceulen on the Determination of π," in: F. Daelemans; J.-M. Duvosquel; R. Halleux; D. Juste (éds.), *Mélanges offerts à Hossam Elkhadem par ses amis et ses élèves*. Bruxelles, Archives et Bibliothèques de Belgique, numéro spécial 83, 189–120.
Kheirandish, Elaheh, 2000: "A Report on Iran's 'Jewel' Codices of Tusi's 'Kutub al-Mutawassitat'," in: N. Pourjavady; Ž. Vesel (éds.), *Nasir al-Din Tusi*. Philosophe et savant du XIIIe siècle. Tehran, Institut Français de Recherche en Iran, 131–144.
King, David A., 1986: *A Survey of the Scientific Manuscripts in the Egyptian National Library*. Publications of the American research center in Egypt, Catalogs, vol. 5. Winona Lake, Indiana, Eisenbrauns.

———, 1999: *World-Maps for Finding the Direction and Distance to Mecca. Innovation and Tradition in Islamic Science*. Leiden and London, Brill/al-Furqan, Islamic Heritage Foundation.

———, 2004: *In Synchrony with the Heavens. Studies in Astronomical Timekeeping and Instrumentation in Medieval Islamic Civilization* (Studies I-IX). Leiden, Brill.

Küçük, Bekir Harun, 2005: *Contexts and Constructions of Ottoman Science with Special Reference to Astronomy*. MA Thesis, Sabancı University, Faculty of Arts and Social Sciences.

Lentz, Thomas W.; Lowry, Glenn D. (eds.), 1989: *Timur and the Princely Vision: Persian Art and Culture in the Fifteenth Century*. Los Angeles, County Museum of Art Washington, Arthur M. Sackler Gallery, Smithsonian Institution.

Lucas, Paul, 1704: *Voyage du Sieur Paul Lucas au Levant. On y trouvera entr'autre une description de la Haute Egypte, suivant le cours du Nil, depuis le Caire jusques aux Cataractes, avec une Carte exacte de ce fleuve, que personne n'avoit donné*, 2 vols. Paris, Guillaume Vandive.

Mar'ašī, Maḥmūd; Ḥusaynī, Aḥmad (eds.), 1352š.–1395 squ.: *Fihrist-i Nusḫahā-yi Ḫaṭṭī-yi Kitāḫāna-yi 'Umūmi-yi Ḥażrat-i Āyatullāh al-'Uẓmā Mar'ashī Najafī*. Qom, Čāpḫāna-yi Mihr-i Ustuvār.

Matvievskaya, Galina Pavlovna; Rozenfel'd, Boris Abramovič, 1985: – *Matematiki i astronomy musul'manskogo srednevekov'ya i ich trudy (VIII-XVII vv.)*, 3 vols, vol. 2, Moscow, Nauka.

Maulavi, Abdul (ed.), 1925: *Catalogue of the Arabic and Persian Manuscripts in the Oriental Public Library at Bankipore*, vol. IX (Persian MSS.), Philology and Sciences. Patna, Government Printing.

——— (ed.), 1927: *Catalogue of the Arabic and Persian Manuscripts in the Oriental Library at Bankipore*, vol XI (Persian MSS). Sciences (continued) and Arts. Government Printing.

——— (ed.), 1937: *Catalogue of the Arabic and Persian Manuscripts in the Oriental Public Library at Bankipore*, vol. XXII (Arabic MSS.), Science. Patna, Government Printing.

Monzavī, Aḥmad (ed.), 1341: *Fihrist-i Nusḫahā-yi Ḫaṭṭī-yi Markaz al-Ma'ārif-i Buzurg-i Islāmī*, 2 vols. Tehrān.

al-Naqšbandī, Usāmā' N.; 'Abbās, Ẓamiyā' M. (eds.), 1982: *Maḫṭūṭāt al-Falak wa-l-Tanjīm fī Maktabat al-Matḥaf al-'Irāqī – Astronomical and Astrological Manuscripts of the Iraq Museum Library*. Baghdad: Wizārat al-Thaqāfa wa-l-I'lām.

Osmanlı Astronomi Literatürü Tarihi (History of Astronomy [sic] Literature During the Ottoman Period). 1997, 2 vols., vol. 1. Istanbul, IRCICA.

Osmanlı Matematik Literatürü Tarihi (History of Mathematical Literature During the Ottoman Period). 1999, 2 vols., vol. 1. Istanbul, IRCICA.

Ragep, F. Jamil, 2005: "'Alī Qušjī and Regiomontanus: Eccentric Transformations and Copernican Revolutions," *Journal for the History of Astronomy*, 36, 359–371.

Rahman, A., 1982: *Science and Technology in Medieval India: A Bibliography of Source Materials in Sanskrit, Arabic and Persian*. In collaboration with M. A. Alvi; S. A. Khan Ghori; K. V. Samba Murthy. New Delhi, Indian National Science Academy.

Richard, Francis, 1997: *Splendeurs persanes. Manuscrits du XIIe au XVIIe siècle*. Paris, Bibliothèque nationale de France.

Roberts, Sean, 2006: *Cartography between Cultures: Francesco Berlinghieri's Geographia of 1482*. Doctoral dissertation, University of Michigan, Ann Arbor.

Šākirī, Ramażān ʻAlī, 1367š.: *Ganj-i Hazār Sāle-yi Kitābḫāna-yi Markazī-yi Āstān-i Quds-i Riḍavī qabla va baʻd az Inqilāb*. Mašhad, Intišārāt-i Kitābḫāna-yi Markazī.

Saliba, George, 1993: "Al-Qušjī's Reform of the Ptolemaic Model for Mercury," *Arabic Science and Philosophy*, 3, 161–203.

———, 2004: "Šams al-Dīn al-Ḫafrī's (d. 1550) Last Work on Theoretical Astronomy," in: N. Pourjavady; Ž. Vesel (eds.), *Sciences, techniques et instruments dans le monde iranien* (Xe XIXe siècles). Tehran, Institut Français de Recherche en Iran, 55–66.

Šānečī, Kāẓim Mudīr; Nūrānī, ʻAbd Allāh; Bīnaš, Taqī (eds.), 1351š.: *Fihrist-i Nusḫehā-yi Ḫattī dar Do Kitābḫāna-yi Mašhad*. Tehran, Čāpḫāna-yi Bahmān.

Sarma, Śreeramula Rajeśwara, 2003: *Astronomical Instruments in the Rampur Raza Library*. Rampur, Raza Library.

Schmitz, Barbara, 1992: *Islamic Manuscripts in the New York Public Library*. With Contributions by L. Khayyat; S. Soucek; M. Pourfarrokh. New York and Oxford, The New York Public Library/Oxford University Press.

Schmitz, Barbara; Desai, Ziyauddin, 2006: *Mughal and Persian Paintings and Illustrated Manuscripts in The Raza Library, Rampur*. New Delhi, Indira Gandhi National Centre for the Art.

Sezgin, Fuat, 1975: *Geschichte des Arabischen Schrifttums*, vol. 5. Leiden, Brill.

Sothebys, 1997: *Oriental Manuscripts and Miniatures*. London, Wednesday 23 April 1997.

Subḥānī, Hāšimpūr; Āq Sū, Ḥussām al-Dīn (eds.), 1374š./1995–1996: *Fihrist-i Nusḫahā-yi Ḫaṭṭī-yi Fārsī-yi Kitābḫāna-yi Dānišgāh-i Istānbūl*. Tehran, Pažūhišgāh-i ʻulūm-i insānī va-muṭālaʻāt-i farhangī.

Ṭāyār, Maḥmūd (ed.), 1378: *Fihrist-i Nusḫahā-yi Ḫaṭṭī-yi Kitābḫāna-yi Imām-i Ṣādiq-i Qazvīn*, vol. 1. Qazvīn, Marā'ī.

Van Dalen, Benno, 2023: "Mathematical Commentary on the Horoscope of Iskandar," ZGAIW.

Velā'ī, Mehdī, 1380: *Fihrist-i Kutub-i Ḫaṭṭī-yi Kitābḫāna-yi Markazī va-Markaz-i Isnād-i Āstān-i Quds-i Riżavī*, vol. 18. Mašhad, Intišārāt-i Kitābḫāna-yi Markazī (*riyāḍiyāt*).

Vesel, Živa, 2001: "Science and Scientific Instruments," in: C. Parham (ed.), *Islamic Period: Applied and Decorative Arts: The Cultural Continuum*. Vol. 3 of *The Splendour of Iran*. N. Pourjavady, General Editor, London, Booth-Clibborn Editions, 260–311.

Waley, Muhammad Isa, 1998: *Supplementary Handlist of Persian Manuscripts, 1966–1998*. London, the British Library.

Welch, Anthony, 1972: *Collection of Islamic Art: Prince Sadruddin Aga Khan*, 4 vols. Geneva, Chateau de Bellerive.

Winter, H. J. J.; Mirza, Ashraf, 1952: "Concerning the Persian Version of Lilavati," *Journal of the Asiatic Society*, 18, 1–10.

Ziriklī, Ḫayr al-Dīn, 1979: *Al-Āʻlam*, vol. 4. Beirut, Dār al-ʻilm li-l-malāyīn.

C Websites

http://facstaff.uindy.edu/~oaks/Biblio/IslamicMathBiblio.htm

APPENDIX

The following three tables present the data on which this paper relied. Table 4.1 surveys dated copies of mathematical, astronomical and astrological works made in the Safavid period. In column one the author of the text if known, the name of an editor if applicable, and the name of an eventual commentator are given. The next column gives the title of each work. Columns three and four list the year and place of the copy. In the last column, all other additionally known information about scribe/s, addressee/s, owner/s etc. is registered. Table 4.2 lists dedicated mathematical works of the Safavid period. Columns one to four follow the usage of Table 4.1. Column five gives the names of people to whom the text was dedicated. Table 4.3 brings together Safavid copies of mathematical works by ancient authors. Column one to four follow the pattern of Table 4.1. Column five contains names of addressees, owners, and copyists.

Table 4.1 Dated Copies Of Mathematical, Astronomical And Astrological Works Made In The Safavid Period

Author	*Title*	*Year of copy*	*Place of copy*	*Additional information*
Anonymous – Ṭūsī	*Šarḥ-i Sī faṣl*	918	Shamakhi	copyist Aḥmad Iṣfahānī
Ptolemy – Naṣīr al-Dīn Ṭūsī	*Tarjama-yi ṯamara-yi Batlamyus*	919		
Abū Muḥammad b. Aḥmad Bihištī Isfarā'aynī	*al-Ḥisāb*	7 Jumāda II 927	Kashan	copyist Ḥasan b. Ġiyāṯ al-Dīn Astarābādī
Niẓām al-Dīn Birjandī	*Šarḥ al-Taḏkira*	929		Muḥammad Amīn
Šams al-Dīn Ḫafrī	*al-Takmila fī l-Taḏkira*	Ṣafar 932		with glosses by Bahā' al-Dīn 'Āmilī

(*Continued*)

Table 4.1 Continued

Author	*Title*	*Year of copy*	*Place of copy*	*Additional information*
Qāżīzāda Rūmī	*Šarḥ Aškāl al-ta'sīs*	937		
anonymous	*Sī bāb dar asṭurlāb*	Ṣafar 941 or 961		copyist Muḥammad Ḥusayn
Ġiyāṯ al-Dīn Manṣūr Daštakī Šīrāzī – Qāżīzāda Rūmī, Čaġmīnī	*Ḥāšiya 'alā šarḥ al-Mulaḫḫaṣ*	954	Shiraz, Madrasa-yi Manṣūriyya	copyist Aḥmad b. Muḥammad Naṣr Allāh
Mīr Ḥusayn b. Mu'īn al-Dīn Maybudī – Qāżīzāda Rūmī, Čaġmīnī	*Ḥāšiya 'alā šarḥ al-Mulaḫḫaṣ*	954	Shiraz, Madrasa-yi Manṣūriyya	copyist Aḥmad b. Muḥammad Naṣr Allāh
Ptolemy – Naṣīr al-Dīn Ṭūsī	*Taḥrīr al-Majisṭī*	Jumāda 956	Qazvin	Muḥammad Ḥusayn b. 'Alī
Abū 'Alā' Bihištī Isfarā'aynī	*Risāla dar ḥisāb va-jabr o muqābala*	956		
Šams al-Dīn Ḫafrī	*al-Takmila fī l-Taḏkira*	960		copyist Muḥammad Ṣādiq
Quṭb al-Dīn Šīrāzī	*al-Tuḥfa al-šāhiyya*	25 Rabī' II 960	Isfahan	copyist Sa'd al-Dīn Muḥammad b. Kamāl al-Dīn Iṣfahānī
'Alī Qušjī	*Hay'a*	961		
Niẓām al-Dīn Birjandī	*Ab'ād wa-ajrām*	961		
Naṣīr al-Dīn Ṭūsī	*Bīst bāb dar aṣṭurlāb*	971	Mašhad	copyist Jalāl al-Dīn Muḥammad; donor to Āqā Zayn al-'Ābidīn, 1166
Niẓām al-Dīn Birjandī	*Bīst bāb dar taqvīm*	971		

Author	*Title*	*Year of copy*	*Place of copy*	*Additional information*
Abū al-Ḫayr Muḥammad Taqī Fārisī	*Ḥall al-taqwīm*	971		
ʿAlī Qušjī	*Faṭḥiyya*	975	Mašhad	copyist Muḥammad Mu'min Aḥmad; donor to Āqā Zayn al-ʿĀbidīn, 1166; same manuscript as the previous text
Qāżīzāda Rūmī	*Šarḥ al-Mulaḫḫaṣ*	975	Mašhad	copyist Muḥammad Bāqir b. ʿAbd al-Qādir b. Hibat Allāh Ḥusaynī
Niẓām al-Dīn Birjandī	*Ajrām-i suflī va avżāʿ-i ajrām-i ʿulvī*	2 Ḏū l-Ḥijja 976		copyist Ġulām ʿAlī b. Darvīš ʿAlī Mašhadī
anonymous	*Misāḥa*	1 Ramażān 981		
anonymous	*Risāla dar ḥisāb*	17 Rajab 982	Shiraz	copyist Ṣadr al-Islām Ibn ʿAlī al-Mawālī; owner Mīrzā Ḥakīm; worked on rules of arithmetic with his teacher ʿAbd al-Raḥīm
Abū al-Ḥasan b. Aḥmad Šarīf Qā'inī	*Muḫtaṣar Rawḍ al-janān*	983	Mašhad	copyist Malik Ḥusayn b. Muḥammad Ṭabasī
Taqī al-Dīn Abū l-Ḫayr Muḥammad b. Muḥammad Fārisī	*Ḥall al-taqwīm*			copyist ʿAlī b. Niʿmat Allāh Ḥusaynī
anonymous	*Taqvīm*	19 Šaʿbān 986		copyist ʿAlī b. Niʿmat Allāh Ḥusaynī; from the same manuscript as previous text

(*Continued*)

Table 4.1 Continued

Author	*Title*	*Year of copy*	*Place of copy*	*Additional information*
Bahā' al-Dīn 'Āmilī	*Tašrīḥ al-aflāk*	11 Rab'ī II 986		copyist 'Abd al-Fattāḥ b. Maṣ'ūd Šarīf, for Sayyid Sa'īd
Muḥammad Ṭabasī	*Misāḥa*	19 Ramażān 986		copyist 'Alī b. Ni'mat Allāh Ḥusaynī; dedicated to Ḫwāja Niẓām al-Dīn Mīrijān; from the same manuscript as the previous text
Niẓām al-Dīn Birjandī – Ṭūsī	*Šarḥ al-Taḏkira*	987		
Jalāl al-Dīn Davvānī (?) – Ṭūsī	*Šarḥ-i Sī faṣl Ṭūsī*	2 Ḏū l-Qa'da 990		
Muḥammad Ašraf b. M. Ja'far Iṣfahānī	*Šarḥ-i Zīj-i Ilḫānī*	Rajab 996	Yazd	
Naṣīr al-Dīn Ṭūsī	*Bīst bāb*	996	Bistam	
Niẓām al-Dīn Nīšābūrī	*al-Šamsiyya*	999		
Šams al-Dīn Ḫafrī	*al-Takmila fī l-Taḏkira*	4 Ḏū l-Qa'da 1001		
anonymous	*Risāla fī ma'rifat al-aṣṭurlāb*	5 Ša'bān 1006		copyist Ismā'īl
Nūrallāh b. Muḥammad Ḥusaynī	*Risāla dar ma'rifat-i aṣṭurlāb*	1006		
'Abd al-Raḥmān Ṣūfī	Kitāb Ṣuwar al-kawākib al-ṯābita	Ḏū l-Ḥijja 1006		
Naṣīr al-Dīn Ṭūsī	*Bīst bāb*	1008	Bistam	
Bahā' al-Dīn 'Āmilī	*Tašrīḥ al-aflāk*	1008		
Euclid – Naṣīr al-Dīn Ṭūsī	*Taḥrīr al-Uṣūl*, Books XIV-XV	Ramażān 1009		
Niẓām al-Dīn Nīšābūrī – Ṭūsī	*Tawḍīḥ al-Taḏkira*	1009		copyist 'Ināyat Allāh Ḥusayn Iṣfahānī
'Alī Qušjī	*Hay'a*	1009		

Author	*Title*	*Year of copy*	*Place of copy*	*Additional information*
Bahā' al-Dīn 'Āmilī	*Ḫulāṣat al-ḥisāb*	1012		copyist 'Alī b. Aḥmad Nabāṭī
Niẓām al-Dīn Birjandī	*Šarḥ-i Zīj-i Uluġ Beg*	Rajab 1014		copyist Muḥammad Zamān b. Ḥusayn Ḫātūnābādī
Bahā' al-Dīn 'Āmilī	*Ḫulāṣat al-ḥisāb*	1012		copyist 'Alī b. Aḥmad Nabāṭī
Niẓām al-Dīn Birjandī	*Šarḥ-i Zīj-i Uluġ Beg*	Rajab 1014		copyist Muḥammad Zamān b. Ḥusayn Ḫātūnābādī
Bahā' al-Dīn 'Āmilī	*al-Ṣafīḥa*	1014	Yazd	copy of autograph by Abū l-Qāsim Yaḥyā Astarābādī
Theodosios	*Taḥrīr Kitāb al-Ukar*, Book III	1014	Yazd	copyist and owner Abū l-Qāsim Yaḥyā Astarābādī; in the same manuscript as the previous text
Theodosios	Fī l-Masākin	1014	Yazd	copyist and owner Abū l-Qāsim Yaḥyā Astarābādī; in the same manuscript as the previous text
Archimedes Theodosios Menelaos – Naṣīr al-Dīn Ṭūsī	*Taḥrīr Kitāb al-Kura wa-l-ustuwāna* *Taḥrīr Kitāb al-Ukar* *Taḥrīr Kitāb Mānālā'ūs fī l-aškāl al-kuriyya*	1015		'Alī Jān b. Ḥaydar 'Alī Haravī

(*Continued*)

Table 4.1 Continued

Author	*Title*	*Year of copy*	*Place of copy*	*Additional information*
Quṭb al-Dīn ʿAbd al-Ḥayy b. ʿIzz al-dīn Ḥusaynī Lārī	*Ḥall va-ʿaqd*	Rabīʿ II 1017	Lar	
Abū l-ʿAlāʾ Muḥammad Bihištī Isfaraʾaynī	*Risāla dar ḥisāb*	1018	Qom	in the same manuscript as the texts copied in 1014 in Lar
ʿAlī Qušjī	*Šarḥ-i Zīj-i jadīd-i sulṭānī*	1018		
Qāżīzāda Rūmī	*Šarḥ al-Mulaḫḫaṣ*	24 Muḥarram 1019	Ganja	copyist Muḥammad Qulī
ʿAlī Qušjī	*Risāla dar ḥisāb*	1020		
Aṯīr al-Dīn al-Abharī	*Ġāyat al-idrāk fī dirāyat al-aflāk*	1021	Mašhad	copyist ʿAbd al-Ġaffūr b. Maṣʿūd al-Ḫāqānī
Jalāl al-Dīn Muḥammad b. ʿAbd Allāh Yazdī	Tuḥfat al-munajjim	1021		
Ḥusayn Kāšifī	*Lavāʾiḥ al-qamar*	1021		
Maḥmūd or Muḥammad Sirāj	*Maʿrifat-i taqvīm*	2 Rabīʿ I 1021	capital of Daylam	annotated by Abū l-Qāsim Istajlū
Theodosios Archimedes Banū Mūsā – Naṣīr al-Dīn Ṭūsī	*Taḥrīr al-Masākin Taḥrīr Kitāb al-Mafrūḍāt Taḥrīr Maʿrifat misāḥat al-aškāl Taḥrīr al-Kura wa-l-ustuwāna Taḥrīr Maqālat Aršimīdīs fī Taksīr al-dāʾira*	Muḥarram-Rabīʿ I 1022		copyist Muḥammad b. Ṭāhir Fāżil Kāšānī

Author	*Title*	*Year of copy*	*Place of copy*	*Additional information*
Quṭb al-dīn Šīrāzī	*Iḫtiyārāt-i Muẓaffarī*	1023		copyist Maṣ'ūd b. Ḥabīb Allāh
Naṣīr al-Dīn Ṭūsī	*Sī faṣl*	1023	Isfahan	copyist 'Alī b. Šāhīd 'Alī Turkumānī
Naṣīr al-Dīn Ṭūsī	*Kašf al-qinā' 'an asrār šakl al-qattā'*	13 Jumāda I 1022; compared in 1024		copyist Muḥammad b. Ṭāhir Fāżil Kāšānī; from the same manuscript as the previous texts
Ptolemy – Ṭūsī	*Taḥrīr al-Majisṭī*	1027	Isfahan	copyist Abū Ja'far Kāfī b. Muḥtašam b. 'Amīd b. Muḥammad Šāhinšāh Qā'inī who studied the work in Isfahan
Ġiyāṯ al-Dīn Kāšī	*Zīj-i Ḫāqānī dar Takmīl-i Zīj-i Ilḫānī*	1028		copyist Ibn Zayn al-'Ābidīn Yazdī
Qāżīzāda Rūmī – Čaġmīnī	*Šarḥ al-Mulaḫḫaṣ*	8 Ḏu l-Ḥijja 1039	Qal'a-yi Aḥmad Bakr	copyist Muḥammad Taqī Mašhadī Ḥusaynī
Muẓaffar b. Muḥammad Qāsim Gunābādī	*Tanbīhāt al-munajjimīn*	1031		
Muḥammad b. Ayyūb Ḥāsib Ṭabarī	*Istiḫrāj*	28 Ṣafar 1031	Dār al-Salṭanat-i Farahabad	copyist Muḥammad Rīżā b. Sulṭān Muḥammad Tabrīzī
Kūšyār b. Labbān	*Tarjama-yi Mujmal al-uṣūl (Rajab 809)*	Jumāda II 1031	Dār al-Salṭanat-i Farahabad	copyist Muḥammad Rīżā b. Ḥajjī Sulṭān Muḥammad Tabrīzī; from the same manuscript as the previous text

(*Continued*)

Table 4.1 Continued

Author	*Title*	*Year of copy*	*Place of copy*	*Additional information*
ʻAlī Qušjī	*Risāla dar ḥisāb*	1031		
Qāżī Kāšif Ardakānī Yazdī	*al-Durūs al-falakiyya*	Rajab 1032		copyist Muḥammad Taqī b. Muḥammad Rīżā Rāzī, student of the author, copied the text from an autograph
Quṭb al-Dīn ʻAbd al-Ḥayy b. ʻIzz al-dīn Ḥusaynī Lārī	*Ḥall va-ʻaqd*	28 D̲ū l-Qaʻda	Isfahan	
anonymous	*al-Hay'a*	13 Jumāda I 1037		
Niẓām al-Dīn Nīšābūrī	*al-Šamsiyya*	1037		Aḥmad Taqī b. Faqīh Muḥammad Aṣmaʻī copied for himself; glosses from *Miftāḥ al-ḥāsib* (sic)
Amīr Muʻīn al-Dīn Muḥammad Ašraf b. Ḥabīb Allāh Šīrāzī	*Šarḥ Ḫulāṣat al-ḥisāb*	9 D̲ū l-Qaʻda 1038	Isfahan	student of Bahā' al-Dīn ʻĀmilī
Kūšyar b. Labbān (?) – Ptolemy	*Mujmal al-uṣūl = Arbaʻ maqālāt*	1038		
Ptolemy	*Arbaʻ maqālāt*	26 D̲ū l-Ḥijja 1038		in the same manuscript as the previous text; manuscript has a note by Muḥammad b. Ḥājjī, secretary of Šāh Ḥusayn; has two seals of Muḥammad Raḥīm Munajjim from 1120 and 1125 (see Table 4.2)

Author	*Title*	*Year of copy*	*Place of copy*	*Additional information*
Qāżīzāda Rūmī – Maḥmūd b. Muḥammad Čaġmīnī	*Šarḥ al-Mulaḫḫaṣ fī l-hay'a*	8 Ḏū l-Ḥijja 1039	Qal'a-yi Aḥmad Bakr	copyist Muḥammad Taqī Mašhadī
Niżām al-Dīn Nīšābūrī	*Tawḍīḥ al-Taḏkira*	4 Rajab 1041		copyist 'Alī b. Maṣ'ūd Ḥusaynī Tafrīšī Qummī
'Alī Qušjī	*Risāla dar ḥisāb*	1041		
Muḥammad Ḫāzinī	*Muḫtaṣar al-Majisṭī*	1041		
Mīrzā Muḥammad Ṭahir b. Ḥusayn Ḫān Vāḥid Qazvīnī (d. 1112)	*al-Ḥisāb*	composed in 1042		
Ġiyāṯ al-Dīn Manṣūr Daštakī Šīrāzī	*Miftāḥ al-munajjimīn*	Rabī' II 1042		
Qāżīzāda Rūmī	*Šarḥ al-Mulaḫḫaṣ*	1043	Dār al-salṭanat Isfahan, Madrasa Luṭf Allāh	copyist Ḥasan Muḥammad Taqī
Ḥasan b. Sa'd Qā'inī – 'Abd al-Raḥmān Ṣūfī	Persian translation of *Kitāb Ṣuwar al-kawākib al-ṯābita*	1043	Mašhad, court of Manūchihr Ḫān	
Ḥasan b. Sa'd Qā'inī – 'Abd al-Raḥmān Ṣūfī	Persian translation of *Kitāb ṣuwar al-kawākib al-ṯābita*	1044	Mašhad, court of Manūchihr Ḫān	
Naṣīr al-Dīn Ṭūsī	*Mu'īniyya*	17 Šawwāl 1049		copyist 'Abd al-Bāqī Muḥammad Akbar

(*Continued*)

Table 4.1 Continued

Author	*Title*	*Year of copy*	*Place of copy*	*Additional information*
Jāmāsb, vizier Guštasb (ascribed)/ translator Mīrzā ʻAbd Allāh Efendī Iṣfahānī (ascribed)	*Jāmāsb-nāma*	1049		
ʻAbd Allāh b. Muḥammad b. ʻAbd al-Razzāq Ḥāsib	*al-Fawā'id al-Bahā'iyya fī al-qawāʻid al-ḥisābiyya*	Ḏū l-Ḥijja 1050		
anonymous	*Ḥāšiya fī l-Fawā'id al-Bahā'iyya*	Ḏū l-Ḥijja 1050	Mašhad	from the same manuscript as the previous text
Bahā' al-Dīn ʻĀmilī	*al-Kur*	Ḏū al-Ḥijja 1050	Mašhad	from the same manuscript as the previous text
anonymous – Ṭūsī	*Šarḥ-i Bīst bāb*	Muḥarram 1051		
Ptolemy – Ṭūsī	*Taḥrīr al-Majisṭī*	Muḥarram 1051	Mašhad	copyist Muḥammad Ṣādiq b. ʻAbd al-ʻAlī Taršīzī; copy is from the manuscript copied by Abū Jaʻfar Qā'inī in Isfahan in 1027
ʻAlī b. ʻĪsā al-Aṣṭurlābī al-Ḥarrānī	*al-Aṣṭurlāb*	1051		copyist Muḥammad Taqī b. Ḥaydar Ḥusaynī; from the same manuscript as the previous text

Author	*Title*	*Year of copy*	*Place of copy*	*Additional information*
anonymous	*al-Rub' al-mujayyab*	1051		copyist Muḥammad Taqī b. Ḥaydar Ḥusaynī; from the same manuscript as the previous text
Abū Ma'šar	*al-Aṣṭurlāb*	1051		copyist Muḥammad Taqī b. Ḥaydar Ḥusaynī; from the same manuscript as the previous text
Ibn Abī Šukr al-Maġribī	*Tasṭīḥ al-aṣṭurlāb*	1051		copyist Muḥammad Taqī b. Ḥaydar Ḥusaynī; from the same manuscript as the previous text
Bahā' al-Dīn 'Āmilī	*Tašrīḥ al-aflāk*	1052	Mašhad	copyist Ġulām Lafīnī
anonymous	*al-Jāmi'*	1052		
Ibn Abī Šukr al-Maġribī	*al-Aḥkām al-nujūmiyya*	1052		from the same manuscript as the previous text
anonymous	*al-Aḥkām al-juz'iyya*	28 Ḏū l-Qa'da 1052		from the same manuscript as the previous text
Bahā' al-Dīn 'Āmilī	*al-Ṣafīḥa*	1053		copyist Ḥusayn b. Ḥājj Muḥammad al-Kāẓimī
Theodosios – Naṣīr al-Dīn Ṭūsī	*Taḥrīr Kitāb al-Ukar*	Muḥarram 1053		
Qāżīzāda Rūmī	*Šarḥ al-Mulaḫḫaṣ*	1054	Isfahan	

(*Continued*)

Table 4.1 Continued

Author	*Title*	*Year of copy*	*Place of copy*	*Additional information*
Sayyid Šarīf Jurjānī – Čaġmīnī	*Šarḥ al-Mulaḫḫaṣ*	Rabī‘ I 1055	Rūstā-yi Jab‘	copyist Muḥammad Ḥasan Ḥurr ‘Āmilī
Ibn Abī Šukr al-Maġribī	*Kitāb yuštamilu ‘alā risāla min ‘ilm al-aḥkām*	1055		copyist Kalb ‘Alī b. Mullā Javād Kāẓimī
Ibn Abī Šukr al-Maġribī	*Risāla fī iqtirānāt al-kawākib wa-mā yadullu bihā*	Ramażān 1055		copyist Kalb ‘Alī b. Mullā Javād Kāẓimī
Uluġ Beg *et al.*	*Zīj-i jadīd-i sulṭānī*	22 Ramażān 1056	Isfahan Masjid-i Jāmi‘-i ‘Abbāsi	copyist M. Rašīd Ḥasanī; owner Qāżī M. Šarīf b. Mullā Musṭafā, Šayḫ al-Islām, Ardelān
Jalāl al-Dīn Muḥammad b. ‘Abd Allāh Yazdī	*Tuḥfat al-munajjimīn*	1056	Mašhad	
Autolykos – Ṭūsī	*Taḥrīr al-Kura al-mutaḥarrika*	1056		
Uluġ Beg *et al.*	*Zīj-i jadīd-i sulṭānī*	6 Ḏū al-Qa‘da 1057		copyist Muḥammad Taqī
Abū l-Rayḥān Bīrūnī	*Kitāb al-isti‘āb li-l-wujūh al-mumkina fī ṣīnā‘at al-aṣṭurlāb*	Ḏū al-Ḥijja 1057		copyist ‘Abd Allāh al-Laṭīf Muḥammad Šarīf b. Ḥājjī Maqṣūd ‘Alī al-Iṣfahānī
anonymous	*Nujūm* (?)	1057		copyist Muḥammad Ḥusayn b. Muḥammad Yūsuf Ṭāliqānī; collection was written for the astrologer Muḥammad Rīżā b. Muḥammad Taqī

Author	*Title*	*Year of copy*	*Place of copy*	*Additional information*
Ibn Abī Šukr al-Maġribī	*Aḥkām taḥāwīl sīnī al-ʿālam*	Ḏū al-Qaʿda 1057		copyist Muḥammad Ḥusayn b. Muḥammad Yūsuf Ṭāliqānī; from the same manuscript as the previous text
anonymous	*Aḥkām-i qirānāt-i kavākib*	1057		copyist Muḥammad Ḥusayn b. Muḥammad Yūsuf Ṭāliqānī; from the same manuscript as the previous text
anonymous	*Aḥkām-i iḥtirāqāt*	1057		copyist Muḥammad Ḥusayn b. Muḥammad Yūsuf Ṭāliqānī; from the same manuscript as the previous text
Euclid – Ṭūsī	*Taḥrīr al-Muʿṭayāt*	1057		
Menelaos – Ṭūsī	*Taḥrīr Manālā'ūs fī l-aškāl al-kuriyya*	1057		
Ḥāfiz Ḥasan b. Šujāʿ b. Muḥammad b. Ḥasan Tūnī	*Dalīl al-munajjimīn*	19 Rajab 1058		copyist Muḥammad Taqī b. Aḥmad ʿAlī Mašhadī
Archimedes – Ṭūsī	*Taḥrīr al-Ma'ḫūḏāt*	1058		
anonymous – Kūšyār b. Labbān	*Tarjama-yi Mujmal uṣūl al-aḥkām*	3 Muḥarram 1059		copyist Muḥammad Taqī b. Aḥmad ʿAlī Mašhadī; from the same manuscript as the previous text

(*Continued*)

Table 4.1 Continued

Author	*Title*	*Year of copy*	*Place of copy*	*Additional information*
Šams al-Dīn Ḫafrī – Ṭūsī	*al-Takmila fī l-Taḏkira*	7 Muḥarram 1059		
Theodosios – Ṭūsī	*Taḥrīr Kitāb al-ukar*	Muḥarram 1059		
Bahā' al-dīn 'Āmilī	*Ḫulāṣat al-ḥisāb*	1059		
Niżām al-Dīn Birjandī – Ṭūsī	*Šarḥ al-Taḏkira*	Rabī' I 1061		copyist Muḥammad b. Sa'īd b. Muḥammad Amīn Kāšānī
Muḥammad Bāqir b. Zayn al-'Ābidīn	*Maṭla' al-anvār*	1061		dedicated to Ṣafī; copyist Muḥammad Sa'īd b. Faḫr al-Dīn
Bahā' al-Dīn 'Āmilī	*Tuḥfa-yi ḥātimiyya*	1061		dedicated to Ḥātim Beg; copyist Muḥammad Sa'īd b. Faḫr al-Dīn; from the same manuscript as previous text
Muḥammad Bāqir b. Zayn al-'Ābidīn – Autolykos	*al-Ḥāšiya 'alā Kitāb al-Ukar*	22 Rajab 1061		copyist Muḥammad Sa'īd b. Faḫr al-Dīn; from the same manuscript as previous text
Niẓām al-Dīn Nīšābūrī – Ṭūsī	*Tawḍīḥ al-Taḏkira*	10 Jumādā II 1062		copyist Muḥammad Qāsim b. Muḥammad Ṭāhir Iṣfahānī
Šams al-Dīn Ḫafrī	*al-Takmila fī l-Taḏkira*	1063		copyist 'Azīz Allāh b. Yūsuf Ṭabātabā'ī
Naṣīr al-Dīn Ṭūsī (ascribed)	*Tarjama-yi Ṣuwar al-kavākib*	1063		
Bahā' al-Dīn 'Āmilī	*Ḥāšiyat Šarḥ al-Mulaḫḫaṣ*	1063		

Author	*Title*	*Year of copy*	*Place of copy*	*Additional information*
Naṣīr al-Dīn Ṭūsī	*Bīst bāb dar aṣṭurlāb*	1065		
Šams al-Dīn Ḫafrī	*al-Takmila fī l-Taḏkira*	Rabīʻ I 1066		copyist ʻAlī b. Ḥājjī Muḥammad Gurgānī Qummī
Muṣliḥ al-Dīn Lārī – ʻAlī Qušjī	*Šarḥ-i Hay'a*	1066		
Bahā' al-Dīn ʻĀmilī	*Ḫulāṣat al-ḥisāb*	1067		copyist Muḥammad b. Amr Allāh Yazdī
anonymous	*Davā'ir-i aʻẓām*	1 Šaʻbān 1067		
Naṣīr al-Dīn Ṭūsī	*Muʻīniyya*	beginning of Šaʻbān 1068	Isfahan	
ʻAlī Qušjī	*Hay'a*	1068		
Naṣīr al-Dīn Ṭūsī	*Ḥall-i Muškilat-i Muʻīniyya*	beginning of Šaʻbān 1068	Isfahan	in the same manuscript as previous text
Niẓām al-Dīn Birjandī	*Bīst bāb dar aṣṭurlāb*	Ḏū al-Qaʻda 1070		
Naṣīr al-Dīn Ṭūsī	*Ḥall-i Muškilat-i Muʻīniyya*	beginning of Šaʻbān 1068	Isfahan	in the same manuscript as previous text
Niẓām al-Dīn Birjandī	*Bīst bāb dar aṣṭurlāb*	Ḏū al-Qaʻda 1070		
Kāšif al-Dīn Muḥammad Ardakānī Yazdī	*Rubʻ-i mujayyab*	Ḏū al-Qaʻda 1070		in the same manuscript as previous text
Muḥammad Bāqir b. Zayn al-ʻĀbidīn	*Mir'āt al-anwār fī maʻrifat sāʻāt al-nahār*	Ḏū al-Qaʻda 1070		in the same manuscript as the previous text
Muṣliḥ al-Dīn Lārī – ʻAlī Qušjī	*Šarḥ-i Hay'a*	1071	Tiflis	dedicated to Humāyūn
ʻAlī Qušjī	*Hay'a*	1071	Tiflis	
Niẓām al-Dīn Birjandī	*Ḥāšiyat Šarḥ al-Mulaḫḫaṣ*	1071		copyist Ḥamza b. Abī Bakr
anonymous	*Ḥāšiya Taḥrīr Uṣūl Uqlīdis*	Ḏū l-Qaʻda 1071		

(*Continued*)

Table 4.1 Continued

Author	*Title*	*Year of copy*	*Place of copy*	*Additional information*
Niẓām al-Dīn Birjandī – Ṭūsī	*Šarḥ al-Taḏkira*	16 Šawwāl 1072 proofread 8 Rabīʿ I 1079		copyist ʿAbd al-Ġanī Ḥusaynī, see copy of ʿAlī b. Riḍwān's commentary on Ptolemy's *Tetrabiblos*, 1079
Niẓām al-Dīn Nīšābūrī – Ṭūsī	*Tawḍīḥ al-Taḏkira*	1072	Tabriz	
Abū Naṣr Aḥmad b. Ibrāhīm b. Fāris b. Ḥasan [Ḥusayn]	*al-Jāmiʿ al-ṯābita*	1073		in the same manuscript as texts by Abū l-Qāsim ʿAlī b. Aḥmad Balḫī, Aristotle and Hermes, copied in 1084
Abū Maʿšar Balḫī	*al-Mudḫal al-kabīr*	1073		in the same manuscript as the previous text
Muḥammad b. Muḥammad Kašġārī	*Šarḥ-i Sī faṣl*	1073		copyist ʿAlī Taqī b. Ḥājjī Muḥammad Amīn Qārī Sāravī; in the same manuscript as the previous text
ʿAlī Qušjī	*Ḥisāb = Mīzān al-ḥisāb = Zubdat al-ḥisāb*	Rabīʿ II 1073		copyist ʿAlī Taqī b. Ḥājjī Muḥammad Amīn Qārī Sāravī; in the same manuscript as the previous text
anonymous	*Ḥisāb al-ḍarb*	1072 (1073?)		in the same manuscript as the previous text

Author	*Title*	*Year of copy*		*Place of copy*	*Additional information*
Bahā' al-Dīn 'Āmilī		*al-Ṣafīḥa*	Ṣafar 1073	Mašhad	
Naṣīr al-Dīn Ṭūsī		*Sī faṣl*	12 Ša'bān 1073		copyist 'Alī Taqī b. Ḥājjī Muḥammad Amīn Qārī Sāravī; in the same manuscript as the previous text
anonymous		*al-Jabr wa-l-muqābala*	17 Ramażān 1074		
'Alī Qušjī		*Ḥisāb*	1074–75	Qariyat-i Ḫunk, near Qa'in	copyist Muḥammad Mu'min b. Niẓām al-Dīn 'Alī
Šams al-Dīn Ḫafrī – Ṭūsī		*al-Takmila fī l-Taḏkira*	1074	Isfahan	copyist Muḥammad Masīḥ Ḥusaynī Šīrāzī
Šams al-dīn Ḫafrī – Ṭūsī		*al-Takmila fī l-Taḏkira*	1074	Shiraz, Madrasa-yi Isma'iliyya	copyist Muḥammad Rīżā Šīrāzī, transcribed work for his teacher Mīrzā Ibrahīm, the head of the madrasa
Theodosios – Ṭūsī		*Taḥrīr Kitāb al-Ukar*	1074	Isfahan	copyist Muḥammad Ṣāliḥ b. Pīrzāde
'Abd Allāh b. Muḥammad b. 'Abd al-Razzāq Ḥāsib		*al-Fawā'id al-Bahā'iyya fī l-qawā'id al-ḥisābiyya*	1074		copyist Ja'far b. Muḥammad Mu'min
Niẓām al-Dīn Nīšābūrī – Ṭūsī		*Šarḥ-i Sī faṣl*	Rajab 1076		

(*Continued*)

Table 4.1 Continued

Author	*Title*	*Year of copy*	*Place of copy*	*Additional information*
ʻAbd al-Qādir Ruyānī Lahījī	*Tuḥfa-yi niẓāmī = Šarḥ-i Sī faṣl*	Rajab 1076		dedicated by author to Sulṭān b. Sulṭān Yaḥya Kiyā of Tabaristan; in the same manuscript as previous text
Niẓām al-Dīn Nīšābūrī – Ṭūsī	*Tawḍīḥ al-Taḏkira*	Ramaẓān 1078 Rabīʻ I 1081		
Abū l-Ḥasan ʻAlī b. Riḍwān – Ptolemy	*Šarḥ al-Arbaʻ Maqālāt li-Baṭlamyūs*	3 Rajab 1079		copyist ʻAbd al-Ġanī Ḥusaynī Yazdī
ʻAbd al-Razzāq b. Muḥammad Muʻīn Kāšānī	*Zāʼiča*	11 Ṣafar 1080		copyist Ḥabīb Allāh Māzandārānī; dedicated to Jalāl al-Dawla ʻAbd al-ʻAzīz b. Uluġ Beg
Qāżīzāda Rūmī	*Šarḥ al-Mulaḫḫaṣ*	1082	Isfahan, Madrasa-yi Nawwab	copyist Qurbān ʻAlī b. Ramażān Šams al-Dīn Ṭabasī
ʻAlī Qušjī	*Hayʼa*	1082	Herat	copyist Šaraf al-Dīn Ḥusayn
Niẓām al-Dīn Birjandī	*Ḥāšiya ʻalā Šarḥ al-Mulaḫḫaṣ fī l-hayʼa*	1083	Isfahan	copyist Sayf al-Dīn Maḥmūd b. Ḥājjī Ibrāhīm
ʻAlī Qušjī	*Hayʼa*	Muḥarram 1084		copyist Muḥammad ʻAlī b. Āḫūnd Mullā Babāyān Damġānī
Abu l-Qāsim ʻAlī b. Aḥmad Balḫī	*Sirr al-asrar = al-Nukat fī al-tasyīr wa-l-istimrār*	Ḏū al-Qaʻda 1084		

Author	*Title*	*Year of copy*	*Place of copy*	*Additional information*
anonymous	*Aḥkām al-kawākib*	1084		in the same manuscript as the previous text
Aristotle (ascribed)	*Asrār al-nujūm*	5 Jumāda II 1084		in the same manuscript as the previous text
anonymous – Abū Ma'šar	*Muḫtaṣar Asrār al-nujūm*	1084		in the same manuscript as the previous text
Hermes (ascribed)	*al-Āsās = al-Ḫamsa wa-ṯamanūn bāban*	1084		in the same manuscript as the previous text
Hermes (ascribed); *Iṣlāḥ* by Ḥasan b. Aḥmad b. 'Abd Allāh Ṣūfī	*al-Fuṣūl*	1084		in the same manuscript as previous text
'Abd al-Raḥmān Ṣūfī	*Kitāb Ṣuwar al-kawākib al-ṯābita*	Ramażān 1085		copyist Qāsim b. Ḥusayn b. Zāl
Euclid – Ṭūsī	*Taḥrīr Uṣūl Uqlīdis*	Ramażān 1085		copyist Muḥammad Zamān b. Kuttāb Allāh Awjānī
Naṣīr al-Dīn Ṭūsī	*Bīst bāb*	25 Jumāda II 1086		copyist Ibn Amīr 'Abd al-Karīm Muḥammad Muḥsin Ḥusaynī
'Alī Qušjī	*Hay'at*	Ḏu al-Ḥijja 1086		
Muẓaffar b. Muḥammad Qāsim Gunābādī	*Tanbīhāt al-munajjimīn*	1086		
Naṣīr al-Dīn Ṭūsī	*Zubdat al-hay'a*	Ṣafar 1087		
Sayyid Muḥammad Masīḥ Ḥusaynī	*A'dād-i mutaḥābba*	beginning of Rajab 1087		copyist Sayyid Muḥammad Masīḥ Ḥusaynī for himself
Rukn al-Dīn Ḥusayn Āmulī	*Panjāh bāb-i sulṭānī*	1088		

(*Continued*)

Table 4.1 Continued

Author	*Title*	*Year of copy*		*Place of copy*	*Additional information*
Muḥammad Bāqir Majlisī		*Iḫtiyārāt al-ayām*	5 Muḥarram 1088		
Bahā' al-dīn 'Āmilī		*al-Kur*	21 Ṣafar 1088		
Naṣīr al-Dīn Ṭūsī (ascribed)		*Madḫal-i manẓūm*	1088		
Bahā' al-Dīn 'Āmilī		*al-Ṣafīḥa*	Ḏū al-Qa'da 1088		copyist Muḥammad Ašraf b. Ḥājjī Muḥammad Yazdī; from the same manuscript as the previous text
Bahā' al-Dīn 'Āmilī		*Taḥqīq jihāt al-qibla*	1088		copyist Muḥammad Ašraf b. Ḥājjī Muḥammad Yazdī; from the same manuscript as the previous text
Bahā' al-Dīn 'Āmilī		*al-Kur*	1088		copyist Muḥammad Ašraf b. Ḥājjī Muḥammad Yazdī; from the same manuscript as the previous text
Abū Zayn al-Ḥasan Fārisī (?) – al-Niẓām al-Dīn Nīšābūrī		*al-Nukat al-Šamsiyya fī l-qawā'id al-ḥisābiyya*	11 Ḏū al-Ḥijja 1088	Isfahan	dedicated to Šams al-Dīn Muḥammad b. Bahā' al-Dīn; copyist Šayḫ 'Alī b. 'Abd al-Wahhāb

Author	*Title*	*Year of copy*	*Place of copy*	*Additional information*
Bahā' al-Dīn 'Āmilī	*Tašrīḥ al-aflāk*	1089		copyist Muḥammad Ašraf b. Ḥājjī Muḥammad Yazdī; from the same manuscript as the other texts by 'Āmilī copied by the same scribe in 1088
Bahā' al-Dīn 'Āmilī	*al-Ṣafīḥa*	1089		copyist 'Alī b. Ḫiẓr
'Alī Qušjī	*Hay'at*	15 Rajab 1089	Lahijan, Madrasa Khayrabad	copyist Ṣadiq b. Maḥmūd Kiyādehī
Muṣliḥ al-Dīn Lārī	*Šarḥ-I Hay'a*	1091	Shiraz	
anonymous	*Šarḥ-i Bīst bāb*	20 Jumāda I 1092		
Abū Naṣr al-Fārābī	*Ibṭāl aḥkām al-nujūm*	10 Ša'bān 1092		copied for Muḥammad b. Muḥsin b. Murtażā Fayż Kāšānī
Bahā' al-Dīn 'Āmilī	*Tašrīḥ al-aflāk*	1092	Isfahan, Madrasa-yi Nawwab	copyist Qurbān 'Alī b. Ramażān Šams al-Dīn Ṭabasī
Ptolemy – Ṭūsī	*Taḥrīr al-Majisṭī*	Ša'bān 1093		
Ptolemy – Ṭūsī	*Taḥrīr al-Majisṭī*	1094		copyist Muḥyī al-Dīn Muḥammad Kātib; was in possession of Rustam b. Šāh Vīrdī
Bahā' al-Dīn 'Āmilī	*Tašrīḥ al-aflāk*	1095	Mašhad-i Ḥusayn	copyist Maḥmūd b. Ibrāhīm b. 'Abd Allāh Najafī

(*Continued*)

Table 4.1 Continued

Author	*Title*	*Year of copy*	*Place of copy*	*Additional information*
Bahā' al-Dīn 'Āmilī	*al-Ṣafīḥa*	1095	Mašhad-i Ḥusayn	copyist Maḥmūd b. Ibrāhīm b. 'Abd Allāh Najafī; from the same manuscript as the previous text
Bahā' al-Dīn 'Āmilī	*Ḫulāṣat al-ḥisāb*	1095		
Bahā' al-Dīn 'Āmilī	*Ḫulāṣat al-ḥisāb*	21 Ḏū al-Ḥijja 1095		
Theodosios – Ṭūsī	*Taḥrīr al-masākin*	1096		
Bahā' al-dīn 'Āmilī	*Risāla dar taḥqīq qibla*	1097		
anonymous	*Aḥkām al-nujūm*	24 Rabī' II 1098		copyist Muḥammad Ibrāhīm b. Muḥammad Taqī Tabrīzī for Āqā Muḥammad Masīḥ Lāhījānī
anonymous	*Ḥisāb*	1099		copyist Ibn Muḥammad Raḥim Muḥammad Ṣāliḥ
Fāżil Javād Kāẓimī – Bahā' al-Dīn 'Āmilī	*Šarḥ Ḫulāṣat al-ḥisāb*	1099		
Muḥammad Bāqir Yazdī – Bahā' al-Dīn 'Āmilī	*Šarḥ Ḫulāṣat al-ḥisāb*			
Quṭb al-Dīn Šīrāzī	*al-Tuḥfa al-šāhiyya*	Ṣafar 1101		copyist Favāris Damāvandī
Abū l-Maḥāmid Muḥammad b. Maṣ'ūd b. Muḥammad Ġaznavī Buḫārī	*Kifāyat al-ta'līm fī ṣīnā'at al-tanjīm*	12 Rabī' II 1101	Shiraz, Madrasa-yi manṣūriyya	copyist Hadī b. Muḥammad 'Alī b. Ḥājjī Malik Qāsim
Muḥsin b. Murtażā Fayż Kāšānī	*Ġunyat al-anām fī ma'rifat al-sā'āt wa-l-ayyām*	Rabī' I 1104		

Author	*Title*	*Year of copy*	*Place of copy*	*Additional information*
Šams al-Dīn Samarqandī	*Aškāl al-ta'sīs*	8 Ḏu al-Ḥijja 1104		Šams al-Dīn Muḥammad b. Ḥusayn Sijistānī
anonymous	*Šarḥ Maqālat Ibn al-Hayṯam fī ma'rifat al-ašḫāṣ al-qā'ima wa- ā'midat al-jibāl wa-irtifā' al-ġuyūm*	Muḥarram 1105		copyist Muḥammad Sa'īd b. Muḥammad Mu'min Tabrīzī; owners Abū l-Ḥasan b. 'Abd al-Qādir al-Šarīf al-Ṭabīb + Muḥammad 'Alī al-Šarīf al-Ṭabīb
Šams al-Dīn 'Alī Ḥusaynī Ḫalḫālī – Bahā' al-Dīn 'Āmilī	*Šarḥ Ḫulāṣat al-ḥisāb*	1105		copyist Muḥammad Rafī' b. Muḥammad Šarīf Ḥusaynī
Luṭf Allāh Ḥusaynī	*Risāle dar ḥisāb*	1105		
Bahā' al-Dīn 'Āmilī	*al-Ṣafīḥa*	1106	Isfahan	
Šams al-Dīn Abū Bakr Muḥammad b. Aḥmad Marwazī	*al-Tabṣira fī 'ilm al-hay'a*	1107		copyist Muḥammad Rīżā b. 'Azīz Allāh Tūnī
Ḥusayn Šabākī	*Ḥisāb al-farā'iḍ*	1107–1108		
Quṭb al-Dīn 'Abd al-Ḥayy b. 'Izz al-Dīn Ḥusaynī Lārī	*Hall va-'aqd*	22 Ṣafar 1108		copyist Muḥammad Taqī b. Muḥammad Aqdaṣ Šarīf 'Āmilī Andijānī
Mīr Ḥusayn Maybudī – Bahā' al-Dīn 'Āmilī	*Šarḥ Ḫulāṣat al-ḥisāb*	1109		

(*Continued*)

Table 4.1 Continued

Author	*Title*	*Year of copy*	*Place of copy*	*Additional information*
Muẓaffar al-Dīn Gunābādī – Birjandī	*Šarḥ-i Bīst bāb*	1111	Isfahan	
Quṭb al-Dīn ʻAbd al-Ḥayy b. ʻIzz al-dīn Ḥusaynī Lārī	*Ḥall va-ʻaqd*	1112		
Ibn Abī Šukr al-Maġribī	*al-Madḫal al-mufīd . . .*	1113		copyist Muḥammad Mu'min
Ibn Abī Šukr al-Maġribī	*Aḥkām taḥāwīl sīnī al-ʻālam*	1113		
Quṭb al-Dīn ʻAbd al-Ḥayy b. ʻIzz al-Dīn Ḥusaynī Lārī	*Ḥall va-ʻaqd*	24 Rajab 1113		
Euclid – Ṭūsī	*Taḥrīr Uṣūl Uqlīdis*	1114		copyist Muḥammad Qāsim b. Bāqir ʻŪsjānī
Ibn Sīnā	*Kitāb al-šifāʼ (all four parts)*	1116		copyist Rafīʻ al-Dīn Muḥammad Ḥusaynī Ṭabasī
Bahāʼ al-Dīn ʻĀmilī	*Risāla dar taḥqīq qibla*	1116		
Bahāʼ al-Dīn ʻĀmilī	*Tašrīḥ al-aflāk*	1118	Kirman	copyist Muḥammad Ṣādiq Iṣfahānī
Ġiyāṯ al-Dīn Manṣūr Daštakī Šīrāzī	*al-Safīr*	1118		copyist Muḥammad Ṣādiq Iṣfahānī
Šams al-Dīn Ḫafrī	*al-Takmila fī l-Taḏkira*	Jumāda I 1120	Shiraz	copyist Muḥammad ʻAlī b. Muḥammad Ḥusayn
Qivām al-Dīn Ḥusayn b. Šams al-Dīn Ḫafrī	*Jaʻfariyya dar masāʼil-i ḥisābiyya*	23 Ḏū al-Qaʻda1120		

Author	*Title*	*Year of copy*	*Place of copy*	*Additional information*
Muẓaffar al-Dīn Gunābādī – Birjandī	*Šarḥ-i Bīst bāb dar taqvīm*	1124		copyist Muḥammad Hadī b. Šayḫ Bahā' al-Dīn Muḥammad Mi'mār Iṣfahānī
Qāżīzāda Rūmī – Čaġmīnī	*Šarḥ al-Mulaḫḫaṣ fī l-hay'a*	1125		copyist Ḥasan b. 'Abd al-Ġaffūr b. Jāmī Ḥusaynī
Niẓām al-Dīn Birjandī	*al-Ḥāšiya 'alā Šarḥ al-Mulaḫḫaṣ*	1125		copyist Ḥasan b. 'Abd al-Ġaffūr b. Jāmī Ḥusaynī; from the same manuscript as the previous text
Sayyid Muḥammad Manṣūr	*Lubāb al-ḥisāb*	1125		
Bahā' al-Dīn 'Āmilī	*Ḫulāṣat al-ḥisāb*	1127	Tiflis?	
Qāżīzāda Rūmī – Čaġmīnī	*Šarḥ al-Mulaḫḫaṣ fī l-hay'a*	1129		copyist 'Alī Taqī b. Mullā Valī Ganja'ī
anonymous	*Taqvīm*	1129		for Šāh Ḥusayn
anonymous	*Taqvīm*	1130		for Šāh Ḥusayn
anonymous	*Taqvīm*	1131		for Šāh Ḥusayn

Table 4.2 Dedicated Mathematical Texts of the Safavid Period

Author	*Title*	*Date of Copy*	*Place of Copy*	*Dedicated to*
Muḥammad Ṭabasī	*Misāḥa*	19 Ramażān 986		Ḫwāja Niẓām al-Dīn Mīrijān
Muḥyī al-Dīn b. Badr al-Dīn al-Avārī	*Jāmi' al-qirānāt*			Ismā'īl
'Abd al-Mun'im 'Āmilī	*Risāla dar ālāt-i raṣdiyya*	973/1563		Ṭahmāsb

(*Continued*)

Table 4.2 Continued

Author	*Title*	*Date of Copy*	*Place of Copy*	*Dedicated to*
Ġiyāṯ al-Dīn Abū Isḥāq Muḥammad ʿĀšiqī Kirmānī	*Risāla dar ḥisāb*			Ṭahmāsb
Bahā' al-Dīn ʿĀmilī	*al-Kur*			Ṭahmāsb
Bahā' al-dīn ʿĀmilī	*Ḫulāṣat al-ḥisāb*			Ṭahmāsb; Mīrzā Ḥamza
Mīrzā Qāżī Ibn Kašīf al-Dīn Muḥammad Ardakānī Yazdī	*Tuḥfa-yi Muḥammadiyya/ Tuḥfa-yi ʿAbbāsiyya*			Muḥammad Ḫudābanda; ʿAbbās I
Bahā' al-Dīn ʿĀmilī	*Tuḥfa-yi ḥātimiyya = Haftād bāb dar aṣṭurlāb*			Ḥātim Beg
anonymous	*Muḫtaṣar Taḥrīr Kitāb Uqlīdis*	1003	Qazvin	a eunuch of the court
al-Muẓaffar Muḥammad b. Qāsim Gunābādī – Birjandī	*Šarḥ-i Bīst bāb dar maʿrifat-i taqvīm*	1005	Isfahan	ʿAbbās I
Muḥammad Ṣādiq al-Tunikabānī; student of Bahā' al-Dīn ʿĀmilī	*Qānūn al-idrāk fī tašrīḥ al-aflāk*	1007	Isfahan	ʿAbbās I
al-Muẓaffar Muḥammad b. Qāsim Gunābādī	*Tanbīhāt al-munajjimīn*	1014 or 1024 (?)	Isfahan	ʿAbbās I
Quṭb al-Dīn ʿAbd al-Ḥayy b. ʿIzz al-Dīn Ḥusaynī Lārī	*Ḥall va-ʿaqd*	Rabīʿ II 1017 (?)	Lar	ʿAbbās I
al-Muẓaffar Muḥammad b. Qāsim Gunābādī	*Tuḥfa-yi ḥātimiyya*	1019	Isfahan	Ḥātim Beg
Bahā' al-Dīn ʿĀmilī	*Tašrīḥ al-aflāk*			ʿAbbās I
Bahā' al-Dīn ʿĀmilī	*al-Ṣafīḥa*			ʿAbbās I
Quṭb al-Dīn ʿAbd al-Ḥayy b. ʿIzz al-Dīn Ḥusaynī Lārī	*Ḥall al-masā'il*	1037	Lar/ Shiraz	Amīr Abū l-Ḥasan, governor of Fars
Muḥammad Bāqir b. Zayn al-ʿĀbidīn	*Maṭlaʿ al-anwār*		Isfahan	Ṣafī
Zamān b. Šaraf al-Dīn Ḥusayn al-Mašhadī	*Tuḥfa-yi sulaymāniyya (Zīj)*	1078	Mašhad	Sulaymān
Mīr Ṣadr al-Dīn Muḥammad b. Muḥammad Ṣādiq Ḥusaynī Qazvīnī	*Tafrīḥ al-idrāk fī taqḍīḥ Tašrīḥ al-aflāk*	1083	Qazvin	Zunīl Ḫān, governor of Qazvin
Muḥammad ʿAlī b. Muḥammad Qāsim	*Mir'āt-i Sulaymān*	1085		Sulaymān

Author	*Title*	*Date of Copy*	*Place of Copy*	*Dedicated to*
Muḥammad Kāfī b. Abī l-Ḥasan Qā'inī	*Taqvīm*	1103		Sulaymān?
Muḥammad Ašraf Munajjim b. Muḥammad Ṣādiq	*Taqvīm*	1103		Sulaymān
Muḥammad Bāqir b. Zayn al-'Ābidīn; translator: Muḥammad Bāqir b. Mīr Ismā'īl Ḥusaynī Ḫātūnābādī	*Tarjama-yi 'Uyūn al-ḥisāb*			Ḥusayn
Muḥammad Ḥusayn Munajjim b. Abū l-Ḥasan Munajjim Gunābādī	*Taqvīm*	1107		Ḥusayn
Muḥammad Bāqir b. M. Taqī Majlisī	*Miftāḥ al-šuhūr*	11 Ša'bān 1109		Ḥusayn
Muḥammad Raḥīm Munajjim	*Taqvīm*	1111		Ḥusayn
Anonymous	*Taqvīm*	1129		Ḥusayn
Anonymous	*Taqvīm*	1130		Ḥusayn
Anonymous	*Taqvīm*	1131		Ḥusayn

Table 4.3 Safavid Copies of Texts by Ancient Authors

Author	Text	*Date of Copy*	*Place of Copy*	Addressee/Owner/ Copyist
Ptolemy London, BL, MS Add 23392	*Almagest*	Muḥarram 955	Qazvīn?	diagrams were added in Qazvīn Jumādā II, 956 by Muḥammad Ḥusayn b. 'Alī
Euclid Theodosios Autolykos Ya'qūb b. Yūsuf al-Kindī Euclid Theodosios Theodosios Hypsikles Hypsikles Abū Sahl Kūhī Ṯābit b. Qurra Qom, Public Library of Great Ayatollah al-Mar'ašī, MS 7580 Qom	*Data* *Spherics* *Spherics* *Optical errors* *Phainomena* *Habitations* *Risings and Settings* *On Days and Nights* *On Risings* *Knowledge of the size . . .* *On Two Mean Proportionals*	28 Muḥarram 960		

(*Continued*)

Table 4.3 Continued

Author	Text	*Date of Copy*	*Place of Copy*	Addressee/Owner/ Copyist
Theodosios Autolykos Theodosios Euclid Euclid Qom, Public Library of Great Ayatollah al-Mar'aši, MS 4009	*Spherics* *Spherics* *Habitations* *Optics* *Data*	18 Ša'bān 1009	Isfahan	addressee: Āqā Jalāl Muḥammad
Euclid Mašhad, Astān-i Quds, MS 12091	*Elements*, Books XIV-XV	Ramażān 1009		

5

THE PRISON OF CATEGORIES – 'DECLINE' AND ITS COMPANY

Historiography of Islamic societies and their secularized successors often suffers under the burden of century-old categories, values, expectations, approaches, and judgments. Despite all the far-reaching changes in the concepts, methodologies, methods, and values introduced by various fields of postmodern engagement, substantial parts of history-writing with regard to Islamic societies have successfully escaped a reflective questioning of the assumptions, tools, values, and goals held or pursued by their practitioners. Categories such as 'decline' and judgments such as the suppression of philosophy and other 'ancient sciences' by religious orthodoxy and worldly rulers have survived until this day either as accepted truth or as statements to be proved wrong by counter-examples.[1] Nineteenth-century inventions such as the death of philosophy after Ibn Rušd caused by al-Ġazālī's sharp accusation of being internally incoherent and fundamentally incompatible with revelation continue to be told and believed within and outside the Arab world.[2] Unanswerable questions such as why there was no scientific revolution in Islamic

1 I use the label 'ancient sciences' as a short-cut for geometry, number theory, astronomy, astrology, theoretical music, philosophy, medicine, alchemy, and other fields of scholarly knowledge appropriated from Greek, Syriac, Pahlavi, and Sanskrit sources.

2 Mohammed Arkoun has expressed his belief that philosophy in the Islamic world disappeared after Ibn Rušd's death in a public lecture at the Institute for the Study of Muslim Civilisations, Aga Khan University International, London, 2006. Research results of the last two decades have discredited, however, this judgment for Islamic societies in Iran and India. Historical sources such as biographical dictionaries, study programs, and historical chronicles leave no doubt that philosophical treatises by Ibn Sīnā (d. 428/1037), Faḫr ad-Dīn ar-Rāzī (d. 606/1209), Naṣīr ad-Dīn at-Ṭūsī (d. 672/1274) or Ǧalāl ad-Dīn ad-Dawwānī (d. 907/1501) were studied at *madrasas* in Cairo, Damascus or even in cities of northern Africa. See, for instance, Seyyed Hossein Nasr and Oliver Leaman (eds.), *History of Islamic Philosophy* (London: Routledge, 1996); Dimitri Gutas, "The Heritage of Avicenna: The Golden Age of Arabic Philosophy, 1000 – ca. 1350," in *Avicenna and His Heritage*, eds. J. Janssens and D. De Smet (Leuven: Peeters, 2002), 81–97. In his entry "Islamic Philosophy" in the *Routledge Encyclopedia of Philosophy* (New York: Routledge, 2000), q.v., Leaman reformulated the old view of philosophy's demise by writing that due to al-Ġazālī's demand to reject philosophy it came to be under a cloud until the nineteenth century. He also affirmed that philosophy and the religious disciplines always had a rather difficult relationship. Access by www.rep.routledge.com/article/H057.

DOI: 10.4324/9781003372493-6

societies are considered relevant and holding the key for today's conflicts and difficulties.[3] Concepts outside of time and independent of concrete space such as 'Arabic,' 'Islamic', or 'Arabic-Islamic' science/s dominate the approaches to the study of history of science in Islamic societies.[4] Creatures of nostalgia such as 'Golden Age' or 'Islamic/Muslim Renaissance' continue to inspire professionals and amateurs alike while glossing over the fact that most localities in the vast realm of the Islamic world in the ninth, tenth, eleventh, etc. centuries were free and empty of any person that practiced the 'ancient sciences'.[5] Oppositions of all sorts such as 'rationality' versus 'superstition' or 'religion', (Catholic and later Protestant and secular) 'Europe' versus the 'Islamic World', 'Christianity' versus 'Islam', or 'progress' versus 'decline' continue their existence as if immutable and outside of history.[6] Rarely anyone asks whether the entities set up against each other for the sake of mobilizing the one or the other set of values and gaining a platform for evaluation did exist in the period under debate and if so in which form and with what meaning. The elementary message of postmodernist critiques that there is no narrative about the past independent from our present and that every set of beliefs about the past, including academic history-writing, is socially constructed has not achieved yet to wash away the categories that dominated throughout the twentieth century the study of Islamic societies in general and their scientific practices, results, practitioners, and sponsors in particular. During the last decade, stimulated largely by the results of 9/11 and the so-called 'War on Terror', the debate on 'Islam and Science', 'The Golden Age of Islam and Science' and the impact of scientific achievements by scholars from Islamic societies, mostly reduced to 'Muslim scientists', on 'modern' science as developed

3 A serious attempt to discuss this question from the perspective of a scientist is Pervez Hoodbhoy, *Islam and Science: Religious Orthodoxy and the Battle for Rationality* (London: Zed Books Ltd, 1991), Chapter 10. Access by www.chowk.com/show_article.cgi?aid=00000104&channel=university%20ave.

4 Roshdi Rashed (ed.), *Encyclopedia of the History of Arabic Science* (Paris: Éditions du Seuil, 1997), 3 vols.; Seyyid Hossein Nasr, *Islamic Science: An Illustrated Study* (London: World of Islam Festival Trust, 1976).

5 www.imarabe.org/temp/expo/sciences-arabes.html.

6 A serious attempt to discuss the issue of the decline of science in Islamic societies from the perspective of a modernist historian of science is Ahmad Y. al-Hassan's, *Factors behind the Decline of Islamic Science*, 2007, access by www.history-science-technology.com/Articles/articles%208.htm, a revised version of his earlier Epilogue to *Science and Technology in Islam*, Part II, eds. Ahmad Y. al-Hassan, Maqbul Ahmad and Albert Zaki Iskandar (Paris: UNESCO, 2001). An earlier version was published in *Islam and the Challenge of Modernity*, ed. Sharifa Shifa Al-Attas (Kuala Lumpur: International Institute of Islamic Thought and Civilization, 1996), 351–389, undertaken with the intention to refute claims made in a series of articles published in *Nature*, 2 November 2006. See the online edition of *Nature*, 1 November 2006, access by http://www.nature.com/news/2006/061030/full/444035a.html. A more propagandistic discussion of the issue of decline and various views held by historians of science is Ziauddin Sardar's, *Islam and Science: Lecture Transcript* (London: Royal Society, s.d.), access by www.royalsoc.ac.uk/page.asp?tip=1&id=5747. It is the transcript of Sardar's lecture "Islam and Science: Beyond the Troubled Relationship," given 12 December 2006, access by www.royalsoc.ac.uk/event.asp?id=5129&month=,2006.

in various Catholic, Protestant, and later secular societies of Europe and their colonial extensions in the 'New World' has become more intense and widespread as even a cursory *googling* of these key terms shows. Ardent defenders of 'a glorious past' populate the Internet as do convinced deniers of any link between Islam and science. Most of the writers have little knowledge of the historical works and little understanding of their scientific content. They rarely ever read a single scientific text of the past, but they are strong believers in their own claims and values whatever they may be.[7] The latest representatives of this kind of apologetic and academically unsound depiction of the sciences in past Islamic societies are books by Jim Al-Khalili, Nidhal Guessoum, or Muzaffar Iqbal. Commercial interests of publishers like Penguin, I. B. Tauris, Ashgate, and others outweigh since many years academic solidity and give authors without qualification for history of science in Islamic societies a platform within an academic framework. Other apologists of a glorious Muslim past in science and technology and its impact on today's world create their own public spaces by lobbying, fund raising, exhibiting, and publishing in print and electronic form. The most successful of them is FSTC (Foundation for Science, Technology and Civilization). The project with the greatest outreach and hence the strongest capacity to influence attitudes and feelings about the past and the present is its exhibition "1001 Inventions – Discover The Muslim Heritage In Our World". This exhibition emphasizes "1000 years of forgotten Muslim inventions and discoveries", forgotten because of "Western academia's Eurocentric history-writing". It replaces this rejected historiography by an equally ideological message of a Muslim invention of 'modern science' which 'Europe' appropriated without admitting its debt to this very day. The sophisticated show started 21 January 2010 in London, Science Museum, moved next to Istanbul (18 August 2010) on invitation by Prime Minister Racıp Erdoğan and then to New York (December 2010).[8] Recently (25 May 2011) it was opened in Los Angeles by Secretary of State Hillary Clinton.[9] Political interests combined with commerce unite here representatives of groups with conflicting beliefs and permit the rise of a new historical myth.

The task of studying the sciences in past Islamic societies according to the interests and need of members of these societies which A. I. Sabra set already more than ten years ago is situated between these two extremes of (steep) 'decline' and

7 These are merely a few examples. Others can be easily added. See www.iht.com/articles/2001/11/13/edlet_ed3 18.php; www.fasebj.org/cgi/content/full/20/10/1581; www.irfi.org/articles/articles_401_450/golden_ age_of_islam.htm.

8 Jim Al-Khalili, *Pathfinders: The Golden Age of Arabic Science* (London: Penguin, 2010); idem, *The House of Wisdom, How Arabic Science Saved Ancient Knowledge and Gave Us the Renaissance* (London: Penguin, 2011); Nidhal Guessoum, *Islam's Quantum Question* (London, New York: I.B. Tauris, 2010); Muzaffar Iqbal, *Islam and Science* (Farnham, Surrey: Ashgate Publishing Limited, 2002); idem, *Science and Islam* (Westport, CT: Greenwood Pub Group Inc, 2007); idem (ed.), *Islam and Science: Historic and Contemporary Perspectives* (Farnham, Surrey: Ashgate Publishing Limited, 2011), 4 vols., access by www.1001inventions.com.

9 www.1001inventions.com/Hillary_Clinton

(breathtaking) 'progress', 'inferiority' and 'superiority'.[10] It continues to be an unfulfilled goal. In the sense that most studies are studies of texts and instruments and are considered to have reached their end when the scientific content is identified, commented upon, and situated within a larger history of ideas, this task has not even been taken seriously. There are very few biographies of major or minor scholars of the 'ancient sciences' in addition to those who have become beacons of cultural identity in North African and Asian countries of today such as al-Kindī, al-Fārābī, Ibn Sīnā, al-Bīrūnī, Ibn al-Hayṯam, ʿUmar al-Ḫayyām, Ibn Rušd, or Naṣīr ad-Dīn at-Ṭūsī. Equally, there are very few studies of the various forms in which the content of the hundreds of thousands of scientific texts and hundreds of scientific instruments was indeed used at different points in time and space across various Islamic societies.

The context of the specific society in which any of these texts were written and any of these instruments were fabricated is often seen as not particularly relevant except for those areas which are today accepted in David A. King's catchphrase as "science in the service of Islam". Outside of these areas, science continues to be regarded as universal and in no need of a deeper analysis of its local entanglements with one exception – the exchange of ideas across disciplines and cultures. As a result, most of the still very limited efforts to contextualize the 'ancient sciences' in Islamic societies focus on their relationship to religion, the *madrasa,* or the court.

1 'Decline' – an inappropriate historiographical category

For a history-writing that aspires to investigate the 'ancient sciences' in Islamic societies within the frameworks of these societies, 'decline' is an inappropriate category. It is conceptually unsuited since a single category does not suffice for capturing the complexity of an entire society, let alone of several societies. 'Decline' as a historiographical category has a place in a cyclical theory of history as the third stage of development of any social organism, culture, or civilization.[11] Within such cyclical theories, the disappearance of the arts and the sciences is part of the normal course of events, not an aberration from the norm. In theories of history that operate with a linear concept of historical time, as most modernist and postmodernist theories do, 'decline' is a negative term, a violation of 'good behavior'. It mobilizes emotions of aggression or defense depending on the overall position of the historian who uses this term. Because of its loss of status as a historiographical category within linear theories of history, most twentieth-century writers about the 'decline' of the 'ancient sciences' in Islamic societies felt compelled either to find causes for the phenomenon or to present counter-examples

10 A. I. Sabra, "The Appropriation and Subsequent Naturalization of Greek Science in Medieval Islam: A Preliminary Statement," *History of Science* 25 (1987), 223–243.

11 http://etext.virginia.edu/cgi-local/DHI/dhi.cgi?id=dv1-74

proving its non-existence altogether.[12] The spatial and temporal localization of 'decline' was not seriously considered and was in a sense unthinkable due to the problematic status of 'context' among the practitioners of the field. The idea of historicizing the term and its use was not seen as an alternative approach due to a lack of interest in historical epistemology and more generally a kind of suspicion towards the application of theories from the humanities and social sciences to a field that was primarily seen as an extension of the sciences. In addition to the theoretical incoherence produced by the import of a category from a cyclical into a linear theory of history, the conceptual inadequacy of 'decline' rests in the silent assumptions underpinning it. 'Decline' as an explanatory category presupposes 'advance', 'progress', 'innovation', 'change', 'movement', 'growth', 'expansion', and similar categories as the 'natural', 'desirable', and 'achievable' goals of societal life. In such a belief system, scientific knowledge and practice can only be lost or destroyed due to some major destructive force or event. Examples are the invasion of a powerful enemy such as the Mongols, the Christian conquest of Muslim Spain, or British and French colonialism; the occurrence of natural and social disaster such as the plague; the spread of a climate hostile to science such as 'superstition' or 'Islamic orthodoxy'; or economic, technological, political, and military marginalization.[13]

The temporal absurdity of the category of 'decline' has been pointed out by other writers on the subject.[14] If a society or culture is seen as being in continuous decline for something between five to ten centuries, it cannot but raise the suspicion that the analysis is based on wrong assumptions and works with faulty methods. And even if this suspicion could be proven wrong, a negative category that is seemingly a valid description of a society or culture for several centuries obviously misses the forces that allowed this society or culture to survive for such a long time. The notion of 'decline' of the 'ancient sciences' in 'Islam' is also unsuited since it pretends validity for a vast region populated by many different peoples who followed diverse patterns of social organization and were ruled by more than four hundred dynasties of different ethnic origins, forms of social

12 See, for instance, Pervez Hoodbhoy, "Why Didn't the Scientific Revolution Happen in Islam?," 23 December 1997, access by www.chowk.com/show_article.cgi?aid=00000104&channel=university%20ave.

13 One among the many examples expressing this kind of presuppositions is the following excerpt from a newspaper article written by Tanvir Ahmad Khan, a former foreign secretary and ambassador of Pakistan: "Since the beginning of the 20th century, Muslim analysts have endlessly repeated a litany of causes that led to the decline of their great civilization that once pioneered modern learning. At the heart of this analysis was recognition that the spirit of inquiry ebbed away as their centers of intellectual excellence were destroyed by the vengeful Christian reconquest of Spain and the Mongol invasion of the Arab heartland. Elsewhere, as in the unrelated but similar rejection of modern printing technology by the Ottomans in the heart of the Caliphate and the Mogul emperors in India, dissemination of knowledge beyond the elite was considered suspect." (Tanvir Ahmad Khan, "Struggles for a Muslim Renaissance," *Arab News*, 25 May 2007).

14 See, for instance, George Saliba, "Seeking the Origins of Early Modern Science?," [review article] *Bulletin of the Royal Institute for Inter-Faith Studies* 1, no. 2 (Autumn 1999), 139–152.

organization, and religious beliefs. To assume that all these different groups and organizations should have followed one and the same standards, norms, directions, and approaches in regard to the 'ancient sciences' presupposes the belief that there must have been similarities strong enough to counterbalance these differences and even outweigh them. Since even today in a much more intertwined world the differences between cultures, forms of social and political organization, economy and scientific activities as well as results remain substantial, it is not very likely that the vast space inhabited by Islamic societies in the past was characterized by a higher degree of harmony and homogeneity.

Moreover, 'decline' as a category of history-writing is culturally flawed since the concept is deeply value-laden. It began its career as a concept of diplomats from Catholic societies serving as an element of political rhetoric in the sixteenth century. The diplomats argued on the basis of the cyclical understanding of history that was prevalent in fifteenth- and sixteenth-century Italian theories of politics and society that the long series of Ottoman sultans proved by virtue of theory that the Ottoman Empire was in the stage of decline and would dissolve within the next few years or decades at most.[15]

But it is not only the project to historicize the history of science in Islamic societies put before us in 1987 by A. I. Sabra that we need to take seriously. We need to go further and abandon J. Needham's project of a universal history of science in which all the separate cultures of past science contribute towards 'modern' science, each in its own way.[16] There is only one way in which a universal history of science can be maintained – by the recognition that all science is culturally constructed. As cultural constructs, the sciences of different societies can be measured and compared in terms of their complexity, degree of difficulty, and explanatory and prognostic capability, but also their positive as well as negative impact on society, nature, and the universe. But they are not Aristotelian forms which reach their full state in our own sciences. They do not grow in a cumulative manner towards us. Our knowledge and our practices do not comprise of all earlier stages of scientific doctrines and activities. Paradigm shifts entail discarding earlier modes of thinking, writing, arguing, or analyzing. Treating such abandoned types of knowledge with the respect and appreciation due to them does not presuppose an Aristotelian belief in teleology. Respecting and appreciating past types of knowledge for their own forms, goals, contents, and contexts independent of our knowledge and its identity-forging powers rather builds on the assumption of the spatial and temporal locality of all knowledge and the contingency of its traces.

15 http://etext.virginia.edu/cgi-local/DHI/dhi.cgi?id=dv1-74

16 For a succinct summary of Needham's position and his rejection of Hossein Nasr's claim that "science in Islam" was "an unfolding of divine wisdom" see Hans Daiber, "Von der Weisheit Gottes zur Wissenschaft," *Evangelium und Wissenschaft* 42 (2003), 3–13; English translation: "The Way from God's Wisdom to Science in Islam: Modern Discussions and Historical Background," in idem, *The Struggle for Knowledge in Islam: Some Historical Aspects* (Sarajevo: Nevad Kahteran, 2004), 52–66. I thank Hans Daiber for reminding me of his paper and providing me with a copy.

Emancipating history of science in Islamic societies successfully from its old and one-sided categories, which in a sense are remnants of an intellectual colonialism constructed over the course of half a millennium, is, however, only one urgent task of today's historians. This process needs to be accompanied by liberating the field from its Middle Eastern nationalistic as well as Islamic categories and their underlying beliefs. It is not merely wrong to argue like Ernest Renan, Ignaz Goldziher, Gustav Edmund von Grunebaum, Bernard Lewis, or Bassam Tibi that Islamic 'orthodoxy' was always hostile towards science and philosophy and thus caused their demise. It is equally wrong to argue like Hossein Nasr, Ziauddin Sardar or Muzaffar Iqbal that 'Islam' was always open to science and philosophy because there was no central religious authority or clergy in Islam and because of Qur'ānic invitations to recite or Prophetic sayings that women and men should always acquire *ʿilm* even if that meant going as far as China. Not only is *ʿilm* in its singular form a specific kind of knowledge, namely the knowledge of *ḥadīṯ* or more broadly speaking religious knowledge, Muḥammad's goal was not to invite his newly converted followers to study the theories of other cultures. But even if the Prophet would have wished his followers to acquire knowledge of a secular kind outside his own cultural realm, attitudes among Muslims towards such a kind of knowledge varied considerably over time and space. Renan, Goldziher, von Grunebaum or Lewis surely interpreted the sources they worked with in a framework of their own prejudices, but they did not invent the hostile statements of Muslim scholars against philosophy, logic, geometry, algebra, or astrology. It is undeniably true that influential Muslim scholars such as Abū Ḥāmid al-Ġazālī (450–505/1058–1111), Ibn al-Ǧawzī (d. 597/1201), Ibn aṣ-Ṣalāḥ aš-Šahrazūrī (577–642/1181–1245), Šams ad-Dīn aḏ-Ḏahabī (672–749/1274–1348), and Ibn Ṭaymīya (661–728/1263–1328) wrote in strong terms against several or even all of the 'ancient sciences'. Moreover, there was no simple, direct connection between Muḥammad's *umma* and later Islamic societies. The changes that occurred after Muḥammad's death were profound. At latest in the tenth century, political, administrative, and economic decentralization facilitated the emergence of regional cultural identities with particular religious beliefs and practices. Tribal migrations, wars, and slave trade changed the ethnic composition of vast parts of the Islamic world and shifted in certain regions the balance between sedentary and nomadic cultures dramatically altering previous kinds of cultural cohesion. New regional blocks emerged which were not necessarily organized according to religious affiliations. Military, commercial, cultural, and familial alliances transcended religious identity to a much greater extent than is generally allowed for. It does not make sense to assume that the 'ancient sciences' and their practitioners would not have been affected by these deep changes nor that the new societies with their profoundly altered set up should have subscribed to the same attitude towards the 'ancient sciences' as did their predecessors.

In the following three sections, I present some of my arguments against various theses of marginalization and 'decline' of the 'ancient sciences' as offered by Goldziher (rejection by Islamic orthodoxy), Makdisi (exclusion from Sunnī *madrasas*), Michot (lack of courtly patronage), and others. In addition, I offer examples for other ways of interrogating the extant source material and reflecting on similarities and differences between some Islamic dynasties in regard to the 'ancient sciences'. My observations and interpretations do not lead yet to new categories. This is an important task for the future. But they point out some themes that stand out in the sources I study.

2 *Ancient sciences at* Madrasas

In 1981, Makdisi claimed in his highly influential book *The Rise of Colleges: Institutions of Learning in Islam and the West* that the 'ancient sciences' were excluded from the newly founded Sunnī *madrasa*s and remained *ante portas* ever after.[17] Two features of this claim are amazing. First, it is in open conflict with sources. Second, it was accepted by a majority of historians as correct. Two examples for the lasting impact of Makdisi's distortion of the relationship between the *madrasa* and the 'ancient sciences' are Halm's description of the Fātimid *Dār al-ʿilm* and Hassan's discussion of the relevance of the *madrasa* for debates of 'decline'. Halm claimed that there was no institution in the Islamic world that offered an encyclopedic knowledge in the classical sense and that the *madrasa*, while being "a cultural institution of high quality and standard, . . . was always limited" in its teaching:

> to religious knowledge. The instruction and study of medicine or astronomy, algebra or geometry, took place elsewhere, in the often private circles of authorities in each of the sciences. There was one exception. The sixth Fātimid caliph and the sixteenth imam of the Ismāʿīlīs, al-Ḥākim (386–412/996–1021), founded the House of Knowledge (Dār al-ʿilm) in Cairo in the year 1005.[18]

Hassan denied that the exclusion of the 'ancient sciences' from the *madrasa* contributed to their 'decline'. By equating the 'ancient sciences' with the 'rational sciences', a position widespread among historians of science, he extended Makdisi's view to a much larger field of disciplines. Hassan insisted:

> that the study of the rational sciences in Islam was always undertaken independently, and the theological studies were not usually undertaken under the same teachers or at the same institutions. Astronomy and mathematics were pursued mostly in the observatories, within a community of mathematicians and astronomers, where a specialized library was available and observational instruments were in constant use. The medical sciences were studied, as they should be, in the medical school

17 George Makdisi, *The Rise of Colleges: Institutions of Learning in Islam and the West* (Edinburgh: Edinburgh University Press, 1981), 77–78.

18 Heinz Halm, *The Fatimids and Their Traditions of Learning* (London: I.B. Tauris in association with The Institute of Ismaili Studies, 1997), 71.

> of a *bīmāristān* (hospital). The other sciences were studied under individual renowned scientists, most often patronized by the rulers, to whom students travelled from the far realms of Islam. The existence of these individual renowned teachers constituted what may be called a college of professors within a certain large city or a region. Let us not forget also the libraries and the academies, like *Dār al-Ḥikma* in Baġdād, which were devoted to research and to the study of the rational sciences. Most of the *madrasas*, on the other hand, were established by persons in power or by pious and wealthy individuals who endowed a part of their wealth to a waqf which supported the school. The purpose was always religious, and the studies were naturally mainly those of law and theology.[19]

The 'ancient sciences' were, however, not identical with the 'rational sciences' nor was either of the two indeed excluded from the *madrasa* or taught by different groups of teachers. It is unclear when the 'ancient sciences' began to move into the *madrasas*, mosques, and *ḫānaqāhs*. At latest in the late twelfth century, some of them can be found in these institutions in Salǧūq Anatolia, ʿAbbāsid Baġdād, Artuqid and Zangid northern Iraq, and Ayyūbid Damascus and Cairo. Other areas such as Iran saw a similar development, if not during Salǧūq rule, then under their successors, the Ilḫānids. This move took three forms – copying manuscripts for the library, teaching and studying either within the physical confinement of such an institute or with a *madrasa* teacher elsewhere, and the donation of special chairs for particular disciplines. While the two first formats are documented in substantial numbers in the sources, sponsoring chairs for 'ancient sciences' was a much less regular event.[20] Chairs for medicine were among the first and most often donated professorships for the 'ancient sciences' and were, in addition to hospitals,

19 Access by www.history-science-technology.com/Articles/articles%208.htm.

20 Scientific texts copied and taught at *madrasas* in Egypt, Syria, Iraq, Anatolia and Iran include works of the Banū Mūsā, Abū l-Wafāʾ, Ibn al-Hayṯam, Nasīr ad-Dīn at-Ṭūsī, Quṭb ad-Dīn Šīrāzī, Ibn Sīnā, Uluġ Beg, Ǧamāl ad-Dīn al-Maridānī, Ibn al-Maǧdī, Ibn al-Hāʾim, Sibṭ al-Maridānī, Bahāʾ ad-Dīn al-ʿĀmilī, Ibn an-Nafīs, Euclid, Ptolemy, Menelaos, Apollonios, Hippocrates, and other authors. Examples for copying are: mathematical texts written by as-Siǧzī (fl. late tenth century) and Ibn al-Hayṯam (d. 432/1041) by a great-great-great-grandson of Niẓām al-Mulk in the *Nizāmīya madrasa* finished in September 9, 1215; Ibn Sīnā's (d. 428/1036) *al-Qānūn fī t-tibb* in the same *madrasa* 68 years later (13 Muḥarram 682/April 13, 1283) and one year later, in the same *madrasa* of Naṣīr ad-Dīn aṭ-Ṭūsī's *Taḏkira fī ʿilm al-hayʾa*; another copy of this text was written in 853/1448 in the *Davud Paşa madrasa* in Istanbul; Quṭb ad-Dīn Šīrāzī's (d. 710/1311) *Nihāyat al-idrāk fī dirāyat al-aflāk* was copied in the *Gök madrasa* of Sivas in 682/1283 and his *at-Tuḥfa aš-šāhīya* in the medical *madrasa* of Sivas in 722/1322 and in the *Sahib Ata madrasa* in Konya in 785/1383; al-Kindī's (d. ca. 256/870) optics in 896/1491 at the *Kāmilīya madrasa* in Cairo; Šams ad-Dīn M. ibn Maḥmūd Āmulī's *Šarḥ al-Qānūn* of Ibn Sīnā's main medical work in 737/1337 at the *Wazīrīya Rāšidīya madrasa* in Sultānīya; Uluġ Beg's (r. 850/1447–1449) *Zīǧ* in the *Masǧid-i Ǧāmiʿ-i Ǧadīd-i ʿAbbāsi* in Isfahan 1056/1646; Qāḍīzāda ar-Rūmī's (d. 836/1432) *Šarḥ Aškāl at-taʾsīs* for Uluġ Beg in the *Ḫatūnīya madrasa* in Erzerum in 1077/1666 and in the *Fatih madrasa* in Istanbul in 1081/1670; ʿAlī Qušçī's (d. 879/1474) *Hayʾa* in 1091/1680 in the *Ḫayrābād madrasa* in Lāhiǧān. This list is by no means exhaustive.

attached to *madrasa*s and mosques in Salǧūq Anatolia, ʿAbbāsid Baġdād, and Mamlūk Cairo.[21] In the fourteenth century, the first chair for *ʿilm al-mīqāt* (science of time keeping) was donated in Mamlūk Cairo.[22] The Ottomans made the official support for *ʿilm al-mīqāt* and the mathematical sciences a regular component of their policies towards the *ʿilmiyye*. Occasionally, chairs were provided for other fields of knowledge, too, such as arithmetic or alchemy.[23] In most cases, the donors came from the ruling military elites. *Madrasa*s specializing in teaching medicine were also donated by physicians in Ayyūbid and Mamlūk Damascus and possibly in Salǧūq Anatolia.[24] In other regions of the Islamic world, teaching medicine remained the prerogative of hospitals and of private tutors.

The move of the 'ancient sciences' into the *madrasa* and cognate institutes brought with it several changes in the status and practice of the 'ancient sciences'. The broadening of the educational space at *madrasa*s, mosques, and *ḫānaqāh*s led to a stabilization of teaching and compiling of educational texts in several disciplines of the 'ancient sciences'. Libraries were created within these institutions that contained treatises on the different fields of these sciences plus those branches that were newly formed like *ʿilm al-mīqāt* or *ʿilm al-hayʾa* (mathematical cosmography). Teachers of these disciplines became well established as members of the educational elite of major cities. They were appointed as *mudarrisūn* at *madrasa*s, mosques, or *ḫānaqāh*s, although their teaching posts were not explicitly named for the 'ancient sciences'. This means that *mudarrisūn*, while officially being teachers of *fiqh*, were free to teach any other subject too. In the case of teachers of the mathematical and astronomical disciplines this often meant that they taught primarily *ʿilm al-farāʾiḍ* followed by arithmetic, *ʿilm al-mīqāt*, and algebra. But they also could teach *fiqh*, Arabic, *ʿilm al-hayʾa*,

21 Emilie Savage-Smith, "Medicine," in *Encyclopaedia of the History of Arabic Science*, ed. Roshdi Rashed (London: Routledge, 1996), vol. 3, 903–962; Gary Leiser, "Medical Education in Islamic Lands from the Seventh to the Fourteenth Century," *Journal for the History of Medicine and Allied Sciences* 38 (1983), 48–75; Doris Behrens-Abouseif, *Islamic Architecture in Cairo* (Leiden: Brill, 1986), 138–140; Mehmet Turgut, "Medieval Medical Schools in the Seljuq and Ottoman Empires," *Child's Nervous System* 26 (2010), 147–148.

22 Jonathan Berkey, *The Transmission of Knowledge in Medieval Cairo: A Social History of Islamic Education* (Princeton: Princeton University Press, 1992), 69; David A. King, "On the Role of the Muezzin and the Muwaqqit in Medieval Islamic Society," in *Tradition, Transmission, Transformation: Proceedings of Two Conferences on Pre-Modern Science Held at the University of Oklahoma*, eds. F. Jamil Ragep and Sally P. Ragep with Steven Livesey (Leiden: Brill, 1996), 285–346; Sonja Brentjes, "Shams al-Dīn al-Sakhāwī on *Muwaqqits, Muʾadhdhins*, and the Teachers of Various Astronomical Disciplines in Mamluk Cities in the Fifteenth Century," in *A Shared Legacy, Islamic Science East and West*, Homage to professor J. M. Millàs Vallicrosa, eds. Emilia Calvo, Mercè Comes, Roser Puig and Mònica Rius (Barcelona: Universitat de Barcelona, Publicacions i Edicions, 2008), 129–150, in particular 133.

23 Carl Petry, *The Civilian Elite of Cairo in the Later Middle Ages: Social Autonomy and Political Adversity in Mamluk Egypt* (Princeton: Princeton University Press, 1982).

24 See footnote 21.

geometry, and philosophy.[25] Others who became influential teachers of *fiqh* or *uṣūl ad-dīn* acquired fame as teachers of *ḥikma*, medicine, theoretical geometry, and astronomy.[26] Sabra has termed this change in professional outlook the move from the philosopher-physician and scientist-physician to the jurist-physician.[27] The change in professional outlook is, however, more adequately described as the emergence of a *mudarris* with a multidisciplinary training covering the 'traditional', 'rational', and mathematical sciences. The 'traditional' disciplines contain most often the study of the Qur'ān and *ḥadīṯ* plus *fiqh*. The 'rational' sciences comprise of the two aṣls (*fiqh* and *dīn*), logic, epistemology and other parts of philosophy, *kalām*, rhetoric, and metric. Some writers include also Arabic and grammar. The mathematical sciences as a rule consist of arithmetic, surveying, number theory, Euclidean geometry, *ʿilm al-mīqāt*, *ʿilm al-hayʾa*, the compilation of astronomical handbooks and ephemerides, and theoretical music. Medicine and alchemy appear either under the 'rational' or the mathematical sciences.[28] These changes in disciplinary perspectives meant that it were not the practitioners of the 'ancient sciences' alone who altered their profile merging it with the study of religious disciplines. The *mutakallimūn* too modified their profiles and integrated parts of mathematics, astronomy, logic, and philosophy into their training and scholarly writing. In the fifteenth century at the latest, most members of the educational elite and several sons of the military elite in Egypt and Syria studied at least some introductory texts of the 'rational' and the mathematical sciences. Copying such texts was a standard educational practice. Some members of the elites even composed their own writings on these disciplines, most often in poetic form as a *naẓm*. In Iran and subsequently in Central Asia and India, members of the educational as well as courtly elites delved more deeply into theory, in particular *ʿilm al-hayʾa*, a process that began in the thirteenth century.[29] A further change that took place with the move of the 'ancient sciences' into the *madrasa* and cognate institutes was the spread of these disciplines from centers into the provinces.[30]

25 Brentjes, "Shams al-Dīn al-Sakhāwī," 144.
26 Ibid., 132.
27 Sabra, "The Appropriation and Subsequent Naturalization of Greek Science in Medieval Islam," 125–127.
28 Sonja Brentjes, "The Location of the Ancient or 'Rational' Sciences in Muslim Educational Landscapes (AH 500–1100)," *Bulletin of the Royal Institute for Inter-Faith Studies*, 4, no. 1, Amman (2002), 47–71.
29 Commentaries on Nasīr ad-Dīn aṭ-Ṭūsī's astronomical works were written by scholars such as Niẓām ad-Dīn Nīsābūri, Sayyid al-Ǧurgānī, Niẓām ad-Dīn Birgāndī, Ǧalāl ad-Dīn Dawwānī and Šams ad-Dīn Ḫafrī who all were leading religious figures. Most of them were closely connected with Ilḫānid, Muẓaffarid, Tīmūrid, and other princes. Several copies of these texts were nicely written and illustrated.
30 Sonja Brentjes, "The Mathematical Sciences in the Safavid Empire: Questions and Perspectives," in *Muslim Cultures in the Indo-Iranian World during the Early-Modern and Modern Periods*, eds. D. Hermann and F. Speziale (Berlin: Klaus Schwarz Verlag, Tehran: Institut Français de Recherche en Iran, 2010), 325–402. See the previous paper in this volume.

The stabilization of teaching the 'ancient sciences' within the *madrasa* and cognate institutes together with the broadening modification of the educational profiles of the formerly more clearly separated practitioners of the 'ancient' and of the religious sciences is a major cultural innovation that took place in several Islamic societies, in particular those with larger territories and more affluent military rulers and urban elites. Many more people acquired at least some knowledge in these sciences than in the centuries of the exclusive location of the 'ancient sciences' in the spheres of courts. The practitioners of these disciplines became part of the *ʿulamāʾ* and partook in their social reputation. Cultural rituals such as marriages with daughters of high ranking *ʿulamāʾ*, employments as ambassadors, attribution of honorific titles and *laqab*s, official mourning ceremonies in mosques, or burials at privileged burial sites in distinguished neighborhoods included all *ʿulamāʾ* independent of their subject of teaching and interest. The price to be paid for this newly available stability and respect was adopting the teaching methods and values of the religious disciplines – memorizing, authority-centered learning versus disciplinary study and fitting into teachers' chains. Good memory, virtuosity in tricky problems, utility for others, social deference, and religious steadfastness are properties that are most often commended in biographical entries. Mathematical texts are listed among those that were learned by heart. Mastering an entire discipline, in contrast, was not seen as the goal of education. Thus, the adoption of the methods and values of teaching religious disciplines by the teachers of the 'ancient sciences' altered not only their format of reproduction, but also the content that was taught and the skills that were advocated. As a consequence, the increase in stability, status, and geographical distribution translated often into a decrease in complexity, degree of difficulty, and comprehensiveness. There were exceptions from this pattern. Some applied to specific disciplines, in particular *ʿilm al-hayʾa* and some parts of *ʿilm al-mīqāt*. Others are represented by individual scholars such as Ibn aš-Šāṭir (703–776/1304–1375), Ibn as-Sarrāǧ (719–748/1319/20–1347/8), Ġiyāṯ ad-Dīn Kāšī (d. 832/1429), or ʿAlī Qušçī (d. 879/1474). Additionally, the tastes and preferences of individual rulers differed in their impact upon the appointment of *mudarrisūn* at *madrasa*s and cognate institutes. But at large, the 'ancient sciences' transformed in outlook and approach into a mirror image of the religious disciplines. It is for the conditions of these fields of knowledge that we have to look in order to understand which forces shaped the practices in the 'ancient sciences'.

3 Courtly patronage for the 'ancient sciences' after 1200

In his paper about Ibn Taymīya, Michot reflects the generally held belief that the fall of the Būyid and Fāṭimid dynasties and the subsequent so-called Sunnī revival led to the abandonment of the practitioners of the philosophical sciences by the elites and in particular the courts when he emphasizes that their fate always depended on this part of Islamic society, and that socio-political conditions led to their demise.[31] Although scholars like ʿUmar al-Ḫayyām (d. 525/1131) and al-Muẓaffar al-Isfizārī (d. before 515/1121) were clearly engaged with the philosophical sciences at Salǧūq courts, they were not mere survivors of a bygone period. Dynasties in Iran in particular sponsored a broad range of disciplines from the 'ancient sciences'. Standard examples are the Ilḫānids and the Tīmūrids. Several well-known scholars from Syria, the Maġrib, and Iran such as Nasīr ad-Dīn aṭ-Ṭūsī, Muḥyī d-Dīn ʿUrḍī (d. 658/1260), Ibn Abī Šukr al-Maġribī (d. 682/1283), or Quṭḇ ad-Dīn Šīrāzī (d. 711/1311) worked as astronomers and astrologers at the Ilḫānid courts of Marāġa, Tabrīz, and Sultānīya, their observatories, and their *madrasa*s. The astronomical and mathematical knowledge of the Tīmūrid prince and ruler Uluġ Beg and his engagement with an observatory and a *madrasa* for teaching mathematics and astrology are equally well-known for more than half a century. But other dynasties, Sunnī and Šīʿī alike, such as the Inǧūʾids, the Muẓaffarids, the Aq Qoyunlu, or the Safavids, sponsored several of the 'ancient sciences', in particular medicine, astrology, *ʿilm al-hayʾa*, geometry, arithmetic, parts of philosophy, and geography. Some of them also patronized the occult sciences, such as alchemy or geomancy. Outside of Iran, the Delhi Sultāns, the Qutbšāhīs, the Muġals, the Ayyūbids, the Mamlūks, the Rasūlids, and the Ottomans all supported several of the 'ancient sciences', often including parts of philosophy, in particular metaphysics and ethics, but occasionally also natural philosophy. The veracity of these observations does not have sufficient power yet to alter the deeply held belief among historians and historians of science that courtly patronage for the 'ancient sciences' disappeared in the post-classical period. Hassan's discussion of why the sciences did not develop in Islamic societies in the same manner as at Catholic and Protestant universities of the early modern period rests on this unquestioned belief:

> But the universities which appeared in the West and which comprised several colleges for theology, law, arts and sciences, and medicine, did not develop in Islam in the same period. This is due to the fact that the madrasas which were supported by the waqf system, and with them the study of law and theology, continued to exist without interruption, whereas the centers for the study of the rational sciences, which were dependent on the strength and the prosperity of the state, deteriorated and ceased to exist with the decline of the Islamic states, and for this reason scientific knowledge did not keep in line with the quick advances of science in Europe after the Scientific Revolution.[32]

31 Yahya Michot, "Vanités intellectuelles . . . L'impasse des rationalismes selon le Rejet de la Contradiction d'Ibn Taymiyyah," *Oriente Moderno* 19 (2001), 597–617, at 602: "Certes, l'espèce des falāsifah proprement dits est alors éteinte, mais c'est parce que son existence était indissolublement liée au soutien et aux salons de princes, d'émirs, de vizirs et d'autres grands notables. Quand donc de tels appuis vinrent à manquer, du fait d'un changement des conditions socio-politiques, ils ne survécurent point."

32 www.history-science-technology.com/Articles/articles%208.htm.

This lack of recognition of the continued support of 'ancient sciences' in the spheres of the courts results most likely from the almost exclusive focus of the historians of science on the study of scientific texts and instruments, while ignoring a more systematic exploration of the larger cultural context of these texts. Historians, on the other hand, tend to ignore the sciences in their studies of politics, war, religion, or the arts. Clues and suggestions from social as well as art historians and dedications in scientific manuscripts suggest that the differences between courtly support for the 'ancient sciences' during the classical period and after the fall of the ʿAbbāsids were less important than previously believed. Courtly patronage of the 'ancient sciences' in post-classical Islamic societies was closely connected with modes of gaining, legitimizing, and maintaining power as was the case in classical societies.[33] The major social group the rulers could turn to for this purpose was that of the local landowners. These landowners were very often religious dignitaries. The most powerful part of this group was the overseers of the major shrines and mosques. They were often *sayyid*s (i.e., descendants of the Prophet and his family), and the heads of this group, the *naqīb*s who controlled and regulated the genealogical documents. A second important part of the local landowners was formed by families who held the offices of a *qāḍī*, a *mudarris*, an *imām*, a *wāʿiz*, or a *muʾaḏḏin* for centuries. The sons of these families usually acquired their reputation through success based on and derived from education – eloquence, familiarity with *fiqh* and *farāʾiḍ*, knowledge of classical Arabic in addition to the spoken language and the local language of literature and poetry, and the capability to debate the fine points of faith.[34] It is here where courtly patronage and *madrasa* education met each other because, as Hoffmann remarked, the respect that princes had to pay to the religious class was measured in terms of their generosity towards the *ʿulamāʾ*.[35] Thus, courtly patronage for the 'ancient sciences' reflects on this level of power sharing the acknowledgement of tendencies and orientations among the teachers and students of *madrasa*s and cognate institutes.

The move of the 'ancient sciences' into these educational institutions secured not only their intellectual continuation but also the continuation of their material existence due to princely funding. As a result, mathematical and astronomical manuscripts with dedications to rulers, princes, amīrs, sons of amīrs, and viziers can be found for most of the wealthier and ambitious dynasties after 1200. An exception is the Mamlūk dynasty who only promoted medicine via

33 Hassan's claim that there was no "demand of the state" for the ancient sciences in later Islamic societies while there had been such a demand in classical societies is not only in conflict with the sources that speak of patronage for medicine, astrology, astronomy, geography, history, arithmetic, and geometry. It is also problematic in a theoretical perspective, because it suggests an identity of court and state. Access by http://www.history-science-technology.com/Articles/articles%208.htm.

34 Birgitt Hoffmann, "Turkmen Princes and Religious Dignitaries: A Sketch in Group Profiles," in *Tīmūrid Art and Culture: Iran and Central Asia in the Fifteenth Century*, eds. Lisa Golombek and Maria Subtelny (Leiden: Brill, 1992), 23–28, in particular p. 25.

35 Hoffmann, "Turkmen Princes," 25.

a courtly *dīwān*. The absence of astrologers from among the courtly offices and the lack of dedications of mathematical and astronomical treatises to specific Mamlūk sultāns and courtiers does not necessarily mean, however, that this dynasty did not support these disciplines at all or even had a hostile attitude towards them. Mamlūk sultans promoted the office of the *muwaqqit* by religious donations as did, for example, Sulṭān Ḥasan (r. 748–752/1347–51, 755–762/1354–61) and by appointing famous scholars of *ʿilm al-mīqāt* to the post of a *mudarris* at *madrasa*s or the head of *ḫānaqāh*s as did Sulṭān Barsbay (r. 825–841/1422–38) in the case of Ibn al-Mağdī (d. 851/1447).[36] Descendants of Mamlūk officials and officers did not look down on the mathematical sciences (*ʿilm al-mīqāt, ʿilm al-hayʾa* or even *ʿilm al-ḥarf*), as the writings of Naṣīr ad-Dīn ibn Qurqmas al-Ḥanafī and Ibn al-Mağdī's career as a *muwaqqit* and *madrasa*/*ḫānqāh* teacher show.[37]

Important and fascinating issues of courtly patronage for the 'ancient sciences' concern the relationships between patron and scholar, the modes of loyalty established in such relationships, the similarities and differences between patronage for scholarship, patronage for art, literature and architecture as well as patronage for military ranks, the types of patrons, the types of remuneration, the function and frequency of dedications and the shift from the 'ancient' to the 'rational sciences'. In a series of papers written in the last three years, I have discussed the languages of patronage and the problems of interpretation, aspects of relationships between princes and scholars of the 'ancient' and 'rational sciences' such as long duration, change of 'patrons', practices of negotiating rights and duties, or types of remuneration.[38] In other papers, I present currently available research and historical evidence for the stimulating impact that the arts and their patronage had upon the continuous production of scientific treatises and their canonization.[39]

36 Šams ad-Dīn as-Saḫāwī, *ad-Dawʾ al-lāmiʿ fī ahl al-qarn at-tāsiʿ*, 12 vols. (Beirut: Dār Maktabat al-Ḥayāt, s.d.), vol. 1, 300.

37 As-Saḫāwī, *al-Dawʾ al-lāmiʿ*, vol. 1, 300; vol. 3, 115 and 235.

38 Sonja Brentjes, "Patronage of the Mathematical Sciences in Islamic Societies: Structure and Rhetoric, Identities and Outcomes," in *The Oxford Handbook of the History of Mathematics*, eds. Eleanor Robson and Jackie Stedall (Oxford: Oxford University Press, 2008), 301–328; idem, "Courtly Patronage of the Ancient Sciences in Post-Classical Islamic Societies," *Al-Qantara* 29 (2008), 403–436; idem, "Ayyubid Princes and Their Scholarly Clients from the Ancient Sciences," in *Court Cultures in the Mus lim World: Seventh to Nineteenth Centuries*, eds. Albrecht Fuess and Jan-Peter Hartung (SOAS/Routledge Studies on the Middle East, London: Routledge, 2010), 326–356.

39 Sonja Brentjes, "The Interplay of Science, Art and Literature in Islamic Societies before 1700," in *Science, Literature and Aesthetics: History of Science, Philosophy and Culture in Indian Civilization*, ed. Amiya Dev, general editor D. P. Chattopadhyaya, Volume 15, Part 3 (New Delhi: PHISPC, Centre for Studies in Civilizations, 2009), 455–486; idem, "The Mathematical Sciences in the Safavid Empire: Questions and Perspectives," 325–402.

4 An internal and an external claim of scientific 'decline'

One Muslim author who is often quoted in discussions about decline is Ibn Ḫaldūn (d. 808/1406). In developing his theory of society in his famous *Muqaddima*, Ibn Ḫaldūn relies on several ancient Greek concepts and theories, among them the idea of history as cycles of rise, climax, and decline. Ibn Ḫaldūn saw a strong link between economic and cultural prosperity of a society and the rise of the disciplines of knowledge as well as between economic crisis and the decline of the scholarly disciplines.[40] The basis for these claims is his view about the relationship between the evolution of sedentary city culture and political power on the one hand and his view on the social status of knowledge on the other. He believed that cities and their prosperity or ruin are the consequence of the existence of dynasties, not their precondition or that these elements were independently coexistent.[41] He regarded scholarly disciplines as crafts due to the idea that skill is the result of habit, habits are corporeal, that is, *sensibilia*, and hence in need of instruction.[42] In addition to these theoretical perspectives, Ibn Ḫaldūn also talked about concrete historical instances in Islamic societies that would back his theory. He claimed that "the tradition of scientific instruction has practically ceased (to be cultivated) among the inhabitants of the Maġrib, because the civilization of the Maġrib has disintegrated and its dynasties have lost their importance."[43] In contrast, scientific instruction continues to flourish in Ḫurāsān and Mā Warā' an-Nahr in the East and Cairo in the West; the early centers of scientific instruction, Baġdād, Basra, and Kūfa, Ibn Ḫaldūn considered as ruined.[44] He emphasized the support for scientific instruction given by the Ayyūbid and Mamlūk dynasties and their officials to it in Cairo through religious donations for *madrasa*s and other buildings due to a fear of confiscation of property and discrimination of their descendants among the officials, that is, the Mamlūks. As a result, people from other regions in the Islamic world streamed to Cairo for studying.[45] Thus, in Ibn Ḫaldūn's view, the Maġrib (and al-Andalus) were the prime examples for the nexus of disappearance of scholarly disciplines and the ruin of dynasties, cities, and civilization, while the East (Egypt, Iran, and Central Asia) demonstrated the continuous flourishing of these disciplines due to the continuous prosperity of their civilization. He saw neither a 'decline' of the 'ancient' nor of the 'rational' and 'traditional sciences' in these Eastern regions in the second half of the fourteenth century. Even Tūnis was still a favorable place for scientific instruction.[46]

40 Ibn Khaldun, *The Muqaddimah: An Introduction to History*, translated from the Arabic by Franz Rosenthal, 3 vols., Bollinger Series XLIII (Princeton: Princeton University Press, 1958), vol. 2, 347–352 and 434–439.

41 Ibid., vol. 2, 235–237 and 270–276.

42 Ibid., vol. 2, 426–433.

43 Ibid., vol. 2, 427.

44 Ibid., vol. 2, 431.

45 Ibid., vol. 2, 435.

46 Ibid., vol. 2, 430.

A look into biographical dictionaries shows that not all of his negative evaluations should be taken at face value. Students and scholars from towns west of Tūnis came to Cairo for study and teaching or passed it as a stop on their way to Medina and Mecca. Some of them had already spent several years with teachers in *madrasa*s or *zāwiya*s exploring the 'rational' and 'traditional' as well as mathematical sciences. The extent of their studies in towns such as Tlemcen and Bejaia, but also in small towns such as Basta, testifies that not all higher learning had disappeared from the Western lands. Muḥammad ibn Muḥammad az-Zawāwī al-Biğāyī (821–c. 864/1418–ca. 1460), for instance, studied in Tlemcen and Bejaia logic, dialectics, philosophy, geometry, mathematical cosmography, mechanics, number theory, theoretical music, optics, burning mirrors, and the theoretical fields of the religious and philological disciplines.[47]

47 As-Saḫāwī, *ad-Daw' al-lāmi'*, vol. 9, 181–182: "Muḥammad ibn Muḥammad . . . Abū l-Faḍl . . . al-Mašdāllī, . . . this refers to a tribe in al-Zawāwa, al-Zawāwī al-Biğāyī . . . and he was known in the East as Abū l-Faḍl and in the West as Ibn Abī l-Qāsim . . . he memorized the two (texts called) *aš-Šātibīya*, al-Ḫirāzī's *Rağaz* on *ar-rasm* and *al-Kāfiya aš-Šāfiya* and *Lāmiyat al-af'āl* by Ibn Mālik on grammar and inflection and *Ġālib at-Tashīl* and his *Alfiya* and Ibn Ḥāğib's *al-Far'ī* and the *Risāla* and at-Tilimsānī's *Urğūza* on *farā'iḍ* and about a quarter of Saḥnūn's *Mudawwana* and *Ṭawāli' al-Anwār* on *usūl ad-dīn* by al-Bayḍāwī and Ibn Ḥāğib's *al-Aslī* and al-Ḫūnağī's *Jumal* and *al-Ḫazrağīya* on metric and *Talḫīṣ* by Ibn al-Bannā' on arithmetic and *Talḫīṣ al-Miftāḥ* and the *Dīwān* of Imrū' al-Qays, an-Nābiġa aḏ-Ḏubyānī, Zuhayr ibn Abī Sulmā, 'Alqama al-Faḥl and Ṭarafa ibn al-'Abd. Then he turned towards comprehension and thus studied inflection and metric with Abū Ya'qūb Yūsuf ar-Rīfī, Arabic, logic, *al-usūl* and *al-mīqāt* with Abū Bakr at-Tilimsānī and he took *al-mīqāt* also with Abū Bakr ibn 'Īsā al-Wanšarīsī. Then he studied grammar with Ya'qūb at-Tīrūnī, grammar and logic with Abū Isḥāq Ibrāhīm ibn Aḥmad ibn Abī Bakr. Then he researched arithmetic with Mūsā ibn Ibrāhīm al-Ḥasnāwī and again arithmetic together with inflection, grammar, the two *aṣl*, rhetoric and the revealed sciences of *tafsīr* and *ḥadīṯ* and *fiqh* with his father. Then he studied the two *asl* with Abū l-Ḥasan 'Alī ibn Ibrāhīm al-Ḥasnāwī, who is the brother of Mūsā, I believe. At the beginning of the year 40 he travelled to Tlemcen and studied there with the famous scholar Muḥammad ibn Marzūq ibn Ḥafīd and with Abū l-Qāsim ibn Sa'īd al-'Uqbānī, Abū l-Faḍl ibn al-Imām, Abū l-'Abbās Aḥmad ibn Zāġū, Abū 'Abd Allāh Muḥammad ibn an-Naǧǧār, known for the strength of his knowledge in *qiyās* through the butcher's knife of concluding by analogy. (He also studied) with Abū Rabī' Sulaymān al-Būzīdī and Abū Ya'qūb Yūsuf ibn Ismā'īl and Abū l-Ḥasan 'Alī ibn Qāsim and Abū 'Abd Allāh Muḥammad al-Būrī and Ibn Afšūš. He (studied) with the first *tafsīr* and *ḥadīṯ* and *fiqh* and the two *aṣl* and *adab* in its branches and logic and dialectic and *falsafiyāt* (the philosophical sciences) and medicine and *handasa*. (He studied) with the second *fiqh* and *usūl ad-dīn* and with the third *tafsīr* and *ḥadīṯ* and medicine and the ancient sciences (*al-'ulūm al-qadīma*) and *tasawwuf*. With the fourth (he studied) rhetoric and arithmetic and *farā'iḍ* and *handasa* and *tasawwuf*. With the fifth (he studied) *usūl al-fiqh* and rhetoric . . . With the sixth (he studied) *fiqh*. With the seventh (he studied) arithmetic and *farā'iḍ*. With the eighth (he studied) arithmetic and algebra and other (things) from its branches and *hay'a* and the movement of weights. And with the eighth (he studied) ephemerides and *mīqāt* in its branches of the arts of the astrolabes and the plates and the sines and *hay'a* and number theory and music and talismans and what resembled them of the science of burning mirrors and optics and the science of the horizons and with the tenth (he studied) medicine. Then he returned to Biğāya."

The external diagnosis of decline of an Islamic society linked to the state of its sciences was first formulated by Catholic and Protestant visitors to the Ottoman Empire.[48] The background to this diagnosis was first the visitors' lack of knowledge of the intellectual spaces within the Ottoman Empire and their inhabitants. Second it was the visitors' inability to navigate between their own cultural forms of scholarship and those in the Ottoman Empire. They were searching for the same, the familiar in a foreign society, and when they did not find it, they declared the other as deficient. Thirdly, political and military motifs stimulated a particularly hostile rhetoric about the Ottomans. Accusing them of neglect of the once flourishing sciences of the Arabs and their own lack of interest in such intellectual matters due to their being barbaric warriors of Scythian origin, while extolling the diligence and studiousness in the 'ancient sciences' of their Šīʿī neighbors in Iran, served as much the elevation of the visitors' own societies as it helped to maintain support for wars against the 'Turk'. Travel literature and diplomatic reports, in particular reports by Venetian ambassadors, abound with remarks or stories about the scientific "underdevelopment" of the "Turks" which are seldom more than prejudices and lack of knowledge. In one case, at least, the author made his theoretical beliefs in a cyclical theory of history explicit. In 1592, Lorenzo Bernardo wrote in his report to the Signoria:

> Three basic qualities have enabled the Turks to make such remarkable conquests, and rise to such importance in a brief period: religion, frugality, and obedience.
>
> From the beginning it was religion that made them zealous, frugality that made them satisfied with little, and obedience that produced men ready for any dangerous campaign.
>
> In an earlier report I discussed at length these three qualities, which were then and always had been typical of the Turks. Now I plan to follow the same order, but to discuss whether any changes have taken place subsequently that might lead us to hope that that empire will eventually decline. For nothing is more certain than that every living thing (including kingdoms and empires) has a beginning, a middle, and an end, or, you might say, a growth, maturity, and decline.[49]

48 See for a more detailed analysis of these claims and their relationship to conflicting descriptions of Ṣafavid Iran the following two papers of mine: Sonja Brentjes, "Pride and Prejudice: The Invention of a 'Historiography of Science' in the Ottoman and Safavid Empires by European Travellers and Writers of the Sixteenth and Seventeenth Centuries," in *Religious Values & the Rise of Science in Europe*, eds. John Brooke and Ekmeleddin Ihsanoğlu (Istanbul: IRCICA, 2005), 229–254; and idem, "Early Modern Western European Travellers in the Middle East and Their Reports about the Sciences," in *Sciences, Techniques et Instruments dans le Monde Iranien (Xe–XIXe Siècle)*, eds. N. Pourjavady and Ž. Vesel (Tehran: Institut Français de Recherches en Iran, Presses Universitaires d'Iran, 2004), 379–420.

49 J. C. Davis, *Pursuit of Power, Venetian Ambassadors' Reports on Spain, Turkey and France in the Age of Philipp II, 1560–1600* (New York: Harper & Row, 1970), 156–157.

While the Safavid elite was portrayed as eager to support the 'ancient sciences' financially and institutionally and the inhabitants of Iran as well versed not only in poetry, but also in astrology and other mathematical sciences, many early modern Catholic and Protestant writers declared the Ottomans and their scholars as merely interested in religion, as lacking books and access to ancient and early modern sources in their respective fields, and as structurally disinclined towards intellectual pursuits. This rhetoric contradicted the scholarly activities in the Ottoman Empire, which did not differ significantly in most of the 'ancient sciences' from those in Safavid Iran. *Madrasa*s here and there taught arithmetic, surveying, algebra, astronomy, geometry, medicine, and parts of philosophy and logic. Students studied Qāżīzāda ar-Rūmī, al-Čaġmīnī, ʿAlī Qušçī, Bahāʾ ad-Dīn al-ʿĀmilī, Naṣīr ad-Dīn aṭ-Ṭūsī, Uluġ Beg, and other authorities. The rhetoric also was in conflict with the activities travelers and envoys from Catholic and Protestant Europe undertook during the seventeenth century in the Ottoman Empire. They met with Jewish, Christian, and Muslim scholars to discuss languages, history, religion, the occult sciences, geography, astronomy, and even occasionally the latest scientific news from Europe. They bought manuscripts on a broad range of subjects including the sciences, hunting for ancient historians like Titus Livius and ancient mathematicians like Apollonios; Arabic philosophers and physicians like Ibn Sīnā; and Arabic, Persian, or Turkish writers, historians, geographers, and philologists like Saʿdī, Mīr Ḫwānd, or Abū l-Fidāʾ. They collected coins, inscriptions, and minerals; observed eclipses, comets, and other heavenly events; measured altitudes; and calculated latitudes. They shopped for medical and other drugs, seeds, fruits, and animals and engaged in trade with almonds and other commodities. Last but not least, they eagerly acquired Russian slaves and Greek temporary wives. They turned the Ottoman Empire into a province of their Republic of Letters and acknowledged it as a reservoir for material sources of knowledge, wealth, and pleasure.[50] In particular during the seventeenth century, Catholic and Protestant scholars were greatly interested in the scholarly resources of the Ottoman Empire as important components for their own scholarly exploits. In certain fields of knowledge such as cartography, geography, and astronomy, this interest continued far into the eighteenth century. The existence of major manuscript libraries in Germany, France, Italy, and Great Britain is a direct outcome of these interests as were the maps, the herbaria, the medical gardens, the cabinets of curiosity, and the museums of the seventeenth and eighteenth centuries.

50 Brentjes, "Pride and Prejudice: The Invention of a 'Historiography of Science';" idem, "'Renegades' and Missionaries as Minorities in the Transfer of Knowledge," in *Multicultural Science in the Ottoman Empire*, eds. Ekmeleddin Ihsanoglu, Kostas Chatzis and Efthymos Nicolaidis (Brepols: Turnhout, 2003), 63–70; idem, "Western European Travelers in the Ottoman Empire and Their Scholarly Endeavors (16th–18th centuries)," in *The Turks*, eds. Hasan Celal Güzel, C. Cem Oğuz and Osman Karatay (Ankara: Yeni Türkiye, 2002), 795–803; idem, "On the Relation between the Ottoman Empire and the West European Republic of Letters (17th–18th centuries)," in *International Congress on Learning & Education in the Ottoman World: Istanbul, April 12–15, 1999*, ed. Ali Çaksu (Istanbul: IRCICA, 2001), 121–148.

Daiber, in his paper on the "way from God's wisdom to science in Islam", points to discussions about the theme of 'decline' which he calls "decadence" in the late nineteenth and the first decades of the twentieth centuries among Arab writers that were nourished decisively, as he argues, by debates among French and English authors about the relative superiority or inferiority of their respective countries and contemporary developments in the theory of evolution. His story strengthens my general point against using this concept as a valid category in today's historiographical practice. A concept with such complex and complicated histories of its own, filled with an immense wealth of mostly negative values impedes any solid investigation of the interests and needs of a concrete Islamic society in any particular scientific practice and the knowledge it created or reproduced. The investigation of these histories is rather an important desideratum to be tackled by such a historiographical approach.

5 Reflections

The major conclusion that I drew some time ago from my encounters with historical sources and their ways of talking about or representing the 'ancient sciences' in various Islamic societies is that it is high time to abandon some of the main ways in which history of science in these societies is conceptualized on its deepest level. We have to stop assuming that all societies in which a majority of people were Muslims had one and the same attitude towards the 'ancient sciences' and hence shared one and one history alone. We rather should assume that different societies had different interests and needs and hence different attitudes. Muslims are as diverse as any other religious group. When it is appropriate to study the sciences and their histories in France, Italy, or any other country, province, capital, city, or town in Europe and look for their specificities rather than assuming that they were all the same all over Europe, why should we look at Islamic societies from a principally different angle? We need to start studying the sciences in specific Islamic societies. We need to historicize the sciences and put them fully into their local cultural context. This does not mean to abandon larger questions and avoid overarching perspectives. But we should not take the generalities as starting points and basis. A second concept that I believe we should abandon is the idea that doing justice to the 'ancient sciences' in Islamic societies means proving the 'greatness' of their practitioners and achievements. Muslims today suffer under an inferiority complex which they try to compensate with, among other things, a constant reference to previous 'greatness' in the 'ancient sciences'. The result of this approach to the past is apology combined with a most serious lack of familiarity with the content as well as the context of these sciences and their practitioners. I do not believe any longer that the race for cultural superiority can be won by pointing to a glorious past which is portrayed as the immediate predecessor or even a fully formed stem cell of 'modern' science, of 'us', of the 'West' today. I do not believe that such a race can be won at all, whatever means will be applied. Pride built on abusing the past is destructive. We need new ways of dealing with the hurt, trauma, and desires of Muslims today. Cultures are different and their peoples believe in different things, see the world with different eyes, pursue different goals, and live their lives differently. Squeezing them all into one long chain of predecessors and successors leads to a substantial loss of history, identity, and individuality. If we should celebrate anything at all, we should celebrate achievements within their own cultural matrix, and we should figure out what people of previous centuries and generations would have celebrated.

The all too easy claim that science is universal and culturally independent overlooks the differences between the cultures and their knowledge. It downplays such differences as irrelevant for the survival of science and its acceptance in different cultures. It is the expectation that science should proceed everywhere in principally the same manner from the lower to the higher, the simpler to the more complex, that has brought us into the quagmire where we are finding ourselves in today – the need to explain why only certain societies and not others created the specific forms of science that today rule supreme across the globe. I am convinced today that it is very difficult to explain things that happened because of the discontinuity and contingency both of historical processes and their transmission. But I continue to hear a nagging voice which says it may be possible. In contrast, I am certain that it is impossible to explain things that did not happen. Things that did not happen did not leave traces and cannot leave traces, not even the contingent and discontinuous traces of things that did happen. Hence, there is no access to things that did not happen. All that can be done is to compare things that happened in one society and culture with things that happened in another and then ask whether the differences between the things that happened can explain the things that did not happen. The problem with such a comparative approach is, however, that it is built on, even if only implicitly, the assumption that certain happenings are normative and hence good or superior, while others are deviations from the norm, for instance 'decline'. Trying to explain things that did not happen through things that happened in a different society inscribes unavoidably the values of the other society unto the society that is studied and seen as lacking. It is a waste of time, time which – as I believe – is better spent in studying the traces that the things which happened in this society have left. The difficulty is to find the questions that are appropriate to these traces and are not merely modified translations of questions asked for the other society. Hence, we need to take serious Sabra's call for a history of science in Islamic societies that asks questions appropriate to these societies rather than sandwiching the history between the history of science in Antiquity and history of science in the Latin Middle Ages. We could, for instance, study the manner through which certain parts of philosophy and logic became cornerstones of certain religious disciplines such as *uṣul ad-dīn* or the ways through which components of Aristotelian cosmology, physics, and meteorology were made acceptable to and accepted by the military as well as administrative and educational elites in certain Islamic societies but not in others. We also could investigate whether differences in social relationships and behaviors expressed by words such as *ḫadama, sanaʿa* or *lāzama* modified the kind of knowledge sought and created within the domain of the 'ancient sciences'. We should take the forms of talking about the 'ancient sciences' as an object of study and find out whether the image presented of these sciences shifted when talked about as a sequence of acts carried out by scholars or by cultural heroes or even those sent to earth by a divine being. Moreover, we should pay serious attention to the religious, literary, and visual elements of scientific treatises and situate them in their specific local context. There are many more aspects both of content and context that could and should be studied, but, above all, we need to take serious Gutas' call for a social and cultural history of the sciences in Islamic societies.[51]

51 Dimitri Gutas, *Greek Thought, Arabic Culture: The Graeco-Arabic Translation Movement in Baghdad and Early ʿAbbāsid Society (2nd–4th/8th–10th C.)* (London: Routledge, 1998).

6

TOWARDS A NEW APPROACH TO MEDIEVAL CROSS-CULTURAL EXCHANGES*

Sonja Brentjes, Alexander Fidora and Matthias M. Tischler

1 The nineteenth century

Medieval cross-cultural exchange of knowledge began its life as an academic research topic in the nineteenth century in debates about the relevance of 'the Arabs' for 'European sciences' and their contribution to 'the sciences' in comparison to 'the Greeks'. Particularly volatile were the intellectual and political controversies in France and Italy in the contexts of modernization, revolutions, republicanism and the political and spiritual powers of the pope and the Catholic Church.[1]

Most of the questions and perspectives formulated during these debates seem to haunt us even now, when our factual knowledge is much greater today than it was 200 years ago. The scholars of the nineteenth century did not propose, however, a theory of cross-cultural exchange. They were much too busy with searching for the progress of the sciences and medieval manuscripts.

2 The impact of Charles Homer Haskins' work

A comprehensive and relatively consistent description of what we call today cross-cultural exchange of knowledge that has had a lasting impact on studies of medieval Catholic Europe was published after the First World War by the US historian Charles Homer Haskins. He regarded this exchange as the transfer of written knowledge through translation. His publications established the undeniable importance of these transfers for the medieval scholarly world and, hence, their high relevance for the intellectual history of the Middle Ages. Methodologically, Haskins focused on texts.

* We thank Rafael Azuar for his contribution to this paper with regard to the importance of material objects, in particular scientific instruments. We also thank Isabelle Draelants, Maribel Fierro, Jens Høyrup, Antoni Malet, Efthymios Nicolaidis, Víctor Pallejà de Bustinza and Anne Tihon for their helpful comments and suggestions.

1 Charette 1995; Vegas González 2005; Brentjes 2016.

DOI: 10.4324/9781003372493-7

Conceptually, he favored the idea that the medieval translators into Latin sought to recover ancient Greek philosophy, medicine and the sciences.

In two works, Haskins presented a series of unknown translations from Greek and Arabic primarily into Latin, in particular texts on astronomy, astrology, geometry, arithmetic, medicine and philosophy.[2] Other domains of knowledge, such as magic, alchemy, divination, history-writing, geography or religious doctrines, remained largely outside his purview. The discovered manuscripts provided Haskins with names of translators, dates and localities. Some of the translations contained prefaces in which the translator legitimated his work and its purpose. In a few cases, the translations were dedicated to a patron, a friend or a colleague. Haskins selected as significant two centuries (the twelfth to thirteenth centuries), four regions (Iberian Peninsula, Southern France, Sicily and Northern Italy) and about a dozen translators (ranging from Petrus Alfonsi, a converted Jew from Huesca, to Michael Scotus, a translator of contested origin who worked in Toledo and Palermo) from among the discovered sources and their information.

Haskins interpreted this material in a teleological and Eurocentric manner characteristic for his time. We will call any such comprehensive, historiographical interpretation an observer narrative. Occasionally we will abbreviate this term by the expression 'model'. We do not suggest with this term that Haskins developed a theoretical schema that he thought was applicable to any situation of translating texts. We simply use it as a label for the longer and more descriptive term 'observer narrative'. A further terminological difficulty concerns the designation of the different societies that came into being, changed, stabilized or disappeared in parts of Europe between approximately 600 and 1500 and whose formalized elite knowledge cultures shared Latin as their language of study and other forms of communication and whose members believed mostly in some form or the other of religious doctrines sanctioned by the Church of Rome. To call them Christian, their language Latin and their territory Europe excludes too many people and communities who also lived in that part of Eurasia that Ptolemy labeled Europe and exchanged knowledge across cultural boundaries. To call them Western is as anachronistic as it is to assume a shared cultural identity. To enumerate them every time we wish to refer to them is cumbersome and inelegant. Such a list of smaller cultural entities certainly has its own conceptual insufficiencies. In a sense, the problem is not solvable in a completely satisfying manner. We have thus decided to call them Occidental Christians and their congregation the Occidental Church. This locates scholars, patrons and craftsmen alike somewhere in the Western regions of Eurasia, sets them apart from their Muslim and Jewish neighbors and highlights their adherence in general terms to what is called today the Catholic Church of Rome.

Haskins' narrative was very successful. It was accepted by many of his successors and dominated research until the end of the twentieth century. Its key concepts are that of a twelfth-century renaissance, the recovery of ancient Greek

2 Haskins 1924, 1927.

philosophy and science and a purposeful intellectual movement. The narrative's actors are Occidental clerics and a few lay people, most of whom travelled to the places where texts to be translated could be found. The intended recipients are archbishops and monks as well as a few royal patrons in the Iberian Peninsula, Sicily, the British Isles and Northern France. They are regarded, however, as of less relevance to the narrative than the translators. They appear as shadowy figures without a clearly elucidated role of their own. The acquisition of languages and other skills remained largely outside of Haskins' narrative framework. The cultural phenomenon of translating from one language into another was primarily understood as a unilateral process. It consisted of mainly accidental acquisitions of texts by members of one single, homogeneous culture identified as Latin, Christian and European.[3] It was imbued with unquestioned positive values ascribed to the work of the translators and, as a kind of projection, to the later users of the translations located primarily at the universities of Bologna, Paris, Oxford and Cambridge, because it allegedly paved the way for the later emergence of modern science in Europe.

Haskins' macro-level interpretation of translations of philosophical, scientific and medical texts from Greek or Arabic into Latin rests on the historiographical beliefs and practices of his time. As a 'model' it is retrospective, restrictive, static and mono-cultural. It operates with the idea of an intentional recovery of ancient Greek knowledge as an unquestioned value that medieval clerics and schoolmen allegedly shared with the humanists of the Renaissance and the neo-humanists of the nineteenth century.[4] Latin sources from the ninth to the twelfth centuries indicate that this was not always the case. Ancient Greek knowledge was often rejected as pagan and Byzantine knowledge was condemned as heretical. Arabic knowledge was propagated because it was presented as well as perceived as rational or because it was seen as an indispensable condition for conversion.[5] Haskins' observer narrative isolates the case of translation from other activities of knowledge acquisition. It separates Latin from other target languages and Greek and Arabic from other source languages of translating. It sets philosophy, medicine and the sciences apart from other fields of knowledge that were subject to translating. It posits texts above instruments, images and maps. It considers elite, written knowledge transfer of greater relevance than oral and manual exchanges of skills and knowledge among merchants, craftsmen or mariners who moved across the Mediterranean

3 Although Haskins recognized the importance of the translations in Salerno and in the Northern regions of the Iberian Peninsula during the tenth century, he separated them from the cultural process, which he saw as a part of the twelfth-century renaissance.

4 Recently, authors like Sylvain Gouguenheim and right-wing Islamophobe propagandists of a 'Western cultural superiority' underlined their adherence to such a simplified and often factually distorted understanding of the complexities of cross-cultural dissemination processes of knowledge.

5 De Libera 2009.

and beyond. Haskins' observer narrative is static because it privileges translation as the only important activity in the transfer of knowledge, largely ignoring the activities necessarily preceding translating as well as those that occur concomitantly and subsequently. As a result, it is mono-cultural and does not recognize the various exchange processes that took place between different communities and cultures. Heterogeneous, multifaceted cultures, many and instable frontiers between them, cross-cultural relationships and shared lives, small-scale local, individually defined projects, personal encounters between members of different communities who pursued different interests and adhered to different beliefs, other actors than translators and patrons and other activities than translating, writing and reading were unthinkable or of minor relevance in Haskins' observer narrative.

Besides Haskins' narrative, the earlier theses of the Roman Catholic physicist and historian Pierre Duhem (1861–1916) had a great impact on how history of science developed in its study of the transformations of the translated texts by scholars at medieval universities in Occidental Christian Europe. He emphasized in particular the contribution of the clergy, a position that was part of his active participation in the struggles of French defenders of the pope and the worldly power of the Roman Catholic Church against republican and secular forces. In his monumental *Système du monde*, Duhem presents medieval science as subordinated under the classification of religious orders.[6] He even advanced the thesis that the condemnations of doctrines taught at the University of Paris in 1277 opened the road to new ideas in natural philosophy and cosmology. Therefore, he argued, 'the Church' not only made medieval science, but also contributed to the renaissance of scientific knowledge. In regard to cross-cultural exchange of knowledge, Duhem represents the positions held in French academia, with variants, during the nineteenth century (i.e., a linear ascent from classical Greece to 'the Arabs' to 'the Latins', who we were in our infancy). Duhem's variation consisted in recognizing Byzantine scientific activities in the early centuries of the East Roman Empire. There were other important discussants on what constituted medieval sciences and who participated in shaping them, such as Lynn Thorndike or George Sarton. But their historiographical positions were rarely taken as a major point of reference during the twentieth century, when issues of cross-cultural exchange of knowledge were discussed. Since we focus in this paper on the changes of Haskins' observer narrative between approximately the 1960s and today, we have decided to abstain from including these other positions in our discussion. Our goal is to stimulate a discussion about concepts, perspectives and methods, not to write a comprehensive survey of the historiographical debates on medieval sciences during the last century.

6 Duhem 1913–1959.

3 From Charles Homer Haskins to our days

The most important successors of Haskins' studies in the twentieth century were Josep Maria Millàs Vallicrosa, Juan Vernet Ginés, Marie-Thérèse d'Alverny, Marshall Clagett and his students, Hubertus L. L. Busard, Menso Folkerts, André Allard and Charles Burnett. While all of them subscribed to basic features of Haskins' 'model' – in particular the centrality of texts, translators, movement; the recovery of ancient Greek philosophy, medicine and sciences; the hierarchy of locations and the temporal focus – each of them also modified it in important ways which reduce in particular its static, restrictive and mono-cultural character.

Millàs Vallicrosa and Vernet incorporated activities of Arabic and Hebrew scholars as well as translations from these two languages into Latin, Catalan and Castilian into the narrative. Due to the specific limitations for academic work on Arabic, Hebrew and Catalan communities during most of the twentieth century in Spain, their methodological approach rested firmly on historical empiricism. They invested most of their research in the discovery, description and publication of scientific texts in Arabic, Hebrew and Catalan, introducing instruments and maps as new objects in the study of medieval knowledge transfer.[7] Millàs Vallicrosa extended Haskins' study of translations and translators to users of translations, investigating in particular the impact that Arabic and Hebrew scholars and their works had exercised upon the intellectual life in the lands of the Crown of Aragon.[8] Vernet, following Millàs Vallicrosa in this line of publication, extended it to the Kingdom of Leon and Castile.[9] Both scholars added to philosophy, medicine and the sciences studies of texts on agriculture, mechanics and other productive or technical arts.[10] They pointed occasionally to the necessity of linking textual studies to investigations of material culture and included magical amulets among the objects of knowledge transfer.[11] Occasionally, and deviating from the otherwise accepted norms among the discussants of knowledge transfer during the twentieth century, they also paid attention to political components of knowledge in the Iberian Peninsula.[12] Conceptually, the most important step towards replacing the uni-directional notion of knowledge transfer by a perspective that identified the cultural phenomenon as cross-cultural exchange of knowledge probably was both scholars' repeated insistence on the cooperative nature of transfer activities.[13]

Marshall Clagett and his students Edward Grant, David Charles Lindberg and John Emery Murdoch, as well as Murdoch's student Edith Dudley Sylla, broadened the material basis for discussing interpretive issues for the mathematical sciences, including mechanics and optics, on the one hand, and Aristotelian natural

7 Millàs Vallicrosa 1927, 1934, 1942; Vernet Ginés 1953, 1986.
8 Millàs Vallicrosa 1931, 1949.
9 Vernet Ginés 1979.
10 Millàs Vallicrosa 1943; Vernet Ginés 1988, 1903.
11 Millàs Vallicrosa 1941.
12 Vernet Ginés 1970.
13 Vernet Ginés 1998, 1999.

philosophy, on the other. Thus, they prepared critical editions of Latin translations of Greek and Arabic texts. They concentrated their efforts on sorting out highly complex and complicated paths of transmission. This work also included the study of the transformations that this new knowledge had undergone in the scholarly circles at the universities of Paris, Cambridge and Oxford.[14] Murdoch and Sylla, together with Michael Rogers McVaugh, contributed substantially to a broadening of the perspectives taken into consideration when interpreting medieval transfer of knowledge: they repeatedly explored issues of context (philosophy, theology, lay public, arts, politics) and reflected on styles, methodological foundations and languages of interpretation.[15]

Hubertus L. L. Busard, Menso Folkerts, Paul Kunitzsch and Richard Lorch insisted, too, on the necessity to uncover textual evidence of transfer, transmission and transformation. They produced critical editions of important mathematical texts derived from late Roman Latin translations or translated from Arabic or Greek.[16] Kunitzsch, partially in cooperation with Lorch, deconstructed age-old myths about individual texts and their authors translated in the late tenth and during the twelfth centuries.[17] He succeeded in determining the type of Arabic manuscript text with which the Italian translator of Ptolemy's *Almagest*, Gerard of Cremona, must have worked in the second half of the twelfth century in Toledo. This enabled him to delineate precisely Gerard's multifaceted working practices in the course of the many years which it took him to finish his translation.[18] The two scholars edited a series of mathematical, astronomical and astrological works from Greek, Arabic, Latin and Hebrew manuscripts, elucidating in this manner the scientific canon broadly shared across time and space among various Mediterranean peoples and communities.[19] Folkerts extended the scope of time and space by paying attention to the diffusion of Gerbert of Aurillac's (tenth century) abacus and its Arabic numerals beyond France and studying mathematical activities at universities and monasteries in Switzerland, Austria, Germany and Poland in the fourteenth and fifteenth centuries.[20] André Allard meticulously studied the translations of Arabic texts on Indian arithmetic and their transformations at the hands of anonymous Latin commentators and readers.[21] He extended Haskins' 'model' to Byzantium by showing that Byzantine monks translated Arabic arithmetical works and integrated them into the intellectual life of Byzantine monasteries.[22] Raymond Mercier, Anne Tihon, Régine Leurquin, Claudy Scheuren and

14 Clagett 1964–1984; Grant 1971; Lindberg 1975; Murdoch/Sylla 1978; Murdoch/Grant 1987.

15 Murdoch/Sylla 1987; Sylla/McVaugh 1997; Murdoch 2003; Sylla/Newman 2009.

16 Busard 1968, 1977, 1983, 1984, 1987, 1996, 2001, 2005; Busard/Folkerts 1992; Folkerts 1970a, 1970b, 1971a, 1972, 1995; Folkerts/Kunitzsch 1997.

17 Kunitzsch 1981, 1989, 1993, 2004.

18 Kunitzsch 1966, 1974, 1975.

19 Kunitzsch/Lorch 2010, 2011; Lorch 1995, 1999, 2001; Lorch/Martínez Gázquez 2005.

20 Folkerts 1971a, 1971b, 1996b, 2002.

21 Allard 1991, 1992.

22 Allard 1973, 1977, 1981.

Paul Kunitzsch also spent much time and effort on studying the cross-cultural exchanges that took place within Byzantium and between Byzantium and its neighbors, focusing primarily on astronomical texts and tables.[23] Their work documents the interest that Byzantine clerical scholars took, in different periods of almost half a millennium, in knowledge produced among their Arabic and Iranian contemporaries. The research of all these scholars shows that translation activities occurred in parallel and independently. It makes clear that the selection of the source as well as the target languages of translation depended largely on local circumstances. It was often not the result of deliberate choice or conscious cultural values as suggested in Haskins' concept of the recovery of Greek philosophy, medicine and sciences. There can be no doubt that in many cases the purpose of medieval cross-cultural exchanges was the acquisition of Arabic, Persian or Hebrew texts on medicine, astrology, astronomy, arithmetic, dreams, alchemy and divination, fields to which one may add philosophical, theological and other kinds of doctrines and writings.

Marie-Thérèse d'Alverny, Charles Burnett and Paul Kunitzsch were perhaps the most ardent supporters of Haskins' observer narrative during the second half of the twentieth century. The great merit of d'Alverny and Burnett consists in their meticulous verification and continuation of the many individual lines of study and suggestions for further research presented by Haskins in his two major publications on the medieval transfer of knowledge. They undertook much archival work in addition to studies of manuscripts in order to gain a clearer understanding of the identities of the translators mentioned by Haskins, the locations where they produced their works, their affiliations to individual churches, ecclesiastic and royal courts, as well as the titles and contents of their translations.[24] They struggled for a better grasp on the social conditions in which the translators lived and worked, in particular their involvement with the so-called 'school of translators of Toledo' or with the Cluniac abbot Petrus Venerabilis.[25] D'Alverny stressed the cooperation between Catholic and Jewish scholars in translating philosophical texts from Arabic into Latin via old Castilian. She included translations of Arabic religious as well as magical texts into Latin into her perspective on the medieval transfer of knowledge.[26] Burnett slowly moved from Haskins' position of the translators' desire to recover ancient Greek philosophy, medicine and sciences to an acknowledgement of the indubitable preeminence of magic, astrology and divination in the translations from Arabic into Latin for many of the Catholic translators working in the Iberian Peninsula and, in the case of Adelard of Bath, working in Norman Sicily, in England and perhaps as a visitor in a few cities of the Latin Crusader

23 Mercier 1994, 1998, 2004; Tihon 1986–1993, 1994, 1994–2001; Tihon e. a. 2001; Tihon/Mercier 1998; Kunitzsch 1964.

24 D'Alverny 1994b; D'Alverny/Hudry 1971; Burnett 1994b, 2008a, 2009.

25 D'Alverny 1994a; Burnett 2006a.

26 D'Alverny 1962, 1982, 1989, 1994a, 1994b.

States in the Levant.[27] Before him, Richard Lemay and Sonja Brentjes already made this point for the Iberian Peninsula.[28] Burnett also argued that the few known translators in the Crusader States of the Levant had made a more substantial contribution to the transfer of knowledge than Haskins had been willing to grant.[29] Kunitzsch's main contributions to Haskins' observer narrative were, on the one hand, his discussion of the relationships between Arabic and Latin scientific texts, tables, nomenclature and instruments from a comparative perspective and, on the other hand, his introduction of images, texts of the so-called folk astronomy and troubadour-poetry, as well as other kinds of literature, into the 'model'.[30]

Transfer of knowledge through oral contact and manual practices, two domains ignored in Haskins' observer narrative, was studied in research programmes headed by Menso Folkerts, Jens Høyrup and Maryvonne Spiesser.[31] They searched for and catalogued Latin and vernacular texts on practical mathematics taught and used outside the universities. Municipalities and private sponsors, merchants and school teachers in cities of Italy, the Provence, Aquitania, Catalonia and, to a lesser degree, Northern Africa were recognized as important actors in this sphere of knowledge transfer. More difficult to trace than is already the case with theoretical texts or scientific instruments, much of the interpretive work remained in the realm of suggestions and speculations.

The trajectories of medical and pharmaceutical texts between various Mediterranean cultures showed similarities with those excavated for scientific or philosophical works, but also clear differences. The study of these fields was undertaken, for instance, in major editorial projects like Avicenna or Averroes latinus. In this manner, the study of medical and pharmaceutical texts was closely related to a major approach pursued in history of philosophy, in particular through the Belgian Aristoteles latinus project. No such large-scale editorial projects were undertaken for scientific texts. This editorial work reflected important features of the accepted interpretive approach. One of these problems consisted in editing original Greek or Arabic texts separately from their Latin or Hebrew translations. Changes in this kind of approach emerged over time, in particular in the last decade of the twentieth century. The project Avicenna latinus at the Royal Academy of Belgium, for instance, began under the leadership of Simone van Riet to consider Arabic and Latin versions together. The Latin translation of Ibn Sīnā's *Physics* was edited in a manner that included a mixed critical apparatus based on an analytical comparison with Arabic manuscripts.[32] The edition of Ibn Sīnā's *al-Qānūn fī l-ṭibb*, however, is considered as too difficult a project for the time being. The difficulties result from the existence of manifold copies of the Arabic

27 Burnett 1996, 2006a.
28 Lemay 1958, 1962, 1987; Brentjes 2000.
29 Burnett 2000, 2003.
30 Kunitzsch 1961, 1977, 1983, 1986, 1996.
31 Folkerts 1981, 1996b; Høyrup 2007; Spiesser 2003.
32 Van Riet/Verbeke 1992; Van Riet e. a. 2006.

original, Latin and vernacular translations of the entire work or of some of its parts and the many different extant commentaries. The study of cross-cultural exchange of knowledge thus confirms that a broad array of historical, linguistic, disciplinary and further technical expertise is required on the side of those who wish to edit and interpret its textual documents.[33]

A further challenge that has emerged more clearly in the last decades of editing texts that represent medieval translations from Arabic or Greek into Latin is the existence of multiple text forms. They can represent several parallel translations, the reshaping of an earlier stage of translation by the same translator or a variety of modifications of different extent and severity executed by users of a translation.[34] This does not only increase the difficulties for the modern editor. It is another feature of the cross-cultural exchange of knowledge that Haskins' narrative does not make room for.

An observer narrative that makes place for the cross-cultural transfer of medical knowledge needs to reconsider periodization and places of transfer. While the importance of Salerno for translations of Arabic medical texts, either themselves translations of older knowledge or compiled on the basis of cross-culturally transmitted theories, pharmaceutics and rules of treatment, has been stressed in Haskins' 'model', the role of North Africa as a place of translation from Arabic into Latin, however, is lost and its role as an intermediary between Baghdad, Cairo, Sicily and al-Andalus is marginalized. By including this region, Jacquart and Micheau took a different stance.[35] They accept, however, the manner in which Haskins interpreted the translators' activities in Southern Italy in the tenth and early eleventh centuries.[36] Despite acknowledging that different possibilities for interpretation exist, they write that it is clear that the goal was to transmit not Arabic, but Greek medicine and, in particular, Galen.[37] Different choices made by the known translators and the critical remarks made at Constantinus Africanus' practice of eliminating all traces of the Arabic authors whose texts he translated call, however, for a more cautious and nuanced evaluation.[38]

Isabelle Draelants' doctoral thesis on Arnold of Saxony's encyclopedia (thirteenth century) reinforces this need to rethink the older positions. Her results support as well as modify Alain de Libera's depiction of the views that such scholars had of ancient authors writing in Greek, Byzantine and Arabic. Arnold of Saxony did not belong to the mendicant orders, although his work shows close links to that of members of the Dominican order.[39] Presenting a long analysis of the texts used and the authorities quoted by Arnold, Draelants concludes that he fused in his

33 Fidora 2012b.
34 Burnett 20008b.
35 Jacquart/Micheau 1990, 91 and 107–118.
36 Ibid., 98–101 and 103–105.
37 Ibid., 101 sq.
38 Ibid., 100 and 107.
39 Draelants 2005, 43 n. 1.

encyclopaedia translations coming from two main areas: Salerno and Toledo.[40] A further important indicator of attitudes held by Arnold and, according to Draelants, also by other authors of encyclopaedias in the thirteenth century, like Bartholomew the Englishman or Vincent of Beauvais, is found in his medical text on causes. There, he used concomitantly texts belonging to the so-called 'old translations' of Aristotelian logic and ethics, new translations of Aristotelian or pseudo-Aristotelian works made from Greek as well as Arabic, translations of treatises by scholars from Islamicate societies and writings by his own contemporaries.[41] Draelants describes this medical text as structurally copying Ibn Sīnā's medical summa (Books III and IV), filling it with details from texts translated by Constantinus Africanus and material from writings of his contemporaries like Albertus Magnus.[42] These observations suggest that Arnold and his contemporaries really did not bother to separate the different translations in any culturally defined grid. They obviously did not share the value hierarchies proposed by Haskins and exaggerated further by Gouguenheim, nor did their practice resemble Duhem's progressist religious readings of the thirteenth century. These men apparently neither looked down on Greek authorities, independent of the period, nor put authors from Islamicate societies on the pedestal of lonely superiority. Their rich tapestry of dialogical textual practice deserves a fresh reflection about their placement in a new observer narrative.

A classical approach to the issues of cooperation and distribution in cross-cultural exchanges of knowledge by Pieter Sjoerd van Koningsveld, namely the codicological study of extant manuscripts from al-Andalus in Arabic or Hebrew, led to impressively different results than those achieved within the prevailing 'model'. Van Koningsveld showed for some 900 manuscripts that the two major groups in twelfth- and thirteenth-century Toledo involved in the selection of scientific texts, their copying, financing and subsequent distribution through booksellers and private owners were wealthy, prominent Jewish families and Muslim prisoners of war held by the Catholic town administration.[43] On the basis of the identified manuscripts, he concluded that after the Christian conquest of Muslim cities, other groups like Mudejars and Mozarabs withdrew rather rapidly from an engagement with advanced sciences, medicine or philosophy. The former focused on the preservation of religious and juridical fundaments of their communal life, adding some interest in practical knowledge such as healing and magic. The latter, van Koningsveld claims, reoriented their cultural outlook partly voluntarily and partly under ecclesiastic pressure towards norms and standards of the Roman rite of the Occidental Church whose representatives dominated life in Toledo throughout the twelfth and in the first decades of the thirteenth centuries.[44] These results from a different field of study present serious challenges to Haskins' perspective and its macro-level approach to knowledge transfer.

40 Draelants 2004, in particular 94–102.
41 Ibid.
42 Ibid., 88 sq.
43 Van Koningsveld 1992.
44 Van Koningsveld 1991.

As a result of all these investigations, Haskins' observer narrative was modified, extended and replaced in so many of its components and details that it ceased to function as the 'leitmotif' of research in the late twentieth century. The time seemed to be ripe to replace it by a new observer narrative that was dynamic, inclusive and receptive. Yet, Haskins' 'model' continued a life of rhetorical veneration, appreciated for its author's broad-scaled approach to the topic. In the 1990s, however, a historiographical shift took place that was unexpected from the perspective of medieval studies because it introduced the early modern period as an important arena to the study of cross-cultural exchanges of knowledge.

In the 1980s, the historiography of the early modern period became a field of intense debate. This debate focused primarily on the narrative of the scientific revolution, on scientific instrument builders and other craftsmen and on the gendered nature of early modern sciences and twenty-century accounts of early modern intellectual history. The other main narrative that connects late medieval with early modern intellectual history, the rise of humanism, its recovery of ancient Greek and Latin knowledge practices and its rejection and devaluation of medieval Latin translations of Arabic scientific, medical and philosophical texts, as well as medieval scholarly authorities from Islamicate societies, was challenged only in some of its components, without being questioned in its totality. Humanist rhetoric continued to be taken at face value and interpreted as an appropriate description of humanist knowledge practices.

The first major challenge of this historiographical narrative came from Nancy Siraisi, who showed that Latin translations of important Arabic medical works, in particular Ibn Sīnā's opus on medicine, continued throughout the early modern period to dominate medical teaching at Italian universities.[45] A decade later, Sonja Brentjes and Dag Nikolaus Hasse challenged two crucial features of the humanist narrative: the rejection of any knowledge coming from Islamicate societies; the rejection and exclusion of all medieval Latin translations of Arabic philosophical, medical or scientific texts. Brentjes demonstrated that many early modern Roman Catholic and Protestant scholars as well as patrons were keenly interested in acquiring materially and intellectually a broad variety of knowledge objects from past as well as contemporary Islamicate societies and invested substantial financial, material, personal and intellectual resources for satisfying their interests.[46] Hasse proved that several outspoken humanist critics of medieval Arabo-Latin translations constructed their supposedly new, humanist translations of ancient Greek texts by presenting major extracts of the rejected medieval translations, at times even only slightly altered, as genuine parts of their own work as translators.[47] Craig Martin, Charles Burnett and Dag Nikolaus Hasse edited or analyzed early modern Latin philosophical texts that relied substantially on the use of medieval

45 Siraisi 1987.
46 Brentjes 2010b.
47 Hasse 2001.

translations of Arabic sources.[48] They argued that the early modern period saw a resurgence of commentaries on Arabo-Latin philosophical texts that, in contrast to the humanist narrative, were not a last flowering of the medieval philosophical debate, but important to early modern revisions of Aristotelian natural philosophy. Brentjes proposed to reinterpret what has been seen since Johann Fück's magisterial book on the rise of Arabic studies in early modern Europe as subjugated to philological and religious studies in terms of a much broader intellectual interest in Oriental intellectual matters that encompassed religions, languages, antiquities, the humanities, medicine and the sciences. She showed how travellers and writers altered their accounts of travelling, reading, talking and investigating in 'Oriental' lands under the impact of norms and customs at home, thereby invalidating their own labors abroad and discarding the foreign knowledge they had spent many hours and much money acquiring.[49] Cross-cultural exchanges of knowledge in the early modern period thus consist of a much greater variety of themes, objects and actors than in the two centuries of Haskins' 'model'. This richness and the greater availability of contextual information demonstrate without doubt that cross-cultural exchanges are difficult, labor-intensive and costly activities that create, transport and legitimize knowledge across cultures, but often entail also acts of dismissing and even destroying knowledge. Furthermore, Kunitzsch even traced the interest of astronomers in various countries in Europe in Arabic texts, images and instruments beyond the early modern period into the eighteenth century.[50]

The first decade of the twenty-first century saw, partly in response to the historiographical changes in regard to the early modern period, an explosion of new proposals for how to interpret various aspects of the phenomenon of cross-cultural exchange of knowledge in the Middle Ages. Senior and junior scholars participated in the intense and varied output of new ideas, illustrating a newly found ambition to classify, evaluate and interpret. Surprisingly, the majority of participants in these theoretical efforts continue to subscribe to several elements of the old 'model': the identification of the transfer activities as a movement; the separation of communities and languages into large cultural blocks as if they were divine immutables; the temporal fixation of the medieval activities in the twelfth and the thirteenth centuries; the focus on scientific, medical and philosophical texts; the privileging of translation. The new perspectives emphasize, in contrast, difference, individuality and interdependence and thus the multiplicity of outlooks and interests; self-representation, legitimization and cultural identities; hybridity and cultural hegemony; social context and new geographical environments. Medieval exchanges are seen as fundamentally different in their social fabric from those of the early modern period, although certain cases, in particular exchanges that involve Jewish communities, are considered as processes of longue durée.

48 Martin 2007; Burnett 1999b; Burnett/Contadini 1999; Hasse 2004a; Hasse 2004b.
49 Brentjes 2009; Brentjes 2010a.
50 Kunitzsch 2006.

Substantially new proposals have been offered in particular by Dimitri Gutas, Dag Nikolaus Hasse, Charles Burnett, Alexander Fidora, Thomas Ricklin, Antoni Malet, Gad Freudenthal and Efthymios Nicolaidis. Gutas and Hasse argue in strong terms for the necessity of a social history of knowledge transfer. Hasse considers the lack of royal patronage and the 'professional' affiliation of the translators to ecclesiastic offices as the two defining features of the medieval phase. In contrast, in the early modern period, physicians affiliated with universities or urban municipalities dominated, in his view, the transfer from Arabic to Latin.[51] Gutas challenges the unified nature of the movement as taken for granted by most of the discussants. He suggests that the translation activities had very different contexts in different periods and locations and hence that different interpretive approaches are necessary. He offers a preliminary proposal for explaining the orientations and inconsistencies of the translations in the twelfth century in the Iberian Peninsula. This proposal rests on the observation that among the different high medieval cultures of the Peninsula, only those in al-Andalus had developed a cultural identity that included introspection and explicit narratives of self-representation. These explicit narratives of self-representation culminate in singling out Andalusian scholarly culture as the best of the contemporary worlds. Gutas finds this ideology of self-aggrandizement – including the specific properties of the intellectual landscapes of al-Andalus as portrayed in the eleventh century by Sa'id al-Andalusi, one of the writers of such narratives of self-representation – reflected in the topics and texts chosen for translation in Toledo and other places of Castile, ignoring largely the activities in Catalonia, Aquitania and the Provence. Gutas suggests that the Catholic translators transferred by and large Andalusian intellectual highlights and trends of the eleventh century into Latin as the consequence of two processes: the counseling by Jewish scholars who had grown up in this culture and identified intellectually with it; the desire of the Catholic conquerors of Andalusian urban centers not merely to acquire the material objects, economic prospects and skilled labor of their Muslim neighbors, but to appropriate also their cultural reputation and splendor. He points out that most of the Catholic translators neither grew up as Arabic native speakers nor received an education in the secular sciences taught in Muslim cities in the Iberian Peninsula. Many of them rather came from other European regions and had to learn Arabic and Romance after their arrival. That is why Gutas suggests that the central actors of the translation movement of the twelfth and early thirteenth centuries were Jewish and Mozarabic mediators who procured the texts and provided the necessary skills, but whose names were written out of history.[52]

In a series of conference papers, Burnett offers a colorful bouquet of ever-changing views on motives, programmes and relationships of Catholic translators. Shedding much of his earlier allegiance to Haskins' narrative, he introduces new

51 Hasse 2006.
52 Gutas 2006.

themes into the debate: advertisement of skills and goals as features of a patronage culture that lacked formal, institutional procedures for raising funds and finding positions; magic, divination and necromancy as lead sciences for interpreting the universe and its symbolic interdependencies and hence as the basis and goal of the activities of translators in the Ebro Valley and Aquitania; dialectic and mathematics as seen by Aḥmad ibn Yūsuf and al-Fārābī's philosophical work *Iḥṣā' al-'ulūm* as models for Gerard of Cremona's programme of translating and teaching; knowledge transfer as a question of intellectual property rights; Arabo-Latin translations between 'humanism and orientalism'; blending knowledge from different cultures versus translating from one language into another.[53]

Thomas Ricklin, who shares Burnett's fascination with participant narratives, disagrees with him in their interpretation, although both read them programmatically. Ricklin's analysis of narrative plots allows him to discover three different types of self-representation as a translator (guided by a teacher; independent selector of texts to be translated; teacher of all Christianity beyond the Iberian Peninsula) and perhaps three different types of constructing a dialogue with a potential readership: mixing private affairs (friendship) with business (translating and selection of topics); making concessions to readers unfamiliar with Arabic knowledge, but presenting the latter as the one to be emulated; mixing Arabic, Greek and Latin styles of presenting and ordering an introduction to a scholarly treatise. This reading of the participant narratives induces Ricklin to speak of far-reaching or even radical proposals for reforming the curricula of the cathedral schools.[54] It might be worthwhile to consider whether his argument needs to be extended to other institutional contexts, since Petrus Alfonsi, for instance, one of the advocates for such a new curriculum, lived at least partially in a monastic environment.

Alexander Fidora and Gad Freudenthal focus in particular on issues related to cross-cultural exchanges between Jewish communities and between Jewish, Muslim and Catholic scholars in philosophy, medicine and the sciences. By studying the appropriation of Latin philosophical works among Jewish scholars of the thirteenth and fourteenth centuries, Fidora adds a new perspective to the study of cross-cultural exchanges in the Middle Ages. Jewish engagement with knowledge from other cultures has been seen so far as located primarily within the Arabic philosophical, scientific and medical intellectual spheres of al-Andalus and as mediating between Arabic and Latin. Alexander Fidora and Mauro Zonta, along with Harvey J. Hames and Yossef Schwartz, have helped to modify this view by studying Hebrew translations of Latin treatises.[55] One of these translations concerns a text by Dominicus Gundisalvi, one of the Toledan translators of Arabic philosophical texts into Latin. Fidora sees here new terrain for reflection. He argues for recognizing multilingual translations and interpretations of

53 Burnett 1994a, 1997, 1999a, 2001, 2002, 2004a, 2004b, 2006b, 2007.

54 Ricklin 2006.

55 Fidora/Zonta 2010; Hames 2012a; See also Zonta 2006.

philosophical doctrines as a new research theme and asking how they have to be presented in relationship to the currently dominating traditions of separate historical narratives about Arabic, Hebrew and Latin philosophical thought in Europe.[56] His concept of multilingualism differs from that introduced by d'Alverny, since he considers the phenomenon of multiple translations of the same or similar material into different target languages.

Analyzing Mauro Zonta's newly compiled survey of all known translations by medieval and early modern Jewish scholars, Freudenthal proposes to differentiate the intellectual interests in Jewish communities in the Provence. He distinguishes between a sustained long-term effort of migrants from the Iberian Peninsula to transfer and integrate Arabic sciences, medicine and, to a lesser degree, philosophy from al-Andalus into their new living spaces, on the one hand, and isolated cases of translating Latin medical texts into Hebrew, on the other. He justifies this schism not merely by numbers (men; books; themes), but by a deep cultural divide that defined in his view the attitudes of Jewish scholars towards the two host cultures of acquirable knowledge: Arabic and Latin. They admired and integrated themselves into the former, but rejected and denounced the latter.[57] He also diagnoses different cultural attitudes between different professional orientations (philosophers versus physicians) and different regional affiliations (Provence versus Italy) concerning the disregard and lack of engagement with the contemporary medical, philosophical or scientific knowledge of the surrounding Latin-speaking Christian communities.[58] He attributes 'cultural isolation' due to 'a deep-seated cultural attitude' to medieval Jewish scholars in the Provence in comparison to a more accommodating practice by Jewish scholars in Spain, Italy or the Ottoman Empire.[59]

Nonetheless, the respective teams led by Fidora and Freudenthal both agree in that the Latin-into-Hebrew translations clearly show that Jewish thought during the European Middle Ages can no longer be described as drawing exclusively on Oriental sources. In their recent two-volume publication *Latin-into-Hebrew*, which contains editions of medieval Hebrew translations of works by Gundisalvi, Albert the Great and the anonymous *Liber de causis*, they arrive at the conclusion that "the perception of the Jewish philosophical tradition as exclusively extraterritorial (i.e. as depending essentially on Oriental sources) will be rectified so as to take into account also Jewish interactions with the majority culture".[60] This confirms the existence of intellectual networks across cultures that were formed among Jewish, Christian and Muslim scholars during the Middle Ages and that account for the striking parallels in the reception of certain philosophical doctrines and traditions. In addition, the studies contained in these volumes raise questions

56 Fidora 2012a.
57 Freudenthal 2011a, in particular 74 sq, 80 and 84.
58 Ibid., 75 and 77–81.
59 Ibid., 84.
60 Fidora e. a. 2013, 17.

which also affect other traditions of translating during the Middle Ages, namely the complex issue of language acquisition (in this case the Jews' learning of Latin) and the use of vernacular intermediaries. Alexander Fidora, Harvey J. Hames, José Martínez Gázquez, Sabine Schmidtke, Matthias M. Tischler and other scholars have also paid renewed attention to the translation of religious texts in the various cultural traditions of the Middle Ages. While Haskins' limitations in this respect were partly outweighed by Marie-Thérèse d'Alverny, recent studies have enlarged considerably our knowledge concerning this phenomenon. Thus, the research group Islamolatina, led by Martínez Gázquez, has started the critical edition of all extant Latin Qur'ān translations.[61] Hames has prepared an edition of the first Hebrew translation of the Gospels,[62] while Schmidtke and her group are studying translations of the Hebrew Bible and the New Testament into Arabic.[63] Several studies on the translation and glossing of religious texts have been recently edited by Tischler and Fidora;[64] the former has also compiled the electronic database 'Bibliotheca Islamo-Christiana Latina',[65] which is based on several studies, dealing with Arabo-Latin biographical, historiographical and religious oral[66] and written text traditions on Muḥammad and Islam, deriving not only from the frontier societies of the Iberian Peninsula[67] but from the Byzantine world as well.[68] Most recently, Fidora and his team have started preparing the critical edition of the Latin Talmud (ca. 1240).

Numerous new results were presented in the last two decades about cross-cultural exchanges in the medieval Eastern Mediterranean with Eastern Europe, Islamicate societies East and South of Byzantium and with communities in Italian colonies along the shores of the Black Sea as well as Italian and Aragonese cities. In his study of the relationship between 'science and the Orthodox church' Efthymios Nicolaidis presents a broadly construed survey on the extant material of translation, migration and transformation of knowledge between Byzantium and its neighbors and enemies.[69] Nicholas de Lange, Anne Tihon and Nicolaidis emphasize, each from the perspectives of their different academic fields, the necessity to investigate the contributions of Jewish scholars inside and outside Byzantium to the exchange of knowledge between the Empire, Latin Europe and Islamicate societies.[70] Tihon insists that the Byzantines continued to believe that they were the heirs of ancient Greek scientific knowledge. But she also demonstrates that until the eleventh century, when the first astronomical text translated

61 E.g., Petrus i Pons 2008. Project description at http://grupsderecerca.uab.cat/islamolatina/.
62 Submitted to Corpus Biblicum Catalanicum: Hames. See also Hames 2012b.
63 Adang e. a. 2013. See also Griffith 2013.
64 Tischler/Fidora 2011.
65 Available at: www.sankt-georgen.de/hugo/forschung/spanien_bicl.php.
66 Tischler 2013.
67 Tischler 2008, 2011a, 2011b.
68 Tischler 2012.
69 Nicolaidis 2011.
70 De Lange 2001b; Nicolaidis 2011, 106–119.

from Arabic (1032) is extant, Byzantine astronomy was rather limited in scope and level.[71] Surveying the academic investigation of Jewish intellectual life in Byzantium, de Lange stresses the long-term interests of Jewish writers in medicine and philosophy and the flourishing of astronomy and astrology in the fourteenth and fifteenth centuries.[72] In this period, Orthodox Byzantine patrons, too, expressed a remarkable interest in Hebrew translations of philosophy, as Stefan Reif shows on the basis of codicological investigations.[73] Nicolaidis draws attention in particular to the relationships between the Karaite communities in the Languedoc, Byzantium and the Crimea from the thirteenth to the fifteenth centuries and their interests especially in the mathematical and medical sciences.[74] New insights into the transfer of knowledge by Jewish communities in Byzantium are to be expected by Daniel J. Lasker and Y. Tzvi Langermann.[75]

In addition to this emergence of data and the diversification of research interests in Byzantine studies, historians of philosophy and science, in particular Gutas and Nicolaidis, have also offered new observer narratives about the motives and inspirations that guided the cross-cultural exchanges of knowledge in the ʻAbbāsid Empire and Byzantium. What is common to both 'models' is their emphasis on socio-cultural factors, in particular the need to legitimize rebellion and rule, to squelch ideologically grounded conflicts, to overcome resistance and to defeat external enemies. Gutas makes a conceptual distinction between translating as an everyday life activity in a multicultural society and translating as a sustained cultural politics of a dynasty or a ruling elite. He argues that only when cross-cultural exchanges of knowledge were part of an official ideology, they attracted sufficient cultural, personal and monetary support and produce new, advanced knowledge.[76] Nicolaidis argues in a similar manner when he speaks of the direct consequences for translating and transferring knowledge between Byzantium, Sicily, Italy or Iran due to Manuel Komnenos' desire during the twelfth century to re-establish the Roman Empire or the newly invented classical Greek identity of Byzantine secular and ecclesiastic elites in the wake of the loss of Constantinople to the papal and Venetian mercenaries of the Fourth Crusade.[77] He adds, however, a further explanatory argument of a different kind: the role of values attached to certain scholarly practices (in the case of Orthodox Byzantine scholars: observations and other manual activities) versus the values attached to an inimical, but knowledgeable foreign culture (depending on the period: the ʻAbbāsid Empire, Ilkhanid Iran or Occidental Europe).[78] This argument is mixed with his own beliefs about the cul-

71 Tihon 2009, in particular 304 sq.
72 De Lange 2001b, 33.
73 Reif 2002, in particular 104.
74 Nicolaidis 2011, 116 sq.
75 Lasker 2011; Langermann 2011.
76 Gutas 1998, 28–106.
77 Nicolaidis 2011, 113–118.
78 Ibid., 107.

tural values of the different players and objects of exchange and leads, at times, to problematic claims that seem to contradict some of the codicological evidence.[79]

Antoni Malet, an outsider to the debates on cross-cultural exchanges of knowledge, proposes an interestingly different conceptual goal for identifying and interpreting the results of such activities. He rejects explicitly the position, held since Haskins by a majority of experts, which posits translation as the main cultural event that has to be investigated and explained. For Malet translation is merely one among many practices that constitute cross-cultural exchange. In his view, the truly important feature of the phenomenon is the creation of hybrid cultures that are open to continued change.[80] This position offers a major challenge not only to the debates among the experts of medieval cross-cultural exchanges, but to current conflicts about the cultural identities in and of Europe. In a case study of fourteenth-century portolan charts, Brentjes has reached a similar conclusion. In her view, the illuminated charts that were produced in Italy and at Majorca can only be adequately understood if a shared cultural space – the so-called middle ground – is assumed that included the entire Mediterranean and the Black Sea with their adjacent cultures and their networks of commercial and cultural brokers.[81]

4 A glance ahead

This fascinating mixture of old and new views on which features are relevant for understanding medieval cross-cultural exchanges of knowledge and the existence of competing observer narratives for cultures in the Eastern Mediterranean and beyond indicate that by now a critical mass of scholars and ideas is available for constructing a new interpretive 'model' that is rich in historical, geographical, cultural and social data, sophisticated in its methodological foundations and representational choices, and reflexive in its values and analytical techniques and methods. The controversial character of some of the new proposals provides another incentive for seriously considering the construction of a new observer narrative. Analyzing the underlying conceptual reasons for the conflicting positions will stimulate the methodological inquiry and holds the chance for enhancing the inclusive character of the new 'model'. Our current positions, which were highlighted in this survey, are dynamic, open, reflexive and in flux. In our view, there is no need for a single narrative for all cross-cultural exchange activities in the medieval Mediterranean, which stresses the concept of movement and linearity for two events (Graeco-Arabic translation under the early 'Abbāsids in the eighth

79 Ibid., in particular his claims about the exclusively practical nature of the knowledge that Byzantine scholars appropriated from Arabic sources and the lack of usage of Arabo-Indian numerals. The latter, while true for a broad class of manuscripts, appear, however, in fly leaves and marginal notes in mathematical manuscripts such as one of the oldest dated extant Greek manuscript of Euclid's *Elements*, Oxford, Bodleian Library, Ms. D'Orville 301.

80 Malet 2000, in particular 195–197.

81 Brentjes 2012.

and ninth centuries; Arabo-Latin and Graeco-Latin translations in the Iberian Peninsula, Sicily and the Latin Crusader States in the twelfth and thirteenth centuries) and dismisses or ignores most of the other activities. In fact, there are a number of small-scale activities among contemporaneous communities whose members interact with each other. The participants in such cross-cultural activities are rarely rigidly bound by their respective communal memberships. On the contrary, cross-cultural exchange of knowledge needs cultural and other brokers. The habits, beliefs and skills of such brokers adapt to the changing conditions and circumstances of their brokerage. Sometimes they are members of a diaspora who have left country, home and family behind, but remained in limbo. At other times they settle comfortably in two or even three culturally different communities. Numerous brokers seem, however, to have stayed in the community of their birth. They shared some parts of their knowledge and skills as goods they were trading. Other parts they safeguarded as assets necessary for defining them as competent mediators for those interested in acquiring objects and knowledge from another community, but disposing only of some of the needed resources. Such brokers and their dealing with clients deserve much more attention than they have received so far.

While translations of scientific, medical, philosophical and theological texts of a formalized content and presentational format continue to be important objects of study and thus a valuable type of source for our goal of creating a new observer narrative, other textual objects of a less formalized content and set-up, such as fly leaves, notes, colophons and ownership marks, or of a different formalization, such as letters, speeches, accounting sheets or recipes, need to receive greater attention. In addition to the written word, visual and material objects were themselves components of cross-cultural exchange and transformation of knowledge in several functions. They served as commercial goods, as objects of royal gift exchange, of cultural copying and code switching, as technical tools and as containers of coveted knowledge and platforms of education. The functions of such textual, visual and instrumental objects of cross-cultural exchange of knowledge have to become an integrated part in a new narrative.

As a tool of research, material objects have the potential to serve as a corrective for a historical narrative exclusively based on scholarly texts. Eight fragments of sundials and astrolabes constructed between the tenth and the thirteenth centuries are known from different places in the Iberian Peninsula. Their study by Millàs Vallicrosa, Barceló, Labarta, King, Samsó, Viladrich, Rius and others signals that knowledge of astronomical theories, parameters and texts arrived in al-Andalus apparently earlier than the preserved texts show.[82] They reveal that their makers and patrons favoured different approaches for choosing among competing data and interpretations: astronomically determined prayer directions, astrological orientations, choices made in Egypt or attributed to companions of Muḥammad.[83]

82 King 1978, 1992, 98 sq., 2002–2003, 115; King/Samsó 2001, 56 sq.; Barceló/Labarta 1988; Labarta 1995; Rius Piniés 2006; Millàs Vallicrosa 1955; Viladrich 1992, 63.

83 King 1978, 370 sq.

Texts do not only seem to have followed instruments in their path from East to West. On the basis of the currently available material, they also seem to represent a later practice among Andalusian experts. These two observations suggest a much more substantial role of practical, instrumental transfers of knowledge as a preparatory framework for a possibly later interest in texts. Thus, we will orient parts of our investigations in this direction.

The observation, on the other hand, that the chronology of the extant astrolabes lags behind the evidence for the movement of their knowledge towards the Latin North presents a further challenge. The Latin texts of Arabic provenance are about half a century older than the oldest extant Andalusian astrolabe. Abstaining from the easiest explanation of material loss and contingency, we think that a comparative inquiry of the relationship between texts and instruments in other cases of cross-cultural exchange of knowledge will result in creating a broader array of interpretive options. A further addition that the study of instruments brings to an observer narrative limited to scholarly texts is the insight that together with bigger socio-cultural processes of cross-cultural exchange of knowledge smaller, individual activities need to be considered. The so-called Carolingian astrolabe, which according to current opinions arrived in the North of the Iberian Peninsula before 980, may have come from somewhere in Italy, where around 900 such instruments seem to have been produced following a Byzantine tradition not yet reflecting the changes introduced by the astrologers in Baghdad.[84]

A further important insight that we shall work on to elaborate in more detail is the embedment of these cross-cultural activities of knowledge exchange in larger cultural processes of confrontation, collaboration, dislocation and settlement. The long-lasting instability, permeability and reinvention of frontiers, dynasties, languages, tribes, sedentary populations, customs, habits and beliefs during those centuries that we label here, as a shorthand, medieval create the multitude of contexts that encouraged mostly men and occasionally women to learn new skills from strangers, buy shiny objects from far away countries and order subordinate subjects to copy or adapt them to local taste and capabilities or collect information about such far away countries from countrymen and foreigners alike. Maps, cloths, paintings, ceramics, manuscript bindings and many other objects testify to this complexity of cross-cultural exchange. These processes bound scholars; nobility; bankers; merchants; priests; healers; sailors; itinerant traders of knowledge of plants, animals, minerals but also of the divine and the supernatural; and the many go-betweens together in a flurry of events, relationships and activities. Seen from this perspective, cross-cultural exchange of knowledge was a way of life, not merely a linear act of translating. We shall try to determine more precisely, if possible, such multifaceted interactions, their loci of contact, transfer, transport and transformation and their participants either as identifiable individuals or as a representative of a social, professional, cultural, linguistic or other group.

84 Viladrich 1992, 61 sq.; Samsó 1992, 2003.

Two major stimulators of cultural exchange of knowledge were marriage and conquest. Marriage necessitated adaptation ideally of the two (or more) sides ceremonially united as a family and if not as a multiple undertaking, then at least of one of the parties involved. Women were here often subject and object of such cross-cultural exchanges, which included the transfer of material objects, religious and other beliefs, the translation of books and the creation of book collections 'seen fit' for the female spirit. Historical chronicles, geographies, prayer books and collections of household recipes were the primary items of such translations from scholarly languages such as Latin into vernaculars or between vernaculars.

Conquest forced adaptation, not only on the side of the conquered, but also on the side of conquerors, because they often lacked recognition as new rulers or experienced resistance by the subdued population and its various civil elites. Here processes of knowledge exchange often took place as elements of the necessary everyday-life adaptation, for instance, in the realm of law where translators continued to participate in court cases long after the conquered population had become bilingual and had learned, voluntarily or not, the new cultural rules. In several cases, such conquests led to cataclysmic changes in the spheres of knowledge and skills. Examples are the conquest of the Umayyad caliphate by the 'Abbāsid family and its allies in the middle of the eighth century, the Almohad dynasty's conquest of al-Andalus between 1146 and 1173, the Aragonese and Castilian conquests of Muslim principalities of al-Andalus between the eleventh and the fifteenth centuries, the creation of the Lascaride dynasty after the conquest of Constantinople by the crusaders in 1204 and others. Destruction of objects, suppression of knowledge and skills, the acquisition of contemporary and the appropriation of 'old' knowledge objects, authorities and teachings for purposes of legitimation of the new rulers, maintenance of irrigation and agricultural production or integration and subjugation of previous intellectual, legal or administrative elites were part and parcel of those turnovers, although each case possessed its specific particularities. In several of these cataclysmic changes, the conquerors used knowledge and its ability to be transported across cultural divides as one of their numerous political strategies for improving their reputation and pacifying parts of the conquered population or to unify the diverse factions of their own elites in struggles against a foreign invader. Thus, reducing such profound changes in the knowledge and skills practiced and valued in such different circumstances of conquering, defending and extending political, economic and military power of one group over others to their intellectual aspects and streamlining these intellectual aspects as stepping stones from 'classical Greece' to the current 'West' leads to a tremendous loss of the complex richness of the past. Hence, multiple methodological platforms and research methods need to be combined with a self-reflexive control of results and interpretations for their dependence on current ideological and political values and commitments in order to break free from essentializing past cultures and their scholarly handling.

Bibliography

Abbreviation

CChr.CM Corpus Christianorum. Continuatio Mediaevalis

Series

Corpus des astronomes byzantins 1–6, Amsterdam 1983–1993; . . . 7–10, Louvain-la-Neuve 1994–2001; general editor: Anne Tihon.

Studies

Abdellatif e. a. 2012: *Acteurs des transferts culturels en Méditerranée médiévale*, ed. by RANIA ABDELLATIF e. a. (Ateliers des Deutschen Historischen Instituts Paris 9), München 2012.

Adamson e. a. 2004 *Philosophy, Science and Exegesis in Greek, Arabic and Latin Commentaries* 1–2, ed. by PETER ADAMSON e. a. (Bulletin of the Institute of Classical Studies of the University of London. Supplement 83, 1–2), London 2004.

Adang e. a. 2013 *The Bible in Arabic among Jews, Christians and Muslims*, ed. by CAMILLA ADANG e. a. (*Intellectual History of the Islamicate World* 1 [2013]), Leiden/Boston 2013.

Aertsen/Pickavé 2004 *'Herbst des Mittelalters'?* Fragen zur Bewertung des 14. und 15. Jahrhunderts, ed. by JAN ADRIANUS AERTSEN/MARTIN PICKAVÉ (Miscellanea Mediaevalia 32), Berlin/New York 2004.

Alberni e. a. 2012 *El saber i les llengües vernacles a l'època de Llull i Eiximenis*. Knowledge and vernacular languages in the age of Llull and Eiximenis, ed. by ANNA ALBERNI e. a. (Textos i estudis de cultura catalana 170), Barcelona 2012.

Allard 1973 ANDRÉ ALLARD: "Les procédés de multiplication des nombres entiers dans le calcul indien à Byzance", in *Bulletin de l'Institut historique belge de Rome* 43 (1973) 111–143.

Allard 1977 ANDRÉ ALLARD: "Le premier traité byzantin de calcul indien. Classement des manuscrits et édition critique du texte", in *Revue d'histoire des textes* 7 (1977) 57–107.

Allard 1981 ANDRÉ ALLARD: *Maximus Planudes, Le grand calcul selon les Indiens*. Histoire du texte, édition critique, traduction et annotations (Centre d'histoire des sciences et des techniques. Sources et travaux 1. Travaux de la Faculté de philosophie et lettres de l'Université catholique de Louvain 27), Louvain-la-Neuve 1981.

Allard 1992 ANDRÉ ALLARD: *Muḥammad ibn Mūsā al-Khwārizmī, Le calcul indien (algorismus)*. Histoire des textes, édition critique, traduction et commentaire des plus anciennes versions latines remaniées du XIIe siècle (Collection sciences dans l'histoire. Collection d'études classiques 5), Paris/Namur 1992.

d'Alverny 1962 MARIE-THÉRÈSE D'ALVERNY: "Survivance de la magie antique", in WILPERT/ECKERT 1962, 154–178 [repr. D'ALVERNY 1994b, n° I].

d'Alverny 1982 MARIE-THÉRÈSE D'ALVERNY: "Translations and translators", in BENSON/CONSTABLE 1982, 421–462 [repr. d'Alverny 1994b, n° II and Addenda et corrigenda, 1].

d'Alverny 1989 MARIE-THÉRÈSE D'ALVERNY: "Les traductions à deux interprètes, d'arabe en langue vernaculaire et de langue vernaculaire en latin", in CONTAMINE 1989, 193–206 [repr. d'Alverny 1994b, n° III].

d'Alverny 1994a MARIE-THÉRÈSE D'ALVERNY (†): *La connaissance de l'Islam dans l'Occident médiéval*, ed. by CHARLES BURNETT. With an appreciation by MARGARET GIBSON (Collected Studies Series 445), Aldershot 1994.

d'Alverny 1994b MARIE-THÉRÈSE D'ALVERNY (†): *La transmission des textes philosophiques et scientifiques au moyen âge*, ed. by CHARLES BURNETT (Collected Studies Series 463), Aldershot 1994.

d'Alverny/Hudry 1971 MARIE-THÉRÈSE D'ALVERNY/FRANÇOISE HUDRY: "Al-Kindi, De radiis", in *Archives d'histoire doctrinale et littéraire du moyen âge* 41 (1974) 139–260.

ANDRÉ ALLARD: "The Arabic origins and development of Latin algorisms in the twelfth century", in *Arabic Sciences and Philosophy* 1 (1991) 233–283.

Barceló/Labarta 1988 CARMEN BARCELÓ/ANA LABARTA: "Ocho relojes de sol hispano-musulmanes", in *al-Qanṭara* 9 (1988) 231–248.

Benson/Constable 1982 *Renaissance and Renewal in the Twelfth Century*, ed. by ROBERT LOUIS BENSON/GILES CONSTABLE, Cambridge (MA) 1982 [also: (Medieval Academy Reprints for Teaching 26), Oxford 1982; repr. Toronto 1991].

Between demonstration and imagination. Essays in the history of science and philosophy presented to John D. North, ed. by LODI NAUTA/ARJO VANDERJAGT (Brill's Studies in Intellectual History 96), Leiden/Boston 1999.

Biard 1999 *Langage, Sciences, philosophie au XIIe siècle.* Actes de la table ronde internationale organisée les 25 et 26 mars 1998, ed. by JOËL BIARD (Sic et non), Paris 1999.

Blum e. a. 1999 *Sapientiam amemus.* Humanismus und Aristotelismus in der Renaissance. Festschrift für Eckhard Keßler zum 60. Geburtstag, ed. by PAUL RICHARD BLUM e. a., München 1999.

Brentjes 2000 SONJA BRENTJES: "Reflexionen zur Bedeutung der im 12. Jahrhundert angefertigten lateinischen Übersetzungen wissenschaftlicher Texte für die europäische Wissenschaftsgeschichte", in COBET e. a. 2000, 269–305.

Brentjes 2009 SONJA BRENTJES: "Immediacy, mediation, and media in early modern Catholic and Protestant representations of Safavid Iran", in *Journal of Early Modern History* 13 (2009) 173–207.

Brentjes 2010a SONJA BRENTJES: "The presence of ancient secular and religious texts in the unpublished and printed writings of Pietro della Valle (1586–1652)", in BRENTJES 2010b, n° III [repr. under the title "The presence of ancient secular and religious texts in Pietro della Valle's (1586–1652) unpublished and printed writings", in FLOOR/HERZIG 2012, 327–345].

Brentjes 2010b SONJA BRENTJES: *Travellers from Europe in the Ottoman and Safavid Empires, 16th–17th Centuries.* Seeking, transforming, discarding knowledge (Collected Studies Series 961), Farnham 2010.

Brentjes 2012 SONJA BRENTJES: "Medieval portolan charts as documents of shared cultural spaces", in ABDELLATIF e. a. 2012, 135–146.

Brentjes 2016 SONJA BRENTJES: "Practicing history of mathematics in Islamicate societies in the 19th century Germany and France", in REMMERT e. a. 2016, 25–52.

Burnett 1994a CHARLES BURNETT: "Advertising the new science of the stars circa 1120–50", in GASPARRI 1994, 147–157.

Burnett 1994b CHARLES BURNETT: "'Magister iohannes hispanus': Towards the identity of a Toledan translator", in TERRASSE 1994, 425–436.

Burnett 1995 CHARLES BURNETT: "The institutional context of Arabic-Latin translations of the Middle Ages: A reassessment of the 'School of Toledo'", in WEIJERS 1995, 214–235.

Burnett 1996 CHARLES BURNETT: *Magic and Divination in the Middle Ages*. Texts and the techniques in the Islamic and Christian worlds (Collected Studies Series 557), Aldershot 1996.

Burnett 1997 CHARLES BURNETT: *The Coherence of the Arabic-Latin Translation Programme in Toledo in the Twelfth Century* (Max-Planck-Institut für Wissenschaftsgeschichte. Preprint 78), Berlin 1997 [revised version: BURNETT 2001].

Burnett 1999a CHARLES BURNETT: "Dialectic and mathematics according to Ahmad ibn Yusuf: A model for Gerard of Cremona's programme of translation and teaching?", in BIARD 1999, 83–92.

Burnett 1999b CHARLES BURNETT: "Al-Kindi in the Renaissance", in BLUM e. a. 1999, 13–30.

Burnett 2000 CHARLES BURNETT: "Antioch as a link between Arabic and Latin culture in the twelfth and thirteenth centuries", in DRAELANTS e. a. 2000, 1–78 [with 9 plates] [repr. BURNETT 2009, n° IV and Addenda and Corrigenda, 2 sq.].

Burnett 2001 CHARLES BURNETT: "The coherence of the Arabic-Latin translation programme in Toledo in the twelfth century", *Science in Context* 14 (2001) 249–288 [repr. BURNETT 2009, n° VII and Addenda and Corrigenda, 4 sq.].

Burnett 2002 CHARLES BURNETT: "The translation of Arabic Science into Latin: A case of alienation of intellectual property?", in *Bulletin of the Royal Institute for Inter-Faith Studies* (Amman) 4 (2002) 145–157.

Burnett 2003 CHARLES BURNETT: "The transmission of Arabic astronomy via Antioch and Pisa in the second quarter of the twelfth century", in HOGENDIJK/SABRA 2003, 23–51.

BURNETT 2004a CHARLES BURNETT: "Arabic and Latin astrology compared in the twelfth century: Firmicus, Adelard of Bath and 'Doctor Elmirethi' ('Aristoteles Milesius')", in BURNETT e. a. 2004, 247–263.

Burnett 2004b CHARLES BURNETT: "The blend of Latin and Arabic sources in the Metaphysics of Adelard of Bath, Hermann of Carinthia, and Gundisalvus", in LUTZ-BACHMANN e. a. 2004, 41–65.

Burnett 2006a CHARLES BURNETT: "A hermetic programme of astrology and divination in mid-twelfth-century Aragon: The hidden preface in the 'Liber novem iudicum', in BURNETT/RYAN 2006, 99–118.

Burnett 2006b CHARLES BURNETT: "Humanism and orientalism in the translations from Arabic into Latin in the Middle Ages", in SPEER/WEGENER 2006, 22–31.

Burnett 2007 CHARLES BURNETT: "Astrology, astronomy and magic as the motivation for the scientific renaissance of the twelfth century", in VOSS/HINSON LALL 2007, 55–61.

Burnett 2008a CHARLES BURNETT: "Royal patronage of the translations from Arabic into Latin in the Iberian Peninsula", in GREBNER/FRIED 2008, 323–333.

Burnett 2008b CHARLES BURNETT: "Scientific translations from Arabic: The question of revision", in GOYENS e. a. 2008, 11–34.

Burnett 2009 CHARLES BURNETT: *Arabic into Latin in the Middle Ages*. The translators and their intellectual and social context (Collected Studies Series 939), Farnham 2009.

Burnett e. a. 2004 *Studies in the History of the Exact Sciences in Honour of David Pingree*, ed. by CHARLES BURNETT e. a. (Islamic Philosophy, Theology and Science. Text and Studies 54), Leiden 2004.

Burnett/Contadini 1999 *Islam and the Italian Renaissance*, ed. by CHARLES BURNETT/ ANNA CONTADINI (Warburg Institute Colloquia5), London 1999.

Burnett/Ryan 2006 *Magic and the Classical Tradition*, ed. by CHARLES BURNETT/ WILLIAM FRANCIS RYAN (Warburg Institute Colloquia 7), London/Torino 2006.

Busard 1968 HUBERTUS LAMBERTUS LUDOVICUS BUSARD: *The Translation of the Elements of Euclid from the Arabic into Latin by Hermann of Carinthia (?)*, Leiden 1968.

Busard 1977 HUBERTUS LAMBERTUS LUDOVICUS BUSARD: *The Translation of the Elements of Euclid from the Arabic into Latin by Hermann of Carinthia (?), Books VII–XII* (Mathematical Centre Tracts 84), Leiden 1977.

Busard 1983 HUBERTUS LAMBERTUS LUDOVICUS BUSARD: *The First Latin Translation of Euclid's Elements Commonly Ascribed to Adelard of Bath.* Books I – VIII and Books X. 36 – XV. 2 (Pontifical Institute of Mediaeval Studies. Studies and Texts), Toronto 1983.

Busard 1984 HUBERTUS LAMBERTUS LUDOVICUS BUSARD: *The Latin Translation of the Arabic Version of Euclid's Elements Commonly Ascribed to Gerard of Cremona.* Introduction, edition and critical apparatus (Asfār 2), Leiden 1984.

Busard 1987 HUBERTUS LAMBERTUS LUDOVICUS BUSARD: *The Mediaeval Latin Translations of Euclid's Elements, Made Directly from the Greek* (Boethius 15), Stuttgart 1987.

Busard 1996 HUBERTUS LAMBERTUS LUDOVICUS BUSARD: *A Thirteenth-Century Adaptation of Robert of Chester's Version of Euclid's Elements* 1–2 (Algorismus 17), München 1996.

Busard 2001 HUBERTUS LAMBERTUS LUDOVICUS BUSARD: *Johannes de Tinemue's Redaction of Euclid's Elements.* The so-called Adelard III version 1–2 (Boethius 45), Stuttgart 2001.

Busard 2005 HUBERTUS LAMBERTUS LUDOVICUS BUSARD: *Campanus of Novara and Euclid's Elements* 1–2 (Boethius 51), Stuttgart 2005.

Busard/Folkerts 1992 HUBERTUS LAMBERTUS LUDOVICUS BUSARD/MENSO FOLKERTS: *Robert of Chester's (?) Redaction of Euclid's Elements, the So-Called Adelard II Version* 1–2 (Science Networks 8–9), Basel 1992.

Büttgen e. a. 2009 *Les Grecs, les Arabes et nous.* Enquête sur l'islamophobie savante, ed. by PHILIPPE BÜTTGEN e. a. (Ouvertures), Paris 2009.

Charette 1995 FRANÇOIS CHARETTE: *Orientalisme et histoire des sciences.* L'historiographie européenne des sciences islamiques et hindoues, 1784–1900, Thèse de M.A., Université de Montréal, 1995.

Clagett 1964–1984 MARSHALL CLAGETT: *Archimedes in the Middle Ages 1: The Arabo-Latin Tradition* (University of Wisconsin. Publications in Medieval Science 6), Madison (Wisc.) 1964; . . . 2: The translations from the Greek by William of Moerbeke (Memoirs of the American Philosophical Society 117 A–B), Philadelphia (Pa.) 1976; . . . 3: The fate of the medieval Archimedes. 1300 to 1565 (. . . 125 A–B and 125), Philadelphia (Pa.) 1978; . . . 4: A supplement on the medieval Latin traditions of Conic sections. 1150–1566 (. . . 137 A and 137), Philadelphia (Pa.) 1980; . . . 5: Quasi-Archimedean geometry in the thirteenth century . . . (. . . 157), Philadelphia (Pa.) 1984.

Cobet e. a. 2000 *Europa.* Die Gegenwärtigkeit der antiken Überlieferung, ed. by JUSTUS COBET e. a. (Essener Beiträge zur Kulturgeschichte 2), Aachen 2000.

Contamine 1989 *Traduction et traducteurs au moyen âge*. Actes du Colloque international du CNRS organisé à Paris, Institut de recherche et d'histoire des textes, les 26–28 mai 1986, ed. by GENEVIÈVE CONTAMINE, Paris 1989.

Curry 1987 *Astrology, Science and Society*. Historical essays, ed. by PATRICK CURRY, Woodbridge 1987.

Draelants 2004 ISABELLE DRAELANTS: "Introduction à l'étude d'Arnoldus Saxo et aux sources du 'De floribus rerum naturalium'", in STAMMEN/WEBER 2004, 85–121.

Draelants 2005 ISABELLE DRAELANTS: "La science naturelle et ses sources chez Barthélemy l'Anglais et les encyclopédistes contemporains", in VAN DEN ABEELE/MEYER 2005, 43–99.

Draelants e. a. 2000 *Occident et Proche-Orient*. Contacts scientifiques au temps des croisades, ed. by ISABELLE DRAELANTS e. a. (Réminisciences 5), Turnhout 2000.

Duhem 1913–1959 PIERRE DUHEM: *Le système du monde*. Histoire des doctrines cosmologiques de Platon à Copernic 1: La cosmologie héllenique, Paris 1913; . . . 2: La cosmologie héllenique, Paris 1914; . . . 3: L'astronomie latine au moyen âge, Paris 1915; . . . 4: L'astronomie latine au moyen âge, Paris 1916; . . . 5: La crue de l'aristotélisme, Paris 1917; . . . 6: Le reflux de l'aristotélisme, Paris 1954; . . . 7–9: La physique parisienne au XIVe siècle, Paris 1956–1958; . . . 10: La cosmologie du XVe siècle. Écoles et universités au XVe siècle, Paris 1959.

Durán Guardeño 2000 *El legado de las matemáticas*. De Euclides a Newton. Los genios a través de sus libros (Exhibition Catalogue), ed. by ANTONIO JOSÉ DURÁN GUARDEÑO, Sevilla 2000.

Fidora 2012a ALEXANDER FIDORA: "The sefer ha-nefesh: A first attempt to translate Aristotle's 'De anima' into Hebrew", in VAN OPPENRAAY/FONTAINE 2012, 159–172.

Fidora 2012b ALEXANDER FIDORA: "La transmissió multilingüe de textos filosòfics a l'Edat Mitjana", *Medievalia. Revista d'Estudis Medievals* 15 (2012) 43–46.

Fidora e. a. 2013 *Latin-into-Hebrew*. Texts and studies 1: Studies, ed. by ALEXANDER FIDORA e. a. (Studies in Jewish History and Culture 39), Leiden/Boston 2013.

Fidora/Zonta 2010 ALEXANDER FIDORA/MAURO ZONTA (ed./trans.): *Vicent Ferrer, Quaestio de unitate universalis – Ma'amar nikhbad ba-kolel*. Latin and Hebrew texts edited together with a Catalan and English translation (Bibliotheca Philosophorum Medii Aevi Cataloniae 1), Santa Coloma de Queralt e. a. 2010.

Floor/Herzig 2012 *Iran and the World in the Safavid Age*, ed. by WILLEM FLOOR/EDMUND HERZIG (International Library of Iranian Studies 2), London e. a. 2012.

Folkerts 1970a MENSO FOLKERTS: *'Boethius' Geometrie II*. Ein mathematisches Lehrbuch des Mittelalters (Boethius. Texte und Abhandlungen zur Geschichte der exakten Wissenschaften 9), Wiesbaden 1970.

Folkerts 1970b MENSO FOLKERTS: *Ein neuer Text des Euclides Latinus*. Faksimiledruck der Handschrift Lüneburg D 4° 48, f. 13r – 17v, Hildesheim 1970.

Folkerts 1971a MENSO FOLKERTS: *Anonyme lateinische Euklidbearbeitungen aus dem 12. Jahrhundert* (Veröffentlichungen der Kommission für Geschichte der Mathematik und Naturwissenschaften 8. Denkschriften der Mathematisch-Naturwissenschaftlichen Klasse der Österreichischen Akademie der Wissenschaften 116, 1), Wien 1971 [repr. Wien 1985].

Folkerts 1971b MENSO FOLKERTS: "Mathematische Aufgabensammlungen aus dem ausgehenden Mittelalter. Ein Beitrag zur Klostermathematik des 14. und 15. Jahrhunderts", in *Sudhoffs Archiv* 55 (1971) 58–75.

Folkerts 1972 MENSO FOLKERTS: "Pseudo-Beda, De arithmeticis propositionibus. Eine mathematische Schrift aus der Karolingerzeit", in *Sudhoffs Archiv* 56 (1972) 22–43 [repr. FOLKERTS 2003, n° III].

Folkerts 1981 MENSO FOLKERTS: "Mittelalterliche mathematische Handschriften in westlichen Sprachen in der Herzog August Bibliothek Wolfenbüttel. Ein vorläufiges Verzeichnis", in *Centaurus* 25 (1981) 1–49.

Folkerts 1989 MENSO FOLKERTS: *Euclid in medieval Europe* (Questio de rerum natura 2), Winnipeg 1989 [repr. FOLKERTS 2006, n° III].

Folkerts 1995 MENSO FOLKERTS: "Die frühesten lateinischen Texte über das Rechnen mit indisch-arabischen Ziffern", in SCHÜTT/WEISS 1995, 157–174.

Folkerts 1996a MENSO FOLKERTS: "Andreas Alexander – Leipziger Universitätslehrer und Cossist", in GEBHARDT/ALBRECHT 1996, 53–61.

Folkerts 1996b MENSO FOLKERTS: "Frühe Darstellungen des Gerbertschen Abakus", in FRANCI e. a. 1996, 223–243.

Folkerts 2002 MENSO FOLKERTS: "Die Handschrift Dresden, C 80 als Quelle der Mathematikgeschichte", in GEBHARDT 2002, 353–378.

Folkerts 2003 MENSO FOLKERTS: *Essays on Early Medieval Mathematics*. The Latin tradition (Collected Studies Series 751), Aldershot 2003.

Folkerts 2006 MENSO FOLKERTS: *The Development of Mathematics in Medieval Europe*. The Arabs, Euclid, Regiomontanus (Collected Studies Series 811), Aldershot 2006.

Folkerts/Kühne 2006 *Astronomy as a Model for the Sciences in Early Modern Times*. Papers from the international symposium Munich, 10–12 march 2003, ed. by MENSO FOLKERTS/ANDREAS KÜHNE (Algorismus 59), Augsburg 2006.

Folkerts/Kunitzsch 1997 MENSO FOLKERTS/PAUL KUNITZSCH: *Die älteste lateinische Schrift über das indische Rechnen nach al-Ḫwārizmī*. Edition, Übersetzung und Kommentar (Abhandlungen der Bayerischen Akademie der Wissenschaften. Philosophisch-Historische Klasse N. F. 113), München 1997.

Forstner 1991 *Festgabe für Hans-Rudolf Singer*. Zum 65. Geburtstag am 6. April 1990 überreicht von seinen Freunden und Kollegen, ed. by MARTIN FORSTNER, Frankfurt am Main e. a. 1991.

Franci e. a. 1996 *Itinera mathematica*. Studi in onore di Gino Arrighi per il suo 90 compleanno, ed. by RAFFAELLA FRANCI e. a., Siena 1996.

Freudenthal 2011a GAD FREUDENTHAL: "Arabic and Latin cultures as resources for the Hebrew translation movement, comparative considerations, both quantitative and qualitative", in FREUDENTHAL 2011b, 74–104.

Freudenthal 2011b *Science in Medieval Jewish Cultures*, ed. by GAD FREUDENTHAL, Cambridge 2011.

Gasparri 1994 *Le XIIe siècle*. Mutations et renouveau en France dans la première moitié du XIIe siècle, ed. by FRANÇOISE GASPARRI (Cahiers du Léopard d'Or 3), Paris 1994.

Gebhardt 2002 *Verfasser und Herausgeber mathematischer Texte der frühen Neuzeit*, ed. by RAINER GEBHARDT (Schriften des Adam-Ries-Bundes Annaberg-Bucholz 14), Annaberg-Buchholz 2002.

Gebhardt/Albrecht 1996 *Rechenmeister und Cossisten der frühen Neuzeit*, ed. by RAINER GEBHARDT/HELMUTH ALBRECHT (Schriften des Adam-Ries-Bundes Annaberg-Buchholz 7), Annaberg-Buchholz 1996.

Goyens e. a. 2008 *Science Translated.* Latin and vernacular translations of scientific treatises in medieval Europe, ed. by MICHÈLE GOYENS e. a. (Mediaevalia Lovaniensia I. Studia 40), Leuven 2008.

Grant 1971 EDWARD GRANT: "Henricus Aristippus, William of Moerbeke and two alleged mediaeval translations of Hero's 'Pneumatica'", in *Speculum* 46 (1971) 656–669.

Grebner/Fried 2008 *Kulturtransfer und Hofgesellschaft im Mittelalter.* Wissenskultur am sizilianischen und kastilischen Hof im 13. Jahrhundert, ed. by GUNDULA GREBNER/JOHANNES FRIED (Wissenskultur und gesellschaftlicher Wandel 15), Berlin 2008.

Griffith 2013 SIDNEY HARRISON GRIFFITH: *The Bible in Arabic.* The scriptures of the 'people of the book' in the language of Islam (Jews, Christians, and Muslims from the Ancient to the Modern World), Princeton (NJ)/Oxford 2013. Gutas 1998.

Gutas 1998 DIMITRI GUTAS: *Greek thought, Arabic Culture.* The Graeco-Arabic translation movement in Baghdad and early 'Abbasid society (2nd–4th/5th–10th c.), London 1998 [repr. London 1999; London 2005].

Gutas 2006 DIMITRI GUTAS: "What was there in Arabic for the Latins to receive? Remarks on the modalities of the twelfth-century translation movement in Spain", in SPEER/WEGENER 2006, 3–21.

Hames 2012a HARVEY J. HAMES (ed./trans.): *Raimundi Lulli Opera Latina.* Supplementum Lullianum 3: Ha-melacha ha-ketzara: A Hebrew translation of Ramon Llull's 'Ars brevis' (CChr.CM 247), Turnhout 2012.

Hames 2012b HARVEY J. HAMES: "Translated from Catalan: Looking at a fifteenth-century Hebrew version of the Gospels", in ALBERNI e. a. 2012, 285–302.

Hames HARVEY J. HAMES: *Els quatre evangelis en hebreu traduïts del català* (Corpus Biblicum Catalanicum 35), Barcelona (submitted).

Haskins 1924 CHARLES HOMER HASKINS: *Studies in the History of Medieval Science*, Cambridge (MA) 1924.

Haskins 1927 CHARLES HOMER HASKINS: *The Renaissance of the Twelfth Century*, Cambridge (MA) 1927.

Hasse 2001 DAG NIKOLAUS HASSE: "Die humanistische Polemik gegen arabische Autoritäten. Grundsätzliches zum Forschungsstand", *Neulateinisches Jahrbuch* 3 (2001) 65–79.

Hasse 2004a DAG NIKOLAUS HASSE: "The attraction of Averroism in the Renaissance: Vernia, Achillini, Prassicio", in ADAMSON e. a. 2004, 2, 131–147.

Hasse 2004b DAG NIKOLAUS HASSE: "Aufstieg und Niedergang des Averroismus in der Renaissance. Niccolò Tignosi, Agostino Nifo, Francesco Vimercato", in AERTSEN/PICKAVÉ 2004, 447–473.

Hasse 2006 DAG NIKOLAUS HASSE: "The social conditions of the Arabic-(Hebrew-)Latin translation movements in medieval Spain and in the Renaissance", in SPEER/WEGENER 2006, 68–86.

Hogendijk/Sabra 2003 *The Enterprise of Science in Islam.* New perspectives, ed. by JAN PIETER HOGENDIJK/ABD AL-HAMID SABRA (Dibner Institute Studies in the History of Science and Technology), Cambridge (MA) 2003.

Holmes/Waring 2002 *Literacy, Education and Manuscript Transmission in Byzantium and Beyond*, ed. by CATHERINE HOLMES/JUDITH WARING (The Medieval Mediterranean 42), Leiden e. a. 2002.

Høyrup 2007 Jens Høyrup: *Jacopo da Firenze's Tractatus algorismi and Early Italian Abbacus Culture* (Science Networks 34), Basel e. a. 2007.

Jacquart/Micheau 1990 DANIELLE JACQUART/FRANÇOISE MICHEAU: *La médecine arabe et l'Occident médiéval* (Islam et Occident 7), Paris 1990 [repr . . . (Références Maisonneuve et Larose), Paris 1996].

King 1978 DAVID ANTHONY KING: "Three sundials from Islamic Andalusia", in *Journal for the History of Arabic Science* 2 (1978) 358–392 [repr. KING 1987, n° XV and Addenda et corrigenda, 372 sq.].

King 1987 DAVID ANTHONY KING: *Islamic Astronomical Instruments* (Collected Studies Series 253), London 1987 [repr. London 1995].

King 1992 DAVID ANTHONY KING: "Los cuadrantes solares andalusíes", in VERNET GINÉS/SAMSÓ 1992, 89–102.

King 2002–2003 DAVID ANTHONY KING: "An astrolabe from 14th-century Christian Spain with inscriptions in Latin, Hebrew and Arabic", in *Suhayl* 3 (2002–2003) 9–156.

King/Samsó 2001 DAVID ANTHONY KING/JULIO SAMSÓ: "Astronomical handbooks and tables from the Islamic world (750–1900): An interim report", in *Suhayl* 2 (2001) 9–105.

van Koningsveld 1991 PIETER SJOERD VAN KONINGSVELD: "Andalusian-Arabic manuscripts from medieval Christian Spain: Some supplementary notes", in FORSTNER 1991, 811–824.

van Koningsveld 1992 PIETER SJOERD VAN KONINGSVELD: "Andalusian-Arabic manuscripts from Christian Spain: A comparative intercultural approach", in *Israel Oriental Studies* 12 (1992) 75–112.

Köpf/Bauer 2011 *Kulturkontakte und Rezeptionsvorgänge in der Theologie des 12. und 13. Jahrhunderts*, ed. by ULRICH KÖPF/DIETER RICHARD BAUER (Archa Verbi. Subsidia 8), Münster in Westfalen 2011.

Kunitzsch 1961 PAUL KUNITZSCH: *Untersuchungen zur Sternnomenklatur der Araber*, Wiesbaden 1961.

Kunitzsch 1964 PAUL KUNITZSCH: "Das Fixsternverzeichnis in der persischen 'Syntaxis' des Georgios Chrysokokkes", in *Byzantinische Zeitschrift* 57 (1964) 382–411.

Kunitzsch 1966 PAUL KUNITZSCH: *Typen von Sternverzeichnissen in astronomischen Handschriften des 10. bis 14. Jahrhunderts*, Wiesbaden 1966.

Kunitzsch 1974 PAUL KUNITZSCH: *Der Almagest.* Die 'Syntaxis mathematica' des Claudius Ptolemäus in arabisch-lateinischer Überlieferung, Wiesbaden 1974.

Kunitzsch 1975 PAUL KUNITZSCH: "Der Almagest des Claudius Ptolemäus – ein Modellfall mehrstufiger Kulturrezeption", in *Jahrbuch der Akademie der Wissenschaften in Göttingen* 1974, Göttingen 1975, 32–36.

Kunitzsch 1977 PAUL KUNITZSCH: *Mittelalterliche astronomisch-astrologische Glossare mit arabischen Fachausdrücken* (Sitzungsberichte der Bayerischen Akademie der Wissenschaften. Philosophisch-Historische Klasse Jg. 1977, Nr. 5), München 1977.

Kunitzsch 1981 PAUL KUNITZSCH: "On the authenticity of the Treatise on the composition and use of the Astrolabe ascribed to Messahalla", in *Archives internationales d'histoire des sciences* 31 (1981) 42–62.

Kunitzsch 1983 PAUL KUNITZSCH: *Über eine anwa'-Tradition mit bisher unbekannten Sternnamen* (Beiträge zur Lexikographie des klassischen Arabisch 4. Sitzungsberichte der Bayerischen Akademie der Wissenschaften. Philosophisch-Historische Klasse Jg. 1983, Nr. 5), München 1983.

Kunitzsch 1986 PAUL KUNITZSCH: *Peter Apian und Azophi.* Arabische (Sitzungsberichte der Bayerischen Akademie der Wissenschaften. Philosophisch-Historische Klasse Jg. 1986, Nr. 3), München 1986.

Kunitzsch 1989 PAUL KUNITZSCH: *The Arabs and the Stars* (Collected Studies Series 307), Northampton 1989.

Kunitzsch 1993 PAUL KUNITZSCH: "Fragments of Ptolemy's 'Planisphaerium' in an early Latin translation", in *Centaurus* 36 (1993) 97–101.

Kunitzsch 1996 PAUL KUNITZSCH: *Reflexe des Orients im Namensgut mittelalterlicher europäischer Literatur*. Gesammelte Aufsätze (Documenta onomastica litteralia medii aevi. Reihe B: Studien 2), Hildesheim e. a. 1996.

Kunitzsch 2004 PAUL KUNITZSCH: *Stars and Numbers*. Astronomy and mathematics in the medieval Arab and Western worlds (Collected Studies Series 791), Aldershot 2004.

Kunitzsch 2006 PAUL KUNITZSCH: "Late traces of Arabic influence in European astronomy (17th–18th centuries)", in FOLKERTS/KÜHNE 2006, 97–102.

Kunitzsch/Lorch 2010 PAUL KUNITZSCH/RICHARD LORCH (ed.): *Theodosius, Sphaerica*. Arabic and medieval Latin translations (Boethius 62), Stuttgart 2010.

Kunitzsch/Lorch 2011 PAUL KUNITZSCH/RICHARD LORCH (ed.): *Theodosius, De habitationibus*. Arabic and medieval Latin translations (Sitzungsberichte der Bayerischen Akademie der Wissenschaften. Philosophisch-Historische Klasse Jg. 2011, Nr. 1), München 2011.

Labarta 1995 ANA LABARTA: "Un nuevo fragmento de reloj de sol andalusí", in *al-Qanṭara* 16 (1995) 147–150.

De Lange 2001a *Hebrew Scholarship and the Medieval World*, ed. by NICHOLAS DE LANGE, Cambridge 2001.

De Lange 2001b NICHOLAS DE LANGE: "Hebrew scholarship in Byzantium", in DE LANGE 2001a, 23–37.

Langermann 2011 Y. TZVI LANGERMANN: "Science in the Jewish communities of the Byzantine cultural orbit: New perspectives", in FREUDENTHAL 2011b, 438–454.

Lasker 2011 DANIEL J. LASKER: "Medieval Karaism and science", in FREUDENTHAL 2011b, 427–438.

Lemay 1958 RICHARD JOSEPH LEMAY: *The 'Introductorium in Astronomiam' of Albumasar and the Reception of Aristotle's Natural Philosophy in the Twelfth Century*, New York 1958.

Lemay 1962 RICHARD [JOSEPH] LEMAY: *Abu Ma'shar and Latin Aristotelianism in the 12th Century*. The recovery of Aristotle's Natural Philosophy through Arabic astrology (American University of Beirut. Publications of the Faculty of Arts and Sciences. Oriental Series 38), Beirut 1962.

Lemay 1987 RICHARD [JOSEPH] LEMAY: "The true place of astrology in medieval science and philosophy: Towards a definition", in CURRY 1987, 57–73.

de Libera 2009 ALAIN DE LIBERA: "Les Latins parlent aux Latins", in BÜTTGEN e. a. 2009, 175–215.

Lindberg 1975 DAVID CHARLES LINDBERG: *A Catalogue of Medieval and Renaissance Optical Manuscripts* (Pontifical Institute of Mediaeval Studies. Subsidia mediaevalia 4), Toronto 1975.

Lindberg 1978 *Science in the Middle Ages*, ed. by DAVID CHARLES LINDBERG (The Chicago History of Science and Medicine), Chicago (IL)/London 1978.

Lorch 1995 RICHARD LORCH: *Arabic Mathematical Sciences*. Instruments, texts, transmission (Collected Studies Series 517). Aldershot 1995.

Lorch 1999 RICHARD LORCH: "The Treatise on the astrolabe by Rudolph of Bruges", in NAUTA/VANDERJAGT 1999, 55–100.

Lorch 2001 RICHARD LORCH: "Greek – Arabic – Latin: The transmission of mathematical texts in the Middle Ages", in *Science in Context* 14 (2001) 313–331.

Lorch/Martínez Gázquez 2005 RICHARD LORCH/JOSÉ MARTÍNEZ GÁZQUEZ: "Qusta ben Luca, 'De sphera uolubili', in *Suhayl* 5 (2005) 9–62.

Lutz-Bachmann e. a. 2004 *Metaphysics in the Twelfth Century*. On the relationship among philosophy, science and theology, ed. by MATTHIAS LUTZ-BACHMANN e. a. (Textes et études du moyen âge 19), Turnhout 2004.

Malet 2000 ANTONI MALET: "Mil años de matemáticas en Iberia", in DURÁN GUARDEÑO 2000, 193–223.

Martin 2007 CRAIG MARTIN: "Rethinking renaissance averroism", in *Intellectual History Review* 17 (2007) 3–19.

Mercier 1994 RAYMOND MERCIER (ed.): *An Almanac for Trebizond for the Year 1336* (Corpus des astronomes byzantins 7), Louvain-la-Neuve 1994.

Mercier 1998 RAYMOND MERCIER: "The astronomical tables of George Gemistus Plethon", in *Journal for the History of Astronomy* 29 (1998) 117–127.

Mercier 2004 Raymond Mercier: *Studies on the Transmission of Medieval Mathematical Astronomy* (Collected Studies Series 787), Aldershot 2004.

Millàs Vallicrosa 1927 JOSEP MARIA MILLÀS I VALLICROSA: *Documents hebraics de jueus catalans* (Institut d'Estudis Catalans. Secció Històrico-Arqueològica. Memòries 1. 3), Barcelona 1927.

../../../../Geek Squad Data Backup 7.27.2020/Users/Paige/Desktop/15031s/15031-5677 Brentjes/03 from CE/15031-5677-Ref Mismatch Report.docx - LStERROR_157Millàs Vallicrosa 1931 JOSÉ MARÍA MILLÀS VALLICROSA: *Assaig d'història de les idees físiques i matemàtiques a la Catalunya medieval* 1 (Estudis universitaris catalans. Série monográfica 1), Barcelona 1931.

Millàs Vallicrosa 1934 JOSÉ MARÍA MILLÀS VALLICROSA: "Manuscritos hebraicos de la Biblioteca Capitular de Toledo", in *al-Andalus* 2 (1934) 395–430.

../../../../Geek Squad Data Backup 7.27.2020/Users/Paige/Desktop/15031s/15031-5677 Brentjes/03 from CE/15031-5677-Ref Mismatch Report.docx - LStERROR_158Millàs Vallicrosa 1941 JOSÉ MARÍA MILLÀS VALLICROSA: "Amuleto musulmán de origen aragonés", in *al-Andalus* 6 (1941) 317–326.

Millàs Vallicrosa 1942 JOSÉ MARÍA MILLÀS VALLICROSA: *Las traducciones orientales en los manuscritos de la Biblioteca Catedral de Toledo*, Madrid 1942.

../../../../Geek Squad Data Backup 7.27.2020/Users/Paige/Desktop/15031s/15031-5677 Brentjes/03 from CE/15031-5677-Ref Mismatch Report.docx - LStERROR_159Millàs Vallicrosa 1943 JOSÉ MARÍA MILLÀS VALLICROSA: "Traducción castellana del 'Tratado de Agricultura' de Ibn Wafid", in *al-Andalus* 8 (1943) 281–332.

Millàs Vallicrosa 1949 JOSÉ MARÍA MILLÀS VALLICROSA: *Estudios sobre historia de la ciencia española* (Instituto 'Luis Vives' de Filosofía y Asociación para la Historia de la Ciencia Española. Sección de Historia de la Filosofía Española. Estudios 2), Barcelona 1949.../../../../Geek Squad Data Backup 7.27.2020/Users/Paige/Desktop/15031s/15031-5677 Brentjes/03 from CE/15031-5677-Ref Mismatch Report.docx - LStERROR_160

Millàs Vallicrosa 1955 JOSÉ MARÍA MILLÀS VALLICROSA: "Los primeros tratados de astrolabio en la España árabe", in *Revista del Instituto egipcio de estudios islámicos en Madrid* 3 (1955) 35–49.

Murdoch 2003 JOHN EMERY MURDOCH: "Transmission into use: The evidence of marginalia in the medieval Euclides Latinus", in *Revue d'histoire des sciences* 56 (2003) 369–382.

Murdoch/Grant 1987 *Mathematics and Its Applications to Science and Natural Philosophy in the Middle Ages*. Essays in honour of Marshall Clagett, ed. by JOHN EMERY MURDOCH/EDWARD GRANT, Cambridge 1987.

Murdoch/Sylla 1978 JOHN EMERY MURDOCH/EDITH DUDLEY SYLLA: "The science of motion", in LINDBERG 1978, 206–264.

Murdoch/Sylla 1987 *The Cultural Context of Medieval Learning*. Proceedings of the First International Colloquium on Philosophy, Science, and Theology in the Middle-Ages, September 1973, ed. by JOHN EMERY MURDOCH/EDITH DUDLEY SYLLA, Dordrecht 1987.

Nauta/Vanderjagt 1999 *Between demonstration and imagination*. Essays in the history of science and philosophy presented to John D. North, ed. by LODI NAUTA/ARJO VANDERJAGT (Brill's Studies in Intellectual History 96), Leiden/Boston 1999.

Nicolaidis 2011 EFTHYMIOS NICOLAIDIS: *Science and Eastern Orthodoxy*. From the Greek fathers to the age of globalization (Medicine, Science, and Religion in Historical Context), Baltimore (MD) 2011.

van Oppenraay/Fontaine 2012 *The Letter before the Spirit*. The importance of text editions for the study of the reception of Aristotle, ed. by AAFKE MARIA ISOLINE VAN OPPENRAAY/RESIANNE FONTAINE (Aristoteles semitico-latinus 22), Leiden/Boston 2012.

Petrus i Pons 2008 NÀDIA PETRUS [I PONS] (ed.): *Alchoranus Latinus, quem transtulit Marcus canonicus Toletanus*. Estudio y edición crítica, Diss. phil. Bellaterra 2008.

Reif 2002 STEFAN REIF: "Some changing trends in the Jewish literary expression of the Byzantine world", in HOLMES/WARING 2002, 88–111.

Remmert e. a. 2016 *Historiography of Mathematics in the 19th and 20th Centuries*, ed. by VOLKER R. REMMERT e. a., Basel 2016.

Ricklin 2006 THOMAS RICKLIN: "'Arabes contigit imitari'. Beobachtungen zum kulturellen Selbstverständnis der iberischen Übersetzer der ersten Hälfte des 12. Jahrhunderts", in SPEER/WEGENER 2006, 47–67.

van Riet e. a. 2006 SIMONE VAN RIET (†) e. a. (ed.): *Avicenna latinus, Liber primus naturalium. Tractatus secundus de motu et de consimilibus* (Avicenna latinus 10), Bruxelles 2006.

van Riet/Verbeke 1992 SIMONE VAN RIET/GERARD VERBEKE (eds.): *Avicenna latinus, Liber primus naturalium. Tractatus primus de causis et principiis naturalium* (Avicenna latinus 1, 8), Louvain-la-Neuve/Leiden 1992.

Rius Piniés 2006 MÒNICA RIUS [PINIÉS]: "La alquibla. ¿Ciencia religiosa o religión científica?", in *Ilu. Revista de ciencias de las religiones*. Anejos 16 (2006) 93–111.

Samsó 1992 JULIO SAMSÓ: "Astronomía teórica en al-Andalus", in VERNET GINÉS/SAMSÓ 1992, 45–52.

Samsó 2003 JULIO SAMSÓ: "El astrolabio carolingio de Marcel Destombes y la introducción del astrolabio en la Catalunya medieval", in *Memorias de la Real Academia de Ciencias y Artes de Barcelona* 60 (2003) 345–353.

Schütt/Weiss 1995 *Brückenschläge*. 25 Jahre Lehrstuhl für Geschichte der exakten Wissenschaften und der Technik an der Technischen Universität Berlin, 1969–1994, ed. by HANS-WERNER SCHÜTT/BURGHARD WEISS, Berlin 1995.

Siraisi 1987 NANCY GILLIAN SIRAISI: *Avicenna in Renaissance Italy*. The canon and medical teaching in Italian universities after 1500, Princeton (NJ) 1987.

Speer/Steinkrüger 2012 *Knotenpunkt Byzanz*. Wissensformen und kulturelle Wechselbeziehungen, ed. by ANDREAS SPEER/PHILIPP STEINKRÜGER (Miscellanea Mediaevalia 36), Berlin/Boston 2012.

Speer/Wegener 2006 *Wissen über Grenzen*. Arabisches Wissen und lateinisches Mittelalter, ed. by ANDREAS SPEER/LYDIA WEGENER (Miscellanea Mediaevalia 33), Berlin/New York 2006.

Spiesser 2003 MARYVONNE SPIESSER: *Une arithmétique commerciale du XVe siècle.* Le 'Compendy de la praticque des nombres' de Barthélemy de Romans (De diversis artibus 70), Turnhout 2003.

Stammen/Weber 2004 *Wissenssicherung, Wissensordnung und Wissensverarbeitung.* Das europäische Modell der Enzyklopädien, ed. by THOMAS STAMMEN/WOLFGANG WEBER (Colloquia Augustana 19), Berlin 2004.

Sylla/McVaugh 1997 *Texts and Contexts in Ancient and Medieval Science.* Studies on the occasion of John E. Murdoch's seventieth birthday, ed. by EDITH DUDLEY SYLLA/MICHAEL ROGERS MCVAUGH (Brill's Studies in Intellectual History 78), Leiden e. a. 1997.

Sylla/Newman 2009 *Evidence and Interpretation.* Studies on early science and medicine in honor of John E. Murdoch, ed. by EDITH DUDLEY SYLLA/WILLIAM ROYALL NEWMAN (Early Science and Medecine 14), Leiden/Boston 2009.

Terrasse 1994 *Comprendre et maîtriser la nature au moyen âge.* Mélanges d'histoire des sciences offerts à Guy Beaujouan, ed. by MICHEL TERRASSE (École Pratique des Hautes Études IVe Séction: Sciences Historiques et Philologiques 5. Hautes études médiévales et modernes 73), Genève 1994.

Tihon 1986–1993 ANNE TIHON (general editor): *Corpus des astronomes byzantins.* Amsterdam.

Tihon 1994 ANNE TIHON: *Études d'astronomie byzantine* (Collected Studies Series 454), Aldershot/Brookfield 1994.

Tihon 1994–2001 ANNE TIHON (general editor): *Corpus des astronomes byzantins.* Louvain-la-neuve.

Tihon 2009 ANNE TIHON: "Les sciences exactes à Byzance", in *Byzantion* 79 (2009) 380–434.

Tihon e. a. 2001 ANNE TIHON e. a. (ed.): *Une version byzantine du traité sur l'astrolabe du Pseudo-Mes-sahalla* (Corpus des astronomes byzantins 10), Louvain-la-Neuve 2001.

Tihon/Mercier 1998 ANNE TIHON/RAYMOND MERCIER (eds.): *Georges Gémiste Pléthon, Manuel d'astronomie* (Corpus des astronomes byzantins 9), Louvain-la-Neuve 1998.

Tischler 2008 MATTHIAS MARTIN TISCHLER: "Orte des Unheiligen. Versuch einer Topographie der dominikanischen Mohammed-Biographik des 13. Jahrhunderts zwischen Textüberlieferung und Missionspraxis", in *Archa Verbi* 5 (2008) 32–62.

Tischler 2011a MATTHIAS MARTIN TISCHLER: "Die Iberische Halbinsel als christlich-muslimischer Begegnungsraum im Spiegel von Transfer- und Transformationsprozessen des 12. – 15. Jahrhunderts", in *Anuario de Historia de la Iglesia* 20 (2011) 117–155.

Tischler 2011b MATTHIAS MARTIN TISCHLER: "Transfer- und Transformationsprozesse im abendländischen Islambild zwischen dem 11. und 13. Jahrhundert", in KÖPF/BAUER 2011, 329–379.

Tischler 2012 MATTHIAS MARTIN TISCHLER: "Eine fast vergessene Gedächtnisspur. Der byzantinisch-lateinische Wissenstransfer zum Islam (8. – 13. Jahrhundert)", in SPEER/STEINKRÜGER 2012, 167–195.

Tischler 2013 MATTHIAS MARTIN TISCHLER: "Lost in translation. Orality as a tricky filter of memory in Arabo-Latin processes of transfer", in *Medievalia. Revista d'Estudis Medievals* 16 (2013) 149–158.

Tischler/Fidora 2011 *Christlicher Norden – Muslimischer Süden.* Ansprüche und Wirklichkeiten von Christen, Juden und Muslimen auf der Iberischen Halbinsel im Hoch- und Spätmittelalter, ed. by MATTHIAS MARTIN TISCHLER/ALEXANDER FIDORA (Erudiri Sapientia. Studien zum Mittelalter und zu seiner Rezeptionsgeschichte 7), Münster in Westfalen 2011.

Van den Abeele/Meyer 2005 *Bartholomaeus Anglicus, De proprietatibus rerum*. Texte latin et réception vernaculaire. Lateinischer Text und volkssprachige Rezeption, ed. by BAUDOUIN VAN DEN ABEELE/HEINZ MEYER (De diversis artibus 74), Turnhout 2005.

Vegas González 2005 SERAFÍN VEGAS GONZÁLEZ: "Significado histórico y significación filosófica en la revisión de los planteamientos concernientes a la escuela de Toledo", in *Revista española de filosofía medieval* 12 (2005) 109–134.

Vernet Ginés 1953 JUAN VERNET GINÉS: "La cartografía náutica, ¿tiene un origen hispano-árabe?", in *Revista del Instituto egipcio de estudios islámicos* 1 (1953) 66–91.

Vernet Ginés 1970 JUAN VERNET GINÉS: "Astrología y política en la Córdoba del siglo X", in *Revista del Instituto egipcio de estudios islámicos* 15 (1970) 91–100.

Vernet Ginés 1979 JUAN VERNET [GINÉS]: *Estudios sobre historia de la ciencia medieval*. Reedición de trabajos dispersos, ofrecida al autor por sus discípulos con ocasión de los veinticinco años de su acceso a la cátedra de la Universidad de Barcelona, Bellaterra 1979.

Vernet Ginés 1986 JUAN VERNET GINÉS: *La ciencia en al-Andalus* (Biblioteca de la cultura andaluza 56), Sevilla 1986.

Vernet Ginés 1988 JUAN VERNET GINÉS: "Alfonso el Sabio y la mecánica", in *Boletín de la Real Academia de la Historia* 185 (1988) 29–38.

Vernet Ginés 1993 JUAN VERNET GINÉS: "Ingeniería mecánica del Islam occidental", in *Investigación y ciencia* 201 (1993) 46–50.

Vernet Ginés 1998 JUAN VERNET GINÉS: *Historia de la ciencia española* (Ad litteram 7), Madrid 1998.

Vernet Ginés 1999 JUAN VERNET [GINÉS]: *Lo que Europa debe al Islam de España* (El acantilado 2), Barcelona 1999.

Vernet Ginés/Samsó 1992 *El legado científico andalusí* (Exhibition Catalogue), ed. by JUAN VERNET GINÉS/JULIO SAMSÓ, Madrid 1992.

Viladrich 1992 MERCÈ VILADRICH: "Astrolabios andalusíes", in VERNET GINÉS/SAMSÓ 1992, 53–65.

Voss/Hinson Lall 2007 *The Imaginal Cosmos*. Astrology, divination and the sacred, ed. by ANGELA VOSS/JEAN HINSON LALL. Introduction by GEOFFREY CORNELIUS, Canterbury 2007.

Weijers 1995 *Vocabulary of Teaching and Research between Middle Ages and Renaissance*, ed. by OLGA WEIJERS (CIVICIMA. Études sur le vocabulaire intellectual du moyen âge 8), Turnhout 1995.

Wilpert/Eckert 1962 *Antike und Orient im Mittelalter*. Vorträge der Kölner Mediävistentagungen 1956–1959, ed. by PAUL WILPERT/WILLEHAD PAUL ECKERT (Miscellanea mediaevalia 1), Berlin 1962.

Zonta 2006 MAURO ZONTA: *Hebrew Scholasticism in the Fifteenth Century*. A history and source book (Amsterdam Studies in Jewish Thought 9), Dordrecht 2006.

7

PRACTICING HISTORY OF MATHEMATICS IN ISLAMICATE SOCIETIES IN 19TH-CENTURY GERMANY AND FRANCE

1 Introduction

Interest in Arabic, Persian, and, at times, also Ottoman Turkish texts of the mathematical sciences existed in different parts of Europe since the Middle Ages. Until the nineteenth century, this interest was mostly of a contemporary scientific, but not of a historical nature. Hence, I will not consider these older activities in this paper. I will limit myself to discussing authors, their texts, and in a more limited manner their contexts that contributed to the formation of research on the history of these sciences in a variety of Islamicate societies. One comment needs to be made nonetheless on those older activities. Early on, the discussions of the value of mostly Latin translations of Arabic texts and of scholarly activities in the Ottoman Empire shaped attitudes and perspectives of first Christian and later also secular scholars in Europe to the intellectual achievements in Islamicate societies as a whole. The values that these discussions promoted were prevalently negative, even when respect was paid to the existence of numerous Arabic translations of Greek mathematical works. Latin translations of Arabic scientific, mathematical, medical, and philosophical texts were denounced from the fourteenth century onwards for their cumbersome Latin and their Arabisms. Arabic scholarly achievements were appreciated in the sixteenth and the first half of the seventeenth centuries, while Ottoman scholarship was talked of in derogatory terms. During the second half of the seventeenth century negative evaluations of Arabic scholarship increased, partly because expectations and attitudes changed as a result of the new forms of scholarly enterprise that developed in this period in parts of Europe and partly because some of the former enthusiasts of Oriental studies became disenchanted with their fruits of reading, traveling, and language learning.

As a consequence of these two strands and further political and cultural changes in Europe, North Africa, and western Asia, two main positions emerged that

DOI: 10.4324/9781003372493-8

affected thinking about the mathematical sciences in Islamicate societies profoundly and for a long time. A cultural and a temporal preference emerged that located worthwhile studies of past mathematical sciences in Arabic texts that were either translations from other languages, mostly ancient Greek, or new compositions executed between the later eighth and the early thirteenth centuries. Among those preferred carriers of knowledge, the subset of texts that preserved ancient Greek works was particularly cherished. The thesis of independent new achievements by scholars from regions ruled by the Abbasid dynasty and its representatives like the Buyids and the Seljuqs, while formulated in clear manner already in the 1830s and supported by various writers in the course of that century, became generally acknowledged, supported, and respected by historians of mathematics only in the last third of the twentieth century. The reasons for this profound shift in evaluation and attitude lie in the rapid increase of professionalism in the second half of the twentieth century, in cultural anti-colonial politics in India, Iran, Turkey, and the Arab world after WWI or WWII, as well as in immense migrations of people from western Asia and North Africa to Europe, Canada, and the US and their demands for cultural scholars from Islamicate societies and their activities and results into their canonical views of their profession. This applies in particular to the development of this field since the 1950s.

The second position that dominated evaluations and attitudes throughout the nineteeth century looked at such Arabic contributions from their potential to mediate between ancient Greece and medieval Latin Europe. This sandwiched position continues to be subscribed to by historians of science of other periods, regions, and cultures, in particular those who work on the Renaissance and the early modern period in parts of central and western Europe. Although historians of the mathematical sciences in Islamicate societies turned their back to it in the course of the second half of the twentieth century, its impact on research orientations remains clearly visible. There was, and is, very little research on past mathematical sciences and their practitioners in regions like eastern Iran, Muslim India, or sub-Saharan Africa. The same phenomenon is expressed by the dearth of studies on relationships between mathematical cultures in different Islamicate societies and their eastern or southern neighbors in Asia and Africa, when compared to such research on Graeco-Arabica or Arabo-Latina.

2 On some of the main contextual components

The contexts of new studies, historiographical positions, and interpretive stances with regard to the mathematical sciences in Islamicate societies during the nineteenth century have to be looked for in political, religious, social, and academic conflicts and changes. The specifics of each of these types of context differ between the three countries where the men came from who provided financial and intellectual support for these studies – Germany, France, and Italy.

In Germany, the Napoleonic wars and their impact on German boundaries, reforms, and political aspirations have to be seen as a major source of change.

One of its consequences was Wilhelm von Humboldt's (1767–1835) reform of Prussian universities and the preference it gave to research (Dauben and Scriba 2002, 114). One outcome of this changed orientation of the university profile and the professional duties of university professors and lecturers was the rise of what was called pure mathematics to the detriment of the so-called applied mathematics. This led to a preference for theory over practice in many books and articles on history of mathematics during the nineteenth century. Folkerts, Scriba, and Wußing see Nesselmann as one of the first authors who embodied "the spirit of the educational reform" within history of mathematics, without, however, explaining why that is the case (Dauben and Scriba 2002, 114–115). One possible argument in favor of such a contextualization of Nesselmann's university career is his substantial involvement with novel research questions and methods.

In France, the French Revolution and its successors shaped the political fate of state and people through the different phases of the constitutional monarchy and its successor, the French Republic. In 1830, at the end of the restoration period, Charles X invaded Algeria and started the second phase of French colonialism. This first French colonial war of the nineteenth century continued until 1848, when the second French revolution of the nineteenth century broke out and introduced the Second Republic. During those years, the first round of a major academic controversy over the value of scholarly contributions to the mathematical sciences from Islamicate societies and their relationships to ancient Greek scholarly achievements took place in Paris. Peiffer rightly highlighted the importance that the French conquest of North Africa has had upon the academic atmosphere of this period (Dauben and Scriba 2002, 14). The first chapter on Algeria of *Histoire pittoresque de l'Afrique française*, published in 1845 during the constitutional reign of Louis Philippe, duke of Orléans, begins with the claim, although the war would continue for another three years, that:

> Algeria is today a French province, or rather the new France. This region (will have) a future, (which is) one of the most prosperous and useful for our fatherland, which soon will draw from this addition incomparably greater advantages than those, (which came) from its distant colonies. The time is not far away, when France will be richly recompensed for all the sacrifices, which it has made for fifteen years in order to retain and organize that precious conquest. Today, the government seems determined not to neglect any means for installing forever the civilization, its laws, its mores, and the French industry in those regions where Turkish dominance and belief in the Muslim religion previously upheld barbarism.[1]

1 "L'Algérie est aujourd'hui une province française, ou plutôt la nouvelle France. Cette contrée, est réservée à un avenir des plus prospères et de plus utiles à notre patrie, qui bientôt, retirera de cette adjonction, des avantages incomparablement plus grands que ceux résultants de ses colonies lointaines. L'époque n'est pas éloignée où la France, sera amplement dédomagée de tous les

Before the anonymous propagator of French colonial glory turned to describing French military success, he wished to discuss the peaceful activities, since it was those, which "honor our arts, our sciences, our civilization . . .".[2] Although the following pages treat more or less military activities, occasional comments clarify the involvement of civil, academic personnel, primarily geographers and archeologists, who drew up maps of the newly conquered terrains and pointed the military commanders to Roman ruins that they considered worthy of protection (Histoire pittoresque 1845, 44–45). This primary attention to historical objects, which were seen as part of the colonizing country's own historical legacy, is also visible in earlier descriptions of the conquest. The anonymous writer had also invested, on the other hand, much effort in reading Arabic geographical and historical sources about North Africa (Histoire pittoresque 1845, 46). This tension between the appreciation of ancient Greek and Roman cultures and their intellectual achievements and interest in Arabic scholarly texts continued during the entire nineteenth contributed to the formation of French oriental studies, on the one hand, and two major public conflicts on the academic evaluation of scholarly contributions from Islamicate societies, on the other.[3]

A further major political and cultural context of the historical study of the mathematical sciences in Islamicate societies concerns the relationship between attitudes towards Catholicism and the Roman Catholic Church, on the one hand, and modernization, secularization, and the formation of a "national state", on the other. In France, the turn towards science as a useful instrument for overcoming the destructions wrought by the French Revolution and the Napoleonic wars favored the rise of positivism as the dominant philosophy (Dauben and Scriba 2002, 14). The rejection of religion in general, Islam and the Roman Catholic Church's Inquisition in particular, was an important factor for Renan's wholly negative evaluation of the sciences in Islamicate societies in a public lecture at the Sorbonne in 1883. His anti-religious position reflects the Republican character of the governments of the Third Republic and their struggle against the Catholic Church and its orders in France. The republicans put part of the blame for the lost war against Prussia in 1870-1871 on the clerics and their political allies (i.e., the monarchists) and their dominant role in the state institutions of the Second Empire, in particular the educational system. Modernization was identified with secular education and the separation between Church and State.

A similar situation prevailed in Italy. Mazzotti argues for Baldassare Boncompagni's (1821–1894) case that his approach to history of mathematics and his

---

sacrifices quelle a faits, depuis quinze ans, pour se conserver et organiser cette précieuse conquète. Aujourd'hui, le gouvernement paraît décidé à ne négliger aucuns des moyens d'installer à jamais la civilisation, ses lois, ses mœurs et l'industrie française, dans ces contrées où la domination turque et la croyance de la religion musulmane, entretenaient auparavant la barbarie." (Histoire pittoresque 1845, 5).

2 ". . . ils honorent nos arts, nos sciences, notre civilisation, . . ." (Histoire pittoresque 1845, 33).

3 See below: the Sédillot-Biot controversy and the Renan-al-Afghani dispute.

various publication projects were shaped by his fierce opposition towards the secular and anti-papal forces of the *Risorgimento* (1815–1870), which led to the end of the Papal state in 1870 with which Boncompagni was closely affiliated by social status, conviction, and academic activities (Mazzotti 2000, 260–269, 272–276). Mazzotti believes that the specific Roman conditions explain Boncompagni's interest in medieval Latin and Arabic mathematical texts and other features of his working practice, which included the generous patronage for Franz Woepcke (Mazzotti 2000, 269–282).[4]

In addition to these major political, religious, and cultural developments in Germany, France, and Italy during the nineteenth century, a number of more specific intellectual and institutional changes shaped the academic landscape in the humanities in this time. In addition to the impact of new trends in philosophy and history, highlighted by Folkerts, Scriba, and Wußing, the search for intellectual, linguistic, and other origins of European languages, peoples, and cultures, for instance in India or in ancient Greece, led to the formation of new academic disciplines, such as Indology or comparative and historical linguistics, and to movements such as neo-humanism. Almost all scholars who worked in France and Germany on Arabic or Persian mathematical and astronomical texts also learned Sanskrit. They believed in a strong Indian background to the mathematical sciences in the Abbasid caliphate in addition to their ancient Greek legacy. The overwhelming dominance of so-called pure mathematics at German universities during the nineteenth century, an already-mentioned outcome of Humboldt's educational reforms, was a further important factor that shaped in particular Woepcke's research and writing practice (Dauben and Scriba 2002, 124). Equally, Cantor's *Vorlesungen über Geschichte der Mathematik* is dedicated to the history of this type of mathematics.

In France, the philosophies of Jean Antoine Caritat de Condorcet (1772–1791) and Auguste Comte (1789–1857) created the contexts for a more diversified working practice (Dauben and Scriba 2002, 9, 14). The two authors propagated an image of the new sciences and their practitioners that demanded encyclopedic knowledge and cherished experimentation. This view asked for a combination of mathematics with the sciences, but included also a desire for the search for, collection, edition, translation, and analysis of ancient, medieval, and early modern works (Dauben and Scriba 2002, 14). Peiffer locates the precursor of French colonial attention to "Oriental" achievements in the mathematical sciences in the activities of early modern antiquarians such as Barthélémy d'Herbelot (1625–1695), astronomers and geographers of the Royal Academy such as Jean-Dominic Cassini (1625–1712) and Joseph-Nicolas Delisle (1688–1768), and the cooperation between these groups. This cooperative atmosphere also characterized much of the work undertaken in the nineteenth century, in particular its first half. As Peiffer shows, it was not limited to interpersonal exchanges, but included institutional

4 Mazzotti's analysis of Boncompagni's *Bullettino di bibliografia e storia delle scienze matematiche e fisiche* is, however, flawed. Moreover, he apparently lacks in understanding of the skills and working practice of a medievalist or Arabist. This lack caused several mistakes in his evaluation.

forms and projects such as the translation of Ibn Yunus' (d. 1009) astronomical handbook by a commission of the *Academy of Sciences*, in which Jean-Baptiste Delambre (1749–1822), Pierre-Simon Laplace (1749–1827), and Jean-Jaques Caussin de Perceval (1759–1835) worked together, or the cooperation of Jean-Jacques Sédillot (1777–1832) with Delambre and Laplace at the *Bureau des Longitudes* (Dauben and Scriba 2002, 15–6). Unfortunately, she does not offer any further investigation of the impact of French colonialism on the study of the mathematical sciences in Islamicate societies, either in France or in the colonies (Dauben and Scriba 2002, 20).

The major specific context that shaped not only the research and debates on Islamicate societies and their sciences in Paris but drew the attention of German and Italian scholars as well is the fierce struggle about the value and content of the contributions of the practitioners of the mathematical sciences in those societies. The elder Sédillot publicly expressed his view of the "Arabs" as innovators and original scholars. His two collaborators Delambre and Laplace defended, despite their growing knowledge of "Oriental" astronomy, the older belief according to which the "Arabs" had merely preserved and transmitted Greek scientific texts without adding anything of value to this knowledge. Jean-Jacques Sédillot's son Louis-Amélie (1808–1875) continued to subscribe to his father's judgment. His life-long efforts to substantiate this view were partly the result of a mistake. In 1835, namely, he misinterpreted a passage in Abu l-Wafa''s (940–998) *Almagest* as "variation" and ascribed to Tycho Brahe (1546–1601) (Dauben and Scriba 2002, 17; Sédillot 1835). His main opponent was Jean-Baptiste Biot (1774–1862), a professor of mathematics and physics and member of the Academy of Sciences. Due to the early death of his son Edouard, who was a sinologist and whose unfinished book he prepared for print, Jean-Baptiste became interested in Chinese mathematics and astronomy and moved from there to Indian and Egyptian topics. His linguistic competence was, however, limited and did not include Arabic. Hence, his severe critique of the younger Sédillot's interpretation of Abu l-Wafa''s text relied on translations and prejudices (Dauben and Scriba 2002, 18).

The esteem of scholarly work in Islamicate societies was not limited to astronomy, algebra, arithmetic, and geometry. Before the two Sédillots, another French historian who investigated, among other themes, the Arabo-Latin translations of Aristotle's works, Amable Jourdain (1788–1818), had already declared that the medieval Occident owed "the Arabs" their knowledge of the most important Aristotelian texts and that "the philosophy of Fakhr-eddin, Algazel, Alfarabi, Avicenne, etc., had obtained a great success in the Orient, despite the censorship of (those doctors, who were most closely) attached to the purity of Islamism".[5]

As I will suggest below, the Biot-Sédillot controversy raised awareness of the issue at stake, namely the question of an adequate and fair interpretation of

5 "La philosophie de Fakhr-eddin, d'Algazel, d'Alfarabi, d'Avicenne, etc., avait obtenu un grand succès en Orient, malgré les censures des docteurs les plus attachés à la pureté de l'islamisme." (Jourdain 1819, 111).

scientific achievements in non-Western cultures, an issue that continues to haunt us today. It also motivated serious research, even if on a quantitatively low scale when compared with other themes in history of mathematics that continued through the entire century. Young scholars of the last third of the nineteenth century carried the torch far into the twentieth century. The controversy in its general features found its repetition in the clash between Ernest Renan (1823–1893) and Jamal al-Din al-Afghani (1838–1897) in the 1880s. Without naming names, Renan referred to the Biot-Sédillot controversy and took sides against the latter.[6] The general tone of his lecture was racist and denigrating. He considered Islam as the most authoritarian of the three monotheistic religions and believed that it had always obstructed scientific and philosophical reflection, that nothing of what was called Arabic science and philosophy owed anything to this language or this religion, but everything to ancient Greece.[7] He argued that "religious orthodoxy" had always and everywhere renounced such rationalism, persecuted its practitioners, and vilified its patrons.[8] In his view Ibn Rushd (d. 1198) was the last Muslim philosopher.[9] Renan's false claims dominated academic interpretive perspectives until the last third of the twentieth century. They can be found even today in

6 "La philosophie de Fakhr-eddin, d'Algazel, d'Alfarabi, d'Avicenne, etc., avait obtenu un grand succès en Orient, malgré les censures des docteurs les plus attachés à la pureté de l'islamisme." (Jourdain 1819, 111).

7 Renan (1883): ". . . ce grand ensemble philosophique, que l'on a coutume d'appeler arabe, parce qu'il est écrit en arabe, mais qui est en réalité gréco-sassanide. Il serait plus exact de dire grec; car l'élément vraiment fécond de tout cela venait de la Grèce. On valait, dans ces temps d'abaissement, en proportion de ce qu'on savait de la vieille Grèce. La Grèce était la source unique du savoir et de la droite pensée. La supériorité de la Syrie et de Bagdad sur l'Occident latin venait uniquement de ce qu'on y touchait de bien plus près la tradition grecque. Il était plus facile d'avoir un Euclide, un Ptolémée, un Aristote à Harran, à Bagdad qu'à Paris. Ah! si les Byzantins avaient voulu être gardiens moins jaloux des trésors qu'à ce moment ils ne lisaient guère; si, dès le huitième ou le neuvième siècle, il y avait eu des Bessarion et des Lascaris! On n'aurait pas eu besoin de ce détour étrange qui fit que la science grecque nous arriva au douzième siècle, en passant par la Syrie, par Bagdad, par Cordoue, par Tolède." http://fr.wikisource.org/wiki/L% 27Islamisme_et_la_science (Accessed June 15, 2014).

8 Renan (1883): "La philosophie avait toujours été persécutée au sein de l'islam, mais d'une façon qui n'avait pas réussi à la supprimer. A partir de 1200, la réaction théologique l'emporte tout à fait. La philosophie est abolie dans les pays musulmans. Les historiens et les polygraphes n'en parlent que comme d'un souvenir, et d'un mauvais souvenir. Les manuscrits philosophiques sont détruits et deviennent rares. L'astronomie n'est tolérée que pour la partie qui sert à déterminer la direction de la prière. Bientôt la race turque prendra l'hégémonie de l'islam, et fera prévaloir partout son manque total d'esprit philosophique et scientifique. A partir de ce moment, à quelques rares exceptions près, comme Ibn-Khaldoun, l'islam ne comptera plus aucun esprit large; il a tué la science et la philosophie dans son sein . . . Entre la disparition de la civilisation antique, au sixième siècle, et la naissance du génie européen au douzième et au treizième, il y a eu ce qu'on peut appeler la période arabe, durant laquelle la tradition de l'esprit humain s'est faite par les régions conquises à l'islam. Cette science dite arabe, qu'a-t-elle d'arabe en réalité? La langue, rien que la langue." http://fr.wikisource.org/wiki/L%27Islamisme_et_la_science (Accessed June 15, 2014).

9 Renan (1883): "Quand la science dite arabe a inoculé son germe de vie à l'Occident latin, elle disparaît. Pendant qu'Averroès arrive dans les écoles latines à une célébrité presque égale à celle d'Aristote, il est oublié chez ses coreligionnaires. Passé l'an 1200 à peu près, il n'y a plus un

writings of non-expert academics as well as popular literature. Al-Afghani, a religious reformer and political activist from Iran, engaged in a public debate with the French scholar when he was in Paris. In stark contrast to his rebuttal of Muslim educational reformers in India and their willingness to adopt British sciences as necessary for young Muslim men to master, in his speech to a French public he presented himself as a champion of modern sciences and philosophy and an opponent of all religion. In an almost ironic form, he repeated many of Renan's claims and opposed them, picking out the most ludicrous of them for his critique. He agreed with Renan about the negative impact of his religion on the sciences and philosophy, but rejected the latter's claim that Arabs (and other Semites) were by nature, language, and social form incapable of scientific, philosophical, or metaphysical thought.[10] He emphasized that Christianity and in particular the Roman Catholic Church had also fought against scientific and philosophical theories and continued to do so in the 1880s.[11] Hence the differences between the two religions and the societies they permeated and controlled were not as great as Renan suggested. Islamicate societies were not incapable of modernization and Muslims could overcome the negative aspects of their religion.[12] Al-Afghani's texts against

seul philosophe arabe de renom." http://fr.wikisource.org/wiki/L%27Islamisme_et_la_science (Accessed June 15, 2014).

10 "In truth, the Muslim religion has tried to stifle science and stop its progress. It has thus succeeded in halting the philosophical or intellectual movement and in turning minds from the search for scientific truth. . . . The Arabs, ignorant and barbaric as they were in origin, took up what had been abandoned by the civilized nations, rekindled the extinguished sciences, developed them and gave them a brilliance they had never had. Is not this the index and proof of their natural love for sciences? It is true that the Arabs took from the Greeks their philosophy as they stripped the Persians of what made their fame in antiquity; but these sciences, which they usurped by right of conquest, they developed, extended, clarified, perfected, completed, and coordinated with a perfect taste and a rare precision and exactitude." (Al-Afghani 1968, 176–177, https://disciplinas.stoa.usp.br/pluginfile.php/2004379/mod_resource/content/1/KEDDIE%2C%20Nikki.pdf) (Accessed Oct. 20, 2016).

11 "A similar attempt, if I am not mistaken, was made by the Christian religion, and the venerated leaders of the Catholic Church have not yet disarmed, so far as I know . . . Besides, the French, the Germans, and the English were not so far from Rome and Byzantium as were the Arabs, whose capital was Baghdad. It was therefore easier for the former to exploit the scientific treasures that were buried in these two great cities. They made no effort in this direction until Arab civilization lit up with its reflections the summits of the Pyrénées and poured its light and riches on the Occident. The Europeans welcomed Aristotle, who had emigrated and become Arab; but they did not think of him at all when he was Greek and their neighbor. Is there not in this another proof, no less evident, of the intellectual superiority of the Arabs and of their natural attachment to philosophy?" (Al-Afghani 1968, 177, https://disciplinas.stoa.usp.br/pluginfile.php/2004379/mod_resource/content/1/KEDDIE%2C%20Nikki.pdf) (Accessed Oct. 20, 2016).

12 "If it is true that the Muslim religion is an obstacle to the development of sciences, can one affirm that this obstacle will not disappear someday? How does the Muslim religion differ on this point from other religions? . . . Religions, by whatever names they are called, all resemble each other. No agreement and no reconciliation are possible between these religions and philosophy. Religion imposes on man its faith and its belief, whereas philosophy frees him of it totally or in part." (Al-Afghani 1968, 177, https://disciplinas.stoa.usp.br/pluginfile.php/2004379/mod_resource/content/1/KEDDIE%2C%20Nikki.pdf) (Accessed Oct. 20, 2016).

the so-called naturalists in India and against Renan's untenable evaluation of the sciences and philosophy in Islamicate societies of the past have seen a revival in the last decades in different academic and non-academic circles, whose method often consists in (very simplified) critiques not merely of the colonial period, but of contemporary policies of Western countries, in particular in the Middle East as well as today's sciences, social sciences, the humanities, and other cultural features of Western societies.

3 Historiographical approaches of nineteenth-century writers to the mathematical sciences in Islamicate societies

The main historiographical approaches to writing history of mathematics in Germany during the nineteenth century, shared by writers on every period and mathematical culture, included the study of primary sources, the search for a periodization of the history of mathematics based on philosophical concepts and principles, and the "kulturhistorische" approach (Dauben and Scriba 2002, 115–117, 122–124). As for Islamicate societies, the representatives of these three approaches were Georg Heinrich Ferdinand Nesselmann (1811–1881), Franz Woepcke (1826–1864), and Moritz Cantor (1829–1920). Nesselmann and Woepcke pursued at least two of them in their research practice, while Cantor employed all three when narrating history of mathematics in a universal style.

The most influential French author of treatises on Arabic astronomy, instruments, and mathematics, although not the first nor the last author of such texts in nineteenth-century France, was the already-mentioned Louis-Amélie Sédillot (Charette 1995, 93–94). He also focused on the study of primary source material and scientific progress as achieved by scholars from Islamicate societies. But, going beyond the themes and methods of Nesselmann and Woepcke, he used codicological methods and wrote about or, better, against Indian and Chinese contributions to scientific knowledge.

3.1 Nesselmann's views on how to do history of algebra

In the dedication of his *Versuch einer kritischen Geschichte der Algebra*, Nesselmann characterized his turn to the history of mathematics, in particular algebra, as the result of the "spirited" teaching of mathematics by his high school teacher C.F. Buchner and one of his two university professors of mathematics, Carl Gustav Jacob Jacobi (1804–1851) (Nesselmann 1842, viii). Folkerts, Scriba, and Wußing identified explicit historiographical statements in the preface to this book as influenced by Johann Gottfried von Herder's (1744–1803) philosophical ideas on culture and the new German school of historicism, in particular the works of Barthold Georg Niebuhr (1776–1831) and Leopold von Ranke (1795–1886) (Dauben and Scriba 2002, 115). The book's title and the emphasis on the chosen critical approach to the past confirm Nesselmann's conceptual as well as rhetorical connection to historicism:

> I intended, as the title says, to write a critical history; I did not wish to teach the history of algebra according to tradition, but as it results from a persevering and careful study of the sources and then present it accordingly with fidelity.[13]

His historiographical commitment in this and other passages to the study of primary sources strengthens the bond with the new historical school in the realm of methods.

The first sentences in Chap. I offer a different kind of vocabulary, one that is indeed strongly reminiscent of Herder's philosophical doctrines:

> At the moment when the first seeds of historical consciousness awake in a people, the desire stirs towards a national history, which will bring this self-consciousness to completion. A people without history is a head that does not see the body on which it rests; it is a present that lacks the consciousness of a true past. In this (situation of an) onerous sentiment of emptiness and unconsciousness of a previous existence the historical element evolves at first in the form of the national epic among peoples, as soon as they reached a certain level of intellectual culture; from there it moves slowly to true history . . . This very same desire for history, which ensouls uprising peoples, can also be discovered among the admirers and propagators of the sciences and the arts. This is all the clearer and more definitive, when the material for further formation of the latter arises from within. Thus, the progress of theology, mathematics, and poetry was first worked out in a historical (form).[14]

13 "Ich wollte, wie der Titel besagt, eine kritische Geschichte schreiben, ich wollte die Geschichte der Algebra, nicht wie die Tradition sie lehrt, sondern wie sie sich aus dem ausdauernden und gewissenhaften Studium der Quellen ergiebt, erforschen und demgemäß treu darstellen." (Nesselmann 1842, x–xi).

14 "Sobald in einem Volke die ersten Keime des Selbstbewußtseins erwachen, regt sich in ihm das Bedürfniß nach einer Nationalgeschichte, welche jenes aufstrebende Selbstbewußtsein zur Vollendung bringe. Ein Volk ohne Geschichte ist ein Kopf, welcher den Körper nicht sieht, auf dem er steht; es ist eine Gegenwart, der das Bewußtsein fehlt, daß sie auf einer thatsächlichen Vergangenheit beruht. In diesem drückenden Gefühle der Leere und der Bewußtlosigkeit früherer Existenz entwickelt sich bei den Völkern, sobald sie auf eine gewisse Stufe der geistigen Cultur getreten sind, zuerst das historische Element in der Form des Nationalepos und geht von da allmählig in wirkliche Geschichte über . . . Dieses nämliche Bedürfniß nach Geschichte, welches emporstrebende Völker beseelt, läßt sich auch unter den Verehrern und Fortbildnern der Wissenschaften und Künste wahrnehmen, und zwar um so deutlicher und bestimmter, je mehr den letzteren der Stoff zu ihrer Fortbildung von innen her zukommt; am frühesten sind daher die Fortschritte in der Theologie, der Mathematik und der Poesie geschichtlich bearbeitet worden." (Nesselmann 1842, 1).

But while positions of historicism permeated Nesselmann's research practice, Herder's philosophy remained a distant point of orientation that gave some rhetorical structure to the mass of details newly acquired from primary sources. The overall structure chosen by Nesselmann for presenting his critical history of algebra, namely, does not follow Herder's ideas of people and historical consciousness, but mathematical topics and methods that are ascribed to either peoples or periods (Nesselmann 1842, 30–34).

In addition to these two methodological and philosophical inspirations of Nesselmann's work a third motivation can be recognized in his desire to define a well-delineated historiographical working methodology with specific methods. The discussion of his predecessors in history of mathematics in Chap. I of *Versuch einer kritischen Geschichte der Algebra* leaves no doubt that Nesselmann perceived severe shortcomings in their works. These issues were of immense relevance to him. He decided to break with their approaches and methods and to make explicit what he considered a reliable and proper research as well as writing practice. He praises, for instance, Proclus (c. 412–485) for

> dealing in detail with a subject, explaining fully the theoretical aspects of the most important theorems and doctrines, and presenting the content of mostly lost books in such precision and detail that the more recent mathematicians were enabled by this information to reconstruct those works,[15]

while chastising his immediate predecessors for mainly offering a catalogue of names and titles. The mistakes committed by the latter are so numerous that Nesselmann saw himself obliged to instruct future writers to learn how to quote correctly, to quote only from a primary source, if one had read it, and to avoid substituting modern forms for older thoughts, methods, or results (Nesselmann 1842, 35, 37–38). Finally, he argued that his predecessors had mistaken three different tasks for one: to write the biography of a mathematician; to compile a bibliography; to write in a scientific manner about the history of mathematics (Nesselmann 1842, 38). His declared intention was to write a scientific history of mathematics. That is why he asked his readers not to expect names, but things and not to demand a complete list of books, but to accept as sufficient the description of the content of those few books that had an impact on the progress of mathematics (Nesselmann 1842, 39). He did not believe to have invented this historiographical approach, a feat he ascribed to Jean-Étienne Montucla (1725–1799), but to have been the first who strictly adhered to these principles and worked hard to complete the entire project (Nesselmann 1842, 18, 39).

15 "(er begnügt sich aber nicht, wie so viele neuere Historiker, bloße Namen und Büchertitel zu nennen,) sondern er geht ausführlich in die Sache ein, führt die wichtigsten Sätze und Lehren theoretisch vollständig aus, und giebt den Inhalt von vielen, jetzt größtentheils verloren gegangenen Büchern so genau und ausführlich an, daß es neueren Mathematikern möglich geworden ist, nach diesen Angaben jene Werke wiederherzustellen." (Nesselmann 1842, 7).

An important point to be raised is whether Nesselmann applied his historiographical ideas to his investigations of algebra and arithmetic in Islamicate societies, and, if so, how. This question can be discussed only with the utmost restriction since Nesselmann never published more than the first volume of his critical history of algebra. His translation of Baha' al-Din al-'Amili's *Essence of Arithmetic* contains only a few brief remarks about his goals in publishing the work, a justification of his use of a printed rather than a manuscript version, and a short biographical sketch of the Safavid author (Nesselmann 1842, iii – v, 74–76). In his discussions of the names of algebra, the various number systems, and their symbolic representations he works indeed with Arabic and Persian primary sources in manuscript as well as printed form and presents in his footnotes the translated quotes in their original language, if they are more than a mere phrase or between parentheses in the text, if they are single words or short expressions (Nesselmann 1843, 42–43, 45–51, 73, 78–79 etc.). The scarcity of his source material though is remarkable. In addition to al-'Amili's *Essence of Arithmetic*, a Persian translation with commentary by Rawshan 'Ali Jawnpuri (nineteenth century), and a didactic poem by some Najm al-Din 'Ali Khan, like Jawnpuri a Muslim scholar from India, Friedrich August Rosen's (1805–1837) publication and translation of al-Khwarizmi's algebra, and apparently Thabit b. Qurra's (d. 901) translation of Nicomachus's (2nd c) *Introduction to Arithmetic*, Nesselmann refers only to a very small number of secondary sources such as Silvestre de Sacy's (1758–1831) Arabic grammar.

Despite the brevity of Nesselmann's own remarks in his reprint and translation of al-'Amili's *Essence of Arithmetic*, its preface is relevant to the discussion of his historiographical stance. It makes clear that he thought that editing a text meant working from manuscripts and using more than a single one, if possible (Nesselmann 1843, iii). The guiding model of classical studies is clearly recognizable. Secondly, Nesselmann justified his publication of the Safavid text primarily because of its information on what scholars in Islamicate societies had done with their intellectual heritage since Muhammad b. Musa al-Khwarizmi had written the first algebraic text in Arabic in the early ninth century (Nesselmann 1843, iii – iv). Only at the end, in his biographical excursus, did he also point to the treatise's function as a textbook and its dominant place in Indian madrasas (Nesselmann 1843, 75). This perspective corresponds well with his goal of producing a scientific history of mathematics that looks for progress and not for socio-cultural roles and functions of knowledge or any other property such knowledge might have. In a sense it also sets the tone that was to dominate research and interpretation of the mathematical sciences in Islamicate societies until very recently.

The reputation that Nesselmann's work acquired among his contemporaries is nicely summarized in the short biography by Moritz Cantor:

> Scientific fame came to him, however, first in a work of the year 1842, his "Algebra of the Greeks". It was the first work in German since Kästner that treated mathematical-historical things on the basis of proper

> research; but in erudition, critical insight, and comparative investigative power it ranks far above all German predecessors. Only Chasles, "History of Geometry" (1837) and Libri, "History of Mathematics in Italy" (1838–1841) can be compared with Nesselmann's "Algebra of the Greeks" and form with it the three exemplary models that all successors on the same domain learned from and were inspired by.[16]

3.2 Louis-Amélie Sédillot

The main interpretive issues that were discussed in the nineteenth century concerned the value of mathematical and astronomical contributions by scholars from Islamicate societies, the relative importance of ancient Greek and ancient or early medieval Sanskrit texts for these contributions, and the sciences in Arabic in general. Contributions in Persian or Ottoman Turkish received much less attention.

The discussions about values and evaluation did not begin in the nineteenth century, but have a history that reaches centuries back. The fierceness of the conflicts that arose in particular in France was, however, more intense than ever, with the exception perhaps of the sixteenth century.

Louis-Amélie Sédillot, like his father, spoke out in favor of Arabic ingenuity and was happy to support his views with the authority of Alexander von Humboldt, with whom he also shared his research results. He expressed with verve a position that found supporters as well as opponents throughout the century (Charette 1995, 105–106, 121–139). Among the opponents were scholars such as Guillaume Libri (1803–1869) and Jean-Baptiste Biot (1774–1862). In general terms, Sédillot described his beliefs as follows in the preface to his *Matériaux pour servir à l'histoire des sciences mathématiques chez les Grecs et les Orientaux*:

> . . . from the ninth to the thirteenth century, one sees the formation of one of the most vast literatures that exist; manifold productions, precious inventions attest to the marvelous (intellectual activities) and making felt their action on Christian Europe, they seem to justify the opinion that *the Arabs were in all matters our masters*.[17]

Given the recent French colonial conquests of Arab territories in North Africa and the strong belief held by many French and other European academics in the

16 "(er begnügt sich aber nicht, wie so viele neuere Historiker, bloße Namen und Büchertitel zu nennen,) sondern er geht ausführlich in die Sache ein, führt die wichtigsten Sätze und Lehren theoretisch vollständig aus, und giebt den Inhalt von vielen, jetzt größtentheils verloren gegangenen Büchern so genau und ausführlich an, daß es neueren Mathematikern möglich geworden ist, nach diesen Angaben jene Werke wiederherzustellen." (Nesselmann 1842, 7).

17 ". . . du neuvième au treizième siècle, on voit se former une des plus vastes littératures qui existent; des productions multipliées, des précieuses inventions attestent l'activité merveilleuse des esprits, et faisant sentir leur action sur l'Europe chrétienne, semblent justifier l'opinion que *les Arabes ont été en tout nos maîtres*." (Sédillot 1845–49, vol. 1, iii).

exclusive superiority of ancient Greek sciences, such a bold statement was bound to encounter scorn and rejection among some factions of academia. While in its generality it was certainly more correct than the opposite position that limited the historical role of "Arabic" scholars to the translation of Greek mathematical and scientific works and their preservation until their translation into Latin in the twelfth and thirteenth centuries, one of Sédillot's main specific arguments, the invention of the lunar variation by Abu l-Wafa', was certainly false.

Convinced of the righteousness of his approach, and based on his father's as well as his own results, Sédillot proposed to prove in his book that ". . . the School of Baghdad was able to surpass the Schools of Athens and Alexandria".[18] The main content of the first volume consists in Sédillot's effort to rebuff those who did not accept his interpretation of Abu l-Wafa''s statement. But much more was at stake than the correct or incorrect interpretation of a few lines in an Arabic text of the tenth century. The quarrel was not limited to a comparative up- or downgrading of ancient Greeks, medieval Arabs, and modern Europeans. It also included ancient and medieval Indians and Chinese. The questions, simply formulated, were who had created a scientific astronomy and who had contributed what to such a science? Sédillot ardently defended the right of the "Arabs" to be considered as major contributors to science, while denying such a right to the Chinese and the Indians, and diminishing that of the Greeks. Biot defended the scientific valor of the Greeks, recognized Chinese astronomy as scientific, and argued vehemently "Arabs", whom he described with a slightly altered and, according to Sédillot, anonymous quote from Cervantes's *Don Quijote* as people of whom one could not expect a single truth, because they were embezzlers, falsifiers, and liars.[19] This usage of literary fantasy, which in addition ignores (of course) the various expressions of appreciation for Islamicate cultures and their people found in Cervantes's works, as a scholarly argument indicates the expanse of what was considered feasible in discussing a specific and limited piece of astronomical knowledge. Emotions obviously ran high in French academia in the middle of the nineteenth century. Doubts and suspicions also characterize Libri's more academic objections against Sédillot's "discovery". He doubted that Abu l-Wafa' had undertaken celestial observations (Sédillot 1845–49, vol. 1, 100). He asked why no later astronomer had mentioned the third inequality, if Abu l-Wafa' truly had discovered it (Sédillot 1845–49, vol. 1, 51).[20] Others wondered why the geometrical methods used by Abu l-Wafa' and Tycho Brahe were so close to each other. They suggested that this either meant that Brahe had known of Abu l-Wafa''s discovery or that Abu l-Wafa''s text had been modified or even falsified by a scribe in the seventeenth century (Sédillot 1845–49, vol. 1, 51, 54, 66–67, 87, 91, 93, 100, 115).

18 ". . . l'École de Bagdad a su dépasser les Écoles d'Athène et d'Alexandrie." (Sédillot 1845–49, vol. 1, v).

19 ". . . de los moros no se puede esperar verdad alguna, porque todos son embelecadores, falsarios y chimeristas." (Sédillot 1845–49, vol. 1, 119).

20 For a rejection of Sédillot's interpretation of the Arabic text see (Carra de Vaux 1892).

Sédillot's answers to these points came from three practices: (1) a careful study of the manuscript he worked with and its para-textual information about copying dates, scribal hands, ownership marks, seals, and material features like binding and paper as well as other manuscripts, which offered elements for rebutting the objections; (2) the study of new publications on various points of Islamicate history and art objects from Islamicate societies, which were related to different features of the manuscript; (3) general historical arguments about the relationship between ancient Greek, medieval Arabic, and early modern European astronomy and geometry (Sédillot 1845–49, vol. 1, 52–58, 69, 72–73, 101–106, 116–117 et al.). Current codicologists and, in a more limited manner, also historians of the mathematical sciences in Islamicate societies check the same kind of items when they wish to determine age, provenance, or pathways of any given manuscript through different libraries. Hence, while Sédillot certainly erred and went too far in his efforts to defend his father's reputation and to establish the recognition of the scholarly value of medieval authors of Arabic astronomical and mathematical treatises, he also was one of the pioneers in applying basic codicological skills to the evaluation of the content of such texts. The various points made in this debate indicate the difficulties the participants faced when trying to interpret and evaluate different kinds of data. Prejudices and presuppositions played an important role in these processes as did the lack of technical skills, codicological and other experiences as well as historiographical methodologies (Sédillot 1845–49, vol. 1, 56–57, 67–69, 71–72, 76, 112, 115). In addition to the efforts undertaken by Sédillot to establish a fundament for a history of the mathematical sciences in Islamicate societies that rested on philology, codicology, history, art history, the sciences, the study of the materiality of manuscripts and other relevant objects, and bibliography, he also formulated research tasks that were taken up in the late nineteenth and in particular in the second half of the twentieth centuries, although some of them still await their students: edition, translation, and analysis of major Arabic astronomical texts and tables, in particular those by al-Battani and Ibn Yunus; edition, translation, and analysis of the mathematical works of "the School of Baghdad" (a term used repeatedly by Sédillot) and the astronomical observations carried out in the Abbasid capital during the ninth and tenth centuries; study of regional contributions to astronomy and mathematics, in particular in al-Andalus, the Maghrib, Egypt, Iran, and Transoxania; study of tables and mechanical, astronomical, and mathematical instruments (Sédillot 1845–49, vol. 1, 134–139, 274–364). On the other hand, he also gave voice, as did other French students of Arabic, Persian, or Ottoman manuscripts, to strong and often declarative (i.e., little investigated, historiographical positions), which dominated approaches during the entire twentieth century. According to him, the imagined interpolation of the third inequality into the eleventh-century copy of Abu l-Waf'a's text could not have been executed by an "Arab of the seventeenth century". His argument was the lack of awareness of the "Turks and the Arabs subjugated by them" of the scientific knowledge of the Europeans that made "most of them even today believe in the immobility of the earth" (Sédillot 1845–49, vol. 1, 56). In Sédillot's view, the end of scientific curiosity and activity occurred in the fifteenth century with the death of Ulugh Beg (Sédillot 1845–49, vol. 1, 270). As for the beginning of the mathematical sciences, in particular astronomy, Sédillot locates it in translations of Sanskrit texts, but believes that the shift to Greek theories, parameters, and methods was a quick and easy process, finished already around 820, due to the inferiority of the Indian material (Sédillot 1845–49, vol. 2, 440). He believed, like many historians of the mathematical sciences and scholars of Oriental matters, in the power of progress and scientific truth.

3.3 Woepcke's goals, claims, and methods

Franz Woepcke pursued a different set of goals and ideas than Nesselmann (Narducci 1869). As I will show below, Woepcke was not interested in progress as such, but in the contributions of scholars from Islamicate societies to the mathematical sciences. This nuance might appear insignificant, but was of importance for Woepcke's historiographical practice. It enabled him to look critically at various interpretations of ancient Greek mathematical achievements proposed by his contemporaries. He introduced new questions, worked consequently with primary sources, and systematically pursued the question of what scholars from Islamicate societies had contributed to the mathematical sciences without trying to equate their achievements with nineteenth-century developments. Woepcke's historiographical approach, as expressed in his texts, reflected three elements: (1) the severe conflicts that shook Paris in the first half of the nineteenth century in regard to the history of astronomy and mathematics; (2) Alexander von Humboldt's interests in the achievements of scholars from Islamicate societies in general and the genesis of number and calculation systems in particular; (3) beliefs and values of mathematicians in Germany and France in the middle of the nineteenth century.

By contrast, Woepcke's texts show no clear or substantial relationship with the historiographical positions of his last major patron, Baldassarre Boncompagni (1821–1894), for whom he wrote several treatises and who published and republished a number of his works. Boncompagni's interests in the history of mathematics had shifted in this period primarily "to reconstructing the chronology and the channels of the transmission of mathematical knowledge from the Arabic world to Christian Europe" (Mazzotti 2000, 260). Woepcke was one of Boncompagni's main collaborators with regard to Arabic mathematical manuscripts and the appropriation of Arabic and Indian contributions. The two exchanged some 164 letters, which so far have not been analyzed for the historiographical similarities and differences between the two men.

Given Mazzotti's ascription of Boncompagni's rejection of a synthetic history of mathematics, of evaluations and judgments, and his conscious limitation to library studies, editions, and translations of mathematical texts and a problem-oriented research to his conservative political and religious convictions, the question arises in how far Woepcke's views and working practice had similar features and were due to comparable political, religious, and academic beliefs.

The obituary written by Hippolyte-Adolphe Taine (1828–1893), according to his own testimony a close friend of Woepcke, paints a picture of Woepcke's historiographical convictions that goes beyond the positions expressed by the scholar himself in his papers. It agrees in several points with the attitudes attributed by Mazzotti to Boncompagni: the rejection of a synthetic history of mathematics as unattainable, the insistence on the need to restrict research for a long time to the study of manuscript texts and the accumulation of factual knowledge, and severe critique of speculations and superficial judgments.[21] Taine situates these points, however, exclusively in Woepcke's character traits without even the smallest hint at any other reason behind these beliefs. Moreover, Taine denies that these beliefs signified a strong rejection of synthesis and theoretical reflection by Woepcke. On the contrary, Taine forcefully stresses Woepcke's interest in and adoration of, as he calls it, metaphysical conceptions of history.[22] Hence, the question must remain unanswered and a task for future research as to what shaped Woepcke's historiographical views beyond the three factors that Woepcke explicitly expressed in his publications.

21 ". . . il ne s'était engagé dans les recherches limitées et dans les questions particulières que par une aversion naturelle pour les considérations vagues, et parce qu'il regardait ces travaux bornés et concentrés comme la meilleure discipline de l'esprit . . . Il entrevoyait dans l'avenir, pour la fin de sa vie, une histoire générale des mathématiques, du moins depuis leurs origines dans l'Inde jusqu'à la renaissance. Mais il n'y comptait guère: "On se donne cette espérance à soi-même, me disait-il, c'est pour s'encourager. Mais c'est là une illusion d'esprit; le travail est trop grand, et la vie d'un homme est sujette (sic) à trop de chances." – "Je ferais bien un système, ajoutait-il une autre fois, il n'y faudrait qu'un peu d'invention, et peut-être en suis-je capable comme un autre; mais à quoi bon, puisque mon système ne serait pas prouvé, et pourquoi perdrais-je mon temps à me duper moi-même avec des phrases?" Il pensait que les jugements d'ensemble sur l'ancienne histoire des mathématiques et sur le passage des sciences anciennes aux sciences modernes doivent demeurer encore en suspens pour un ou deux siècles. Il comparait les connaissances que nous avons aujourd'hui sur la science et la civilization arabes aux celles que nous avions au seizième siècle sur la science et la civilization grecques, et croyait que pendant bien longtemps tout travail fructueux doit se réduire comme au siècle de Casaubon et de Scaliger, à la publication des manuscrits . . . Le trait le plus marquant de son esprit était la haine du charlatanisme; il ne devenait moqueur et caustique que sur ce point; et quand il mettait le doigt sur les prétensions et l'insuffisance de quelques contemporains, ses petits exposés de faits, si exacts et d'apparence si sèche, arrivaient au plus haut comique. Pour ce qui est de lui-même, il était toujours prêt à se réduire, même à se rebaisser . . . Son plus vif désir était de n'être jamais dupe de lui-même; il tenait toujours dans sa main une balance pour peser ses opinions; il ne voulait rien admettre que de vrai et de prouvé, et préférait l'ignorance aux conjectures. Il avait un sentiment profond de l'imperfection de nos sciences, des limites de chaque esprit, des bornes du sien entre tous les autres." (Taine 1866, 385–387, 389).

22 "Quoiqu'il eût aimé passionnément la métaphysique, il l'avait laissée derrière lui et la considérait seulement comme une façon commode de grouper les faits, comme un système provisoire, utile pour tirer l'esprit des recherches spéciales et pour le diriger vers les vues d'ensemble . . . Non qu'il fût sèchement positiviste ; il suivait avec intérêt et sympathie les hautes constructions idéales que l'on essaye d'élever sur ces rares soutiens; et il estimait que chacun doit essayer ou esquisser la sienne, et il jugeait qu'après tout le plus noble emploi de science est de fournir matières à ses divinations grandioses par lesquelles, en dépit de nos erreurs et de nos doutes, nous prennons part aux contentements et à l'œuvre des siècles qui nous suivront." (Taine 1866, 390).

3.3.1 The impact of Louis-Amélie Sédillot's appreciation of Abu l-Wafa' and other scholars from Islamicate societies on Woepcke's methodological stance

Woepcke's historiographical practice cannot be isolated from the positions formulated and contested in Paris during the first half of the nineteenth century. The impact of the harsh debates on whether "the Arabs" had contributed anything to human scientific progress beyond the "preservation" of ancient Greek texts cannot be missed in the publications of French scholars during Woepcke's stay in Paris. As his first publication shows, Woepcke's approach was a continuation and extension of that proposed by Sédillot. In this work on 'Umar Khayyam's (d. ca. 1123) algebra, Woepcke states:

> The works of the illustrious mathematicians, whom Greece produced during a period of six centuries, have been, practically without interruption, the object of scholarly works. Since the beginning of the Middle Ages until our days, they have been translated, commented on, published, often by geometers who themselves were very famous. It suffices here to remember the names of Nassir eddin al Thusi, of Bachet de Mezériac, of Halley. The important discoveries during a similarly long period through which the Arab genius has enriched the same science have not been as fortunate in attracting an equal attention. Some have even gone so far as to claim that the Arabs in general have invented nothing or almost nothing beyond that which they scooped from Greek authors translated into Arabic since the time of the caliphs Haroun Alrachid and Almamoun. Careful and extended research will probably lead to markedly different results.[23]

Woepcke hoped to offer clear evidence for his claim. Comparing Khayyam's work to those of Muhammad b. Musa al-Khwarizmi's algebra and Baha' al-Din al-'Amili's arithmetic, the only two Arabic treatises with chapters on algebra published in a European language during the first half of the nineteenth century,

23 "Les œuvres des illustres mathématiciens, que la Grèce a produits pendant l'espace de six siècles, ont été presque continuellement l'objet de travaux savans. Dès le commencement du moyen age, jusqu'à nos jours, elles ont été traduites, commentées, publiées, souvent par des géomètres, qui eux-mêmes avaient une haute célébrité. Il suffira ici de rappeler les noms de Nassir eddin al Thusi, de Bachet de Mezériac, d'Halley, Les découvertes importantes par lesquelles le génie Arabe pendant une période d'une semblable durée, a enrichi la même science, n'ont pas été assez heureuses pour s'attirer une pareille attention. On est même allé jusqu'à soutenir, que les Arabes n'avaient en général rien inventé, ou presque rien au-delà de ce qu'ils avaient puisé des auteurs Grécs, traduits en Arabe depuis le temps des khalifes Haroun Alrachid et Almamoun. Des recherches soigneuses et étendues meneront probablement à des résultats fort différens." (Woepcke 1850, 160).

Woepcke left no doubt as to Khayyam's more advanced mathematical level.[24] In his view, Khayyam's higher scholarly rank is not only expressed in the more complicated mathematical problems and methods, but also in his reliance on Aristotelian philosophy in the definitions of his new theory and in his methodology.[25] The latter Woepcke summarizes as focusing on true difficulties, while ignoring questions of lesser relevance and quoting contemporary scholars whose errors he occasionally corrected.[26] Woepcke's main praise for Khayyam's treatise consists in acknowledging that it pushed back the boundaries of a science.[27] Here we find major themes of research that have continued to stimulate research interests until the end of the twentieth century.

Woepcke also spoke out against premature and inappropriate evaluations of the intellectual achievements of scholars in Islamicate societies during those centuries offered by his contemporaries.

> By comparing the treatises of Mohammed Ben Moûçâ and Behâ Eddìn, Colebrooke arrived at the conclusion (Algebra of the Hindus, Dissertation, p. LXXIX) that algebra had remained almost stationary in the hands of the Musulmans. Would it not be the same to put into doubt the discoveries of Apollonius, Archimedes, Diophant, because neither Euclid's Elements nor Marcianus Capella's "Nuptials of philology and Mercury" gives us knowledge of the most beautiful monuments, which Greek geometry has left (to us)? No, the mathematical sciences did not remain stationary in the Orient in the time from Mohammed Ben Moûçâ until Behâ Eddìn; they took, in an intermediary epoch, an upsurge and a development worthy of true admiration.[28]

24 "Je remarque qu'en général on doit accorder à Alkhàyàmî un rang supérieur à celui de Mohammed Ben Mousa ou de Behà-Eddìn, vue que les triatès de ceux-ci n'ont pour but que l'instruction des commençans, tadnis que celui d'Akhayàmî porte un caractère plus élevé, . . ." (Woepcke 1850, 161). Compare also (Woepcke 1851, xix).

25 ". . . contenant les définitions des notions fondamentales de cette science. Ces définitions sont assez intéressantes, parcequ'elles font voir combien la philosophie d'*Aristote* a influé sur la science Arabe; . . ." (Woepcke 1850, 161).

26 ". . . effleurant seulement les questions d'une portée inférieure; appuyant sur les difficultés réelles, citans les travaux contemporains, corrigeant patfois leurs erreurs, . . ." (Woepcke 1850, 161–162).

27 "Ce n'est pas un des livres, qui reproduisent ce qu'on sait dans une science, mais un de ceux qui en reculent les bornes." (Woepcke 1850, 162).

28 "En comparant entre eux les traités de Mohammed Ben Moûçâ et de Behâ Eddìn, Colebrook était arrivé à la conclusion (Algebra of the Hindus. Dissertation. P. LXXIX), que l'algèbre était resté à peu près stationnaire entre les mains des musulmans. Ne serait-on pas également fondé à mettre en doute les découvertes d'Apollonius, d'Archimède, de Diophante, parce que ni les *Éléments* d'Euclide, ni les "Noces de la philologie et de Mercure" de Marcianus Capella, ne nous font connaître les plus beaux monuments qu'ait laissés la géométrie grecque? Non, les mathématiques ne sont pas restées stationnaires en Orient depuis Mohammed Ben Moûçâ jusqu'à Behâ Eddìn; elles ont pris, à une époque intermédiaire, un essor et un développment dignes d'une véritable admiration." (Woepcke 1851, xix).

3.3.2 Humboldt's questions as sources of inspiration for Sédillot and Woepcke

In Louis-Amélie Sédillot's view, Humboldt had acknowledged explicitly in the *Kosmos*, "a book that France, in a period of political agitation, has not yet fully appreciated, . . . the services that the Arabs have rendered to civilization and at the same time gave an indication of all that which one should expect from future studies (if they) are capably undertaken".[29] Humboldt encouraged Woepcke to study Arabic in order to bring to light these expected achievements. Summarizing the most important ideas, problems, methods, and results was one of the strategies by which Woepcke tried to fulfill this task in several of his most extensive articles. He stressed their programmatic interconnectedness and his goal of proving that scholars from Islamicate societies went beyond the ancient Greek writers in their mathematical activities.

Woepcke expressed his appreciation for and devotion to Humboldt very early in his career by translating the latter's essay *Über die bei den verschiedenen Völkern üblichen Systeme von Zahlzeichen und über den Ursprung des Stellenwertes in den indischen Zahlen* (1829) into French (Woepcke 1851). Six of Woepcke's articles discuss subjects related to Humboldt's treatise, although Boncompagni's patronage was essential for at least half of them (Woepcke 1853, 1854, 1855, 1863, 1865–66).

Humboldt's patronage of Woepcke and his demands for uncovering innovative results in Arabic texts were of an undeniably great stimulus for the young researcher. However, these demands also produced undesirable side effects. In particular the refusal to edit and translate Arabic or Persian texts as a whole and preference for extracts and summaries led to a focus on the mathematical content only and its quick identification with "modern" formulas. Woepcke's working practice was oriented towards excavating highlights to the detriment of any effort to study the many texts available to him in the Imperial Library from at least a limited historical perspective, for instance as groups of interrelated works and authors.

3.3.4 The impact of values and beliefs of nineteenth-century mathematicians on Woepcke's approach to Arabic texts

Woepcke considered mathematicians and perhaps school teachers as a main part of his public. It is to such readers that he offers his discussion of Arabic texts. He tries to please them, adapts his writing style and explanations to them, and apologizes when he thinks they might be displeased. This is not limited to his

29 ". . . a livre que la France, à une époque d'agitations politiques, n'a pas encore suffisament apprécié, . . . (le *Cosmos* de M. de Humboldt trace un tableau impartial des) services que les Arabes ont rendus à la civilisation, et laisse en même temps pressentir tout ce qu'on doit attendre de recherches ultérieures habilement dirigées." (Sédillot 1845–49, vol. 2, ii–iii).

publications in mathematical journals, but also in those of Oriental matters like the *Journal Asiatique*.[30] Typical of Woepcke's argumentation is his reliance on mathematical issues for explaining or criticizing al-Khayyam's concepts, methods, and failures.[31] In his comparative studies, Woepcke exclusively focuses on the mathematical content of a problem or the kind of a method in order to determine whether they are of the same style, similar, or profoundly different. Hence, comparisons of texts, even if they are of a later period than the Arabic or Persian ones he analyzed, are at the heart of his working practice. Questions of how such problems or methods were disseminated, in case they resembled each other, and which historical conditions favored such a transfer of knowledge, to name only two issues of relevance to his comparative studies, played no role in his work.

This relative narrow set up of historical comparison does not mean, however, that Woepcke drew facile, superficial conclusions from his mathematical observations, although he always stressed the need to recognize the intellectual productivity of the scholars whose texts he analyzed.[32] His main argument for the Indian origin of two of Abu l-Wafa''s problems and solutions is that "they are quite noticeably detached from the spirit of the Arabic geometry, which is always faithful, with regard to the form, to its Greek models".[33] His second argument is the judgment of Chasles, who had declared in 1837 that a certain method in Bhaskara's arithmetic was "indeed of Indian origin", but without any comparison with Abu l-Wafa''s text, which was not yet made accessible.[34] Woepcke's praise for Chasles leaves no doubt about his devotion to the French mathematician. It also shows that both found it completely acceptable to make sweeping evaluations on a very thin basis of evidence.

This kind of macro-historical judgment does not necessarily yield wrong results. But it ignores too many questions that need to be studied if we wish to understand

30 An example for this apologetic style is the following: "Comme un reproduction de ces démonstrations aurait décuplé l'étendue de cette notice, j'ai dû me borner à ne donner que les énoncés des propositions, vu le peu d'espace que ce Journal peut accorder à des publications de ce genre. Mais pour satisfaire les géomètres, j'ai placé en note des démonstrations de ces propositions en me servant de la notation algébrique moderne, où le plus souvent la démonstration se réduit à la simple inspection d'une identité." (Woepcke 1852, 422).

31 "Toutefois, il est très-suprenant qu'Alkhayyâmi, en construisant les équations du troisième degré, n'ait pas remarqué l'existence des racines négatives . . . C'est la vicieuse habitude de ne tracer que des demi-cercles, des demi-paraboles, et une seule branche des hyperboles, qui a fait manquer au géomètre arabe cette belle découverte . . . Les Arabes savaient déjà qu'il existait une certaine équation du second degré à deux racines . . . ; si donc Alkhayyâmî avair remarqué que pareillement une équation cubique admettait, en certains cas, trois solutions, il est difficile à croire que cette coincidence entre le degré du problème et le nombre des solutions ne l'eût pas frappé et conduit à des réflexions, et peut-être à des découvertes, ultérieures." (Woepcke 1851, xvi).

32 An example is Woepcke's comparison between several constructions of Abu l-Wafa' and Bhaskara (Woepcke 1855, 219–220).

33 ". . . elles s'éloignent très-sensiblement de l'esprit de la géométrie arabe, toujours fidèle, sous le rapport de la forme, à ses modèles grecs, . . ." (Woepcke 1855, 230).

34 "La seconde est tout-à-fait d'origine indienne; . . ." (Chasles 1837, 454).

processes of knowledge transfer beyond the simple comparison of forms. First and foremost, such fast and facile evaluations reflect the lack of historical training of this generation of writers on the history of mathematics. This interpretive approach is on a par with the dominance of mathematical evaluations of works of the past in Woepcke's research practice. Both features highlight the manner in which he understood his obligations as a historian of mathematics. Being very honest about his beliefs and working methods, Woepcke formulated his position clearly:

> Namely, in the discussion of scientific borrowings made by one people from another one, the criterion, which needs to stand in the first place and which contributes much beyond all the others, is the conformity or the difference of the spirit of the methods; in the actual case, this criterion decides, as we have seen, in favor of the Indian origin of the two constructions of Abu l-Wafa'.[35]

These two core practices had a tremendous impact on the practice of historians of mathematics until recently. Macro-historical judgments and abstention from historicizing mathematical practices and their products continued to dominate the practice of historians of mathematics in Islamicate societies during the entire twentieth century.

4 Cantor's strategies for representing the cultures of the mathematical sciences

Cantor's general history of mathematics follows other historiographical principles and goals than those favored by the specialists. First, Cantor wrote explicitly for only one audience – mathematicians. Second, he had no knowledge of Arabic, Persian, or Ottoman Turkish and could thus only work with translations or research done by specialists. It is here where his main relevance lies for a discussion on historiographical practices in regard to the mathematical sciences in Islamicate societies. Third, as Folkerts, Scriba, and Wußing explained, he was influenced in his approaches to the history of mathematics by the philosophical ideas of Arthur Arneth (1802–1858), who was a high school teacher in Heidelberg and a lecturer (Privatdozent) at the city's university (Dauben and Scriba 2002, 116, 123). Folkerts, Scriba, and Wußing considered two points in Arneth's publications as being of particular relevance to the way in which Cantor conceived of how to do history of mathematics. Arneth's belief that the development of mathematical knowledge

35 "Or, dans la discussion des emprunts scientifiques faits d'un peuple à un autre, le critérium, qui doit figurer en première ligne, et qui l'emporte de beaucoup sur tous les autres, est la conformité ou la différence de l'esprit des méthodes, et dans le cas actuel, ce critérium décide, comme nous venons de le voir, en faveur de l'origine indienne des deux constructions d'Aboûl Wafâ." (Woepcke 1855, 237).

was closely intertwined with general intellectual and cultural history, or as he wrote, the history of the human mind, furnished a strong stimulus for Cantor to look beyond the confines of mathematics proper. Second, Arneth's engagement with the debate on the relationship between Greek and Indian mathematics that prevailed in France during much of the nineteenth century as mentioned already in the discussion of Woepcke's studies motivated a special attention to this issue (Dauben and Scriba 2002, 112, 116).

Moritz Cantor published the first volume of his *Vorlesungen über Geschichte der Mathematik,* which covered the period "from the oldest times until the year 1200", in 1880. No expert of the mathematical sciences in any Islamicate society himself, he could rely on a body of studies in German and French published by authors mentioned repeatedly in this paper and a few others like Jean Jacques Emmanuel Sédillot (1777–1832), Friedrich August Rosen (1805–1837), or Adolf Hochheim (1840–1898). He also worked with other secondary literature, in particular surveys. The most important books of this type for Cantor were Hermann Hankel's (1839–1873) *Zur Geschichte der Mathematik in Althertum und Mittelalter*, Gustav Weil's (1808–1889) *Geschichte der islamitischen Völker von Mohammed bis zur Zeit des Sultan Selim übersichtlich dargestellt* (Weil 1866), and Alfred von Kremer's (1828–1889) *Culturgeschichte des Orients unter den Chalifen* (von Kremer 1875–77). Moreover, he read biographical and bibliographical material as provided, for instance, by Johann G. Wenrich (1787–1847) in his *De auctorum Graecorum versionibus et commentariis Syriacis, Arabicis, Armeniacis Persicisque* or Ferdinand Wüstenfeld (1808–1899) with his *Geschichte der arabischen Aerzte und Naturforscher* (Cantor 1880, vol. 1, 593–700). In addition, Cantor worked with literature from the early nineteenth century, publications of primary sources as early as the seventeenth century, and the most recent publications on history of ancient and medieval, European and non-European mathematics, Arabic literary and political history, and catalogues of cross-cultural translations, commentaries, and pseudepigraphic works. Here his preferred authors in addition to Woepcke, Sédillot, and Hankel were Moritz Steinschneider (1816–1907), Adolf Hochheim (1840–1898), Aristide Marre (1823–1918), Joseph von Hammer-Purgstall (1774–1856), and Friedrich Hultsch (1833–1906).

As his reading practice went beyond the mere mathematical and bio-bibliographical, his writing practice aimed at combining the political, cultural, and scientific in a manner that did not separate too strictly the one from the other. Although Cantor devoted many pages to the description of purely mathematical content, his introduction to Chap. XXXII clearly aims at explaining the emergence of highly successful scholars interested in the mathematical sciences as well as the particularities of the scholarly communities and their specific interests on the basis of what he extracted from the surveys on Islamicate history and culture. Similar efforts characterize Chaps. XXXIV – XXXVII (Cantor 1880, vol. 1, 629–630, 633–634, 636–637, 642, 649–650, 653, 668–669, 680–682).

Cantor follows two different pathways by discussing political and cultural history in its most general forms on the one hand and treating individual scholars not as mere repositories of old or new mathematical knowledge, but as people with a life and stories to be told. In this sense, his "history of mathematics of the Arabs" was conceptually more than the "scientific history of mathematics" of Nesselmann and the study of primary sources for discovering the Arabic achievements by Sédillot or Woepcke. Cantor certainly gave these approaches prominence in his survey, but embedded them in a political and cultural history, which he mainly borrowed from his Orientalist colleagues in Austria and Germany. In my view this effort possesses two impressive features. One is its existence; the second is its relative comprehensiveness, at least for the so-called classical period. Almost all major themes for explaining what today is called the translation movement can already be found in Cantor's summary: the role of Syriac-speaking Christian communities and their centers in Antioch, Edessa, Emesa, and Nisibis; the function of Christian communities in Sasanian Iran and of Middle Persian texts; the impact of Sanskrit texts and the (alleged?) visit of an Indian embassy to the Abbasid court in 773; the enormous thematic breadth, cultural diversity, and number of the translated texts; the role of the administrators, in particular the Barmakids; the relevance of trade, finances, and patronage (Cantor 1880, vol. 1, 594–602). Major translators and scholars of the mathematical sciences of the ninth and tenth centuries are mentioned by name and activities (Cantor 1880, vol. 1, 602–603, 627, 629). In addition, Cantor points to two further themes as if unproblematic: the division of Islamicate societies and their mathematical sciences in Eastern (Morgenland) and Western (afrikanische Nordküste, Spanien und Sicilien) and the emergence of a somehow impermeable separation (Scheidewand) between these two regions; and the fact that translating was often carried out in combination with commenting, although Cantor discusses these two activities separately from each other, since to him only the latter included "original thinking" (Cantor 1880, vol. 1, 605–606).[36] All the themes that Cantor raised shaped the historical practices of scholars in the second half of the twentieth century and their historiographical debates. By contrast, the function of diagrams and diagram letters as carriers of historical information about practices and cross-cultural transformations, which Cantor emphasizes in his reflections on how to determine the cultural origins of al-Khwarizmi's algebra, has received solid attention among recent historians of mathematics only in the last decade or two (Cantor 1880, vol. 1, 621, 630.). Cantor is also conscious of two important methodological points that should guide the practice of a historian of mathematics (or any other field), but did not do so during the twentieth century: historical events or products are rarely the result of a single cause; speculations about possible motivations, factors, or causes that brought forth such events or products need to be based on in-depth and broadly conceived investigations of texts and other historical material. Hence, in sometimes perhaps

36 The last point is expressed in the following statement: "Die Uebersetzungsthätigkeit war auch von einer vielfach commentirenden begleitet, auf die wir aber, da sie immerhin einige Ansprüche an das Selbstdenken des Commentators erhebt, bei den Originalarbeiten zu reden kommen." (Cantor 1880, vol. 1, 605).

tedious excursions, he discusses all possible reasons or interpretations of a method, a text, or an approach, including not only the standard search for whether something is of Greek or Indian origin, but also regional customs, consequences of theological debates, and conflicts with regard to mathematical themes and methods and their evaluation by medieval scholars (Cantor 1880, vol. 1, 616–626, 653–657).

Finally, it is not surprising that the sharp debates about interpretation left their traces in Cantor's depiction of Arabic mathematical texts. He sided with Sédillot and Woepcke by acknowledging original achievements made by Arabic writing scholars from Islamicate societies. He also sided with Biot, Hankel, and others by privileging the impact of Greek scientific works on the mathematical sciences in those later societies and by recognizing as the only other relevant source of inspiration Indian texts.

Despite refusing to follow Hankel in his mostly negative evaluations of the contributions to mathematical progress by Arabic writing scholars, Cantor takes up his predecessor's historiographical perspective in one point: decline. Chapter XXXVI is titled: "The Decline of the Eastern Arabic Mathematics. Egyptian Mathematicians".[37] The men we encounter in the first part of this chapter are: Nasir al-Din Tusi (1201–1274), Ghiyath al-Din Kashi (d. perhaps in 1429) and Baha' al-Din al-'Amili from Iran, Qadizade al-Rumi (d. after 1440) from Anatolia, and Ulugh Beg (d. 1449) from Transoxania. They all are portrayed as the last mathematicians and astronomers and somehow the representatives of decline (Cantor 1880, vol. 1, 668–672). Cantor presents as his understanding of decline the notion "that each progress ends".[38] The men listed as examples do not fit such a claim, as we know today.

A second of Cantor's criterion for decline is the disappearance of "receptivity in the domain of mathematics".[39] Here, his only example is Baha' al-Din al-'Amili's *Essence of Arithmetic* published by Nesselmann (Cantor 1880, vol. 1, 672–675). With this formulation, he did not mean to say that there was no appropriation of new mathematical knowledge from other than Islamicate societies. This theme does not belong to those that Cantor addressed in a serious manner. He simply means the fact that the compiler himself does not offer anything that Cantor could recognize as relevant for the progress of the mathematical sciences (Cantor 1880, vol. 1, 674–675). Cantor's cultural and biographical approach to the history of mathematics in Islamicate societies was limited to macro-history. He had no strong interest in understanding local contexts and their impact on the use

37 "Der Niedergang der ostarabischen Mathematik. Aegyptische Mathematiker." (Cantor 1880, vol. 1, 668).

38 "Mit den Männern, welche wir zuletzt genannt haben, hört jeder Fortschritt bei den Einen auf, während er bei den Anderen zu immer rascherer Gangart sich gestaltet." (Cantor 1880, vol. 1, 672).

39 "Und auch die Empfänglichkeit der Araber auf mathematischem Gebiet war dahin." (Cantor 1880, vol. 1, 672).

and representation of mathematical knowledge. After all, his question was that of most historians of mathematics of the nineteenth and twentieth centuries whose education had focused on mathematics and the sciences: who had contributed what to mathematical progress?

5 Concluding remarks

These limited explorations in the historiography of mathematics in Islamicate societies as practiced in the nineteenth century in France and Germany illustrate the profoundly ideological underpinnings of all these efforts, independent of their methodological orientations and technical tools in philology, mathematics, philosophy, or historiography. This should encourage us more than ever to reflect about our own ideological commitments and their impact upon our approaches to primary sources and their interpretation. The second result of this brief dip into some of the publications of nineteenth-century writers on subjects of the history of mathematics is the recognition that these writers formulated the basic approaches that have dominated our own research during the last decades. Their methodological perspectives and technical arsenal were, despite its factual limitations, broader than what we accept as respectable tools and questions. This came as an unexpected surprise to me and shows how productive an analysis of working practices of earlier generations of historians of mathematics can be. An important conclusion from my interrogations of the works discussed and the practices they represent is thus to encourage further studies of past working practices, their theoretical underpinnings and analytical tools. Such studies, if done with a sympathetic, but critical eye, can only enrich our own practices and improve our own capabilities of investigating and interpreting past mathematical works and their historical meanings. A third point has to be made with regard to interpretive practices. The authors of the nineteenth century were willing, too readily as I think, to generalize on a very thin basis of evidence and to create pictures of centuries of mathematical activities in a broad range of societies and scholarly communities that by now are so deeply anchored in our collective view of Islamicate societies that it takes great efforts to question, check, or deconstruct them. Macro-historical reconstructions with their immanent judgments have proven dangerous, if not detrimental, to a fair understanding of past processes and products. It is a major trap from which we have not yet liberated ourselves in our own practices.

The last comment to be made is not based directly on the material discussed in this paper. It results from a conversation I had recently with Lorraine Daston (Max Planck Institute for the History of Science, Berlin) on the topic of when and why the sciences and technology became identifiers of cultural progress of civilizations. She pointed out that a major shift occurred during the nineteenth century, which is not well studied yet and thus only badly understood. It replaced the concept of people by the idea that different peoples constituted civilizations apart from other such groupings and that such big clusters possessed an identifying culture whose values in a universal history of humanity could be measured in terms of their scientific and technological progress. Colonialism and its accompanying "civilizational mission", which was invented for legitimizing conquest and appropriation of large territories with everything that existed and lived there and the subjugation of "nature and culture" to the unconstrained exploitation by the conquerors, explain parts of this profound shift in perspective and explanatory approach to culture and its values. But other changes in those societies that struggled for global domination must have occurred to call for this new grand narrative of civilizational superiority. One further social change that most likely contributed to its emergence were the internal struggles for or against modernization, secularization, and republicanism. Some of the consequences of those struggles were pointed out in this paper in the description of Boncompagni's patronage of history of mathematics and the questions and methods he favored. His conservative approach to society and politics was certainly not the only reflection of the impact that the fierce struggles within Italian and French societies for their further sociopolitical directions exercised on the practices of historians of mathematics. Hence, this larger perspective on the interpretive frameworks used by nineteenth-century writers about historical subjects also calls for more studies of the practices of our predecessors, their motivations and engagements with past mathematical cultures not only qua historians of mathematics, but as members of a living society with a broad spectrum of problems that necessitated them to take a position and stance through their choices of research topics, approaches, and tools.

References

Al-Afghani, J.-D. (1968). *An Islamic response to imperialism* (pp. 175–187) (N.R. Keddie, Trans.). Berkeley: University of California Press.

Cantor, M. (1880). *Vorlesungen über Geschichte der Mathematik* (Vol. 1). Leipzig.

Carra de Vaux, B. (1892). L'Almageste d'Abū 'lWéfa alBūzdjāni. *Journal Asiatique, sér., 8*(19), 408–471.

Charette, F. (1995). *Orientalisme et histoire des sciences: l'historiographie européenne des sciences islamiques et hindoues, 1784–1900*. Thèse de M.A: Université de Montréal.

Chasles, M. (1837). *Aperçu historique sur l'origine et le développement des méthodes en géométrie*. Brussels: M. Hayez.

Dauben, J. W., & Scriba, C. J. (Eds.). (2002). *Writing the history of mathematics: Its historical development*. Basel: Birkhäuser.

Jourdain, A. (1819). *Recherches critiques sur l'âge et l'origine des traductions latines d'Aristote, et sur des commentaires grecs ou arabes employés par les docteurs scholastiques*. Couronné par l'académie royale des inscriptions et belles-lettres. Paris: Imprimerie de Chapelet. New edition 1843. https://ia600909.us.archive.org/33/items/recherchescritiq00jouruoft/recherchescritiq00jouruoft.pdf

Kremer, A. von (1875–77). *Culturgeschiche des Orients unter den Chalifen* (2 vols). Vienna: Wilhelm Braumüller.

Mazzotti, M. (2000). For science and for the Pope-King: Writing the history of the exact sciences in nineteenth-century Rome. *BJHS, 33*, 257–282.

Narducci, E. (1869). Intorno alla vita ed agli scritti di Francesco Woepcke. *Bullettino di bibliografia e storia delle scienze matematiche e fisiche, 2*, 119–152.

Nesselmann, G.H.F. (1842). *Versuch einer kritischen Geschichte der Algebra.* Berlin: G. Reimer. (Reprint Frankfurt am Main 1969).

Nesselmann, G.H.F. (1843). *Essenz der Rechenkunst von Mohammed Beha-eddin ben Alhossain aus Amul, arabisch und deutsch.* Berlin: G. Reimer.

Renan, E. (1883). *L'Islamisme et la science.* 29 mars Conférence. Université de Sorbonne, Paris.

Sédillot, L. A. (1835). Découverte de la variation, par Aboul-Wafâ, astronome du Xe siècle. *Journal Asiatique, sér.*, *2*(16), 420–438.

Sédillot, L.P.E.A. (1845–1849). *Matériaux pour servir à l'histoire comparée des Sciences mathématiques chez les Grecs et les Orientaux* (Vol. 1). Paris: Firmin Didot Frères.

Taine, H.-A. (1866). *Nouveaux essais de critique d'histoire* (2e éd.). Paris: Librairie de L. Hachette et Cie.

Weil, G. (1866). *Geschichte der islamitischen Völker von Mohammed bis zur Zeit des Sultan Selim übersichtlich dargestellt.* Stuttgart: Rieger.

Woepcke, F. (1850). Notice sur un manuscrit Arabe d'un traité d'algèbre par Aboul Fath Omar Ben Ibrāhīm Alkhayyāmī, contenant la construction géomètrique des équations cubiques. *Journal für die reine und angewandte Mathematik*, *40*, 160–172.

Woepcke, F. (1851). *L'algèbre d'Omar Alkhayyāmī, publiée, traduite et accompagnée d'extraits de manuscrits inédits.* Paris: Benjamin Duprat.

Woepcke, F. (1851). Des systèmes de chiffres en usage chez différents peuples, et de l'origine de la valeur de position des chiffres indiens. *Nouvelles Annales des Mathématiques. Journal des Candidats aux Ecoles Polytechnique et Normale*, *10*, 372207.

Woepcke, F. (1852). Notice sur une théorie ajoutée par Thābit Ben Korrah à l'arithmétique spéculative des Grecs. *Journal Asiatique*, sér., *4*(20), 420–429.

Woepcke, F. (1853). *Extrait du Fakhrī, traité d'algèbre par Aboū Bekr Mohammed Ben Alhaçan Alkarkhī, précédé d'un mémoire sur l'algèbre indéterminée chez les Arabes.* Paris: Imprimerie Impériale.

Woepcke, F. (1854). Recherches sur l'histoire des sciences mathématiques chez les orientaux d'après des traités inédits arabes et persans. Premier article. Notice sur des notations algébriques employées par les Arabes. *Journal Asiatique*, sér., *5*(4), 348–384.

Woepcke, F. (1855). Recherches sur l'histoire des sciences mathé4matiques chez le4s oientaux. D'après des traités inédits arabes et persans. Deuxième article. Analyse et extrait d#un recueil de constructions géometrique s par Aboûl Wafâ, *Journal Asiatique*, 5e série, 5, 218–256, 309–359.

Woepcke, F. (1863). Mémoire sur la propagation des chiffres indiens. *Journal Asiatique*, sér., *6*(1), 27–79, 234–290, 442–529.

Woepcke, F. (1865–66). Introduction au calcul gobārī et hawāī. *Atti dell'accademia de' Nuovi Lincei*, *19*, 365–383.

8

WAS THERE A SHIFT FROM FAITH-NEUTRAL TO FAITH-BASED SCHOLARLY COMMUNITIES IN ISLAMIC SOCIETIES FROM THE CLASSICAL TO THE POST-CLASSICAL PERIOD?[1]

In 2002, Bernard Goldstein stressed the cooperation between scholars of different faiths as one of the most salient features of the early years of science in Islamic societies. Without this cooperation, this crossing of religious boundaries, he believed, the translation movement and its concomitant development of scientific ideas, methods and practices would not have been possible.

One major historiographical problem in history of science in Islamic societies during the twentieth century was the observation that in later centuries scholarly products of the Islamic world fell short of the level reached between the eighth and the eleventh centuries. Many different causes have been declared responsible for this apparent development – the growing impact of religion, the invasion of nomads, natural and health catastrophes, the disappearance of courtly patronage for the sciences or the exclusion of the sciences from the *madrasa*. While the dating of the so-called decline has been challenged and altered time and again and although I consider the concept of decline as inappropriate for understanding the processes of knowledge production in post-classical Islamic societies, the observation that many of the thousands of mathematical

1 This talk given at a conference on convivencia in Madrid was translated in a slightly abbreviated form into Indonesian Bahasa in the frame of a publication project of Syamsuddin Arif in Jakarta: "Sans Islam dan kerjasama lintas agama," in *Islamic Science. Paradigma, Fakta dan Agenda*, ed. Symseuddein Arif, Jakarta: Institute for the Study of Islamic Thought and Civilizations (INSISTS), 2016, 138–155.

DOI: 10.4324/9781003372493-9

texts produced in the centuries after 1300 are elementary introductions into the basics of arithmetic, algebra, surveying or themes of astronomy cannot but be confirmed. Even if we find a growing number of texts debating problems of higher mathematical or astronomical levels, they always will remain a minority of the entire output after 1300. Hence, we continue to have to search for explanations for the phenomenon.

A question to ask is whether shifts in composition of the scientific communities, in particular shifts in religious affiliation, could have played an important role in altering orientation and quality of the texts produced by teachers and students at *madrasa*s as well as scholars at courts and in other socio-cultural contexts. The biobibliographical material available to me, however, does not allow me to engage fully with this question. Thus, I focus in this paper on outlining what needs to be considered for an entrance into the issue of the relationship between scholars of different religious persuasions in different localities and periods and its representation in biographical literature. Before I enter into this discussion, I will first briefly summarize my general outlook on the cultural contexts of what we often label 'the sciences' meaning fields such as geometry, number theory, arithmetic, optics, medicine, theoretical music, philosophy, logic, astrology, time keeping, magic squares or algebra.

1 On the cultural contexts of the sciences

I see four socio-cultural spaces that existed in the classical period as well as the post-classical one where scholars of 'the sciences' worked with, for or side by side with colleagues and patrons – the courts (caliph, local dynasty, vizier, governor), the private household, the hospital and religious institutions. I do not see the observatory or the library as further independent socio-cultural spaces of science since either of the two was part of a court or a private household or, in the case of the library, of a religious institution. I exclude the marketplace because it seems almost inaccessible due to a lack of sources. The boundaries in the four spaces are of course fluid. Major *madrasa*s or mosques were donated by rulers and their relatives and can thus be considered as part of the courts as can many of the important hospitals in Baghdad, Damascus, Cairo, Istanbul or other cities. Courtiers acted in the courtly sphere as patrons and clients, but also had often their own fairly complex households with clients. Two major changes occurred between the eleventh and the fourteenth centuries that transformed these four spaces, their inhabitants and the kind of scholarly practices as well as the fields that were sponsored and appreciated. One change concerns the quantitative distribution patterns of different religious faith communities as a result of conversion, expansion and migration. The ensuing predominance of Islamic institutions and the transformation of other faith communities into minorities across the entire Islamic world west of India with some exceptions in certain parts of the Ottoman Empire altered dramatically the composition of those communities that engaged seriously with 'the sciences'. The second major change consists in the emergence and spread of specifically Muslim spaces and institutions of education and scholarly practices. *Madrasa*s and mosques became primarily populated by Muslim teachers and students as well as their families. Christian, Jewish, Sabian, Zoroastrian and other students and scholars retreated into the spaces of their own communities and found only rarely their way into the classrooms of the Muslim teachers. They could not obtain a teaching position in such institutions nor did their communities apparently set up any analogous kind of institution for higher education although this is admittedly a point difficult to make due to the kind of sources I work with. Cross-denominational encounters between scholars of different religious creeds did not disappear though completely. Several major as well as minor dynasties such as the Ilkhanids, Artuqids, Ayyubids, Ottomans, Mughals or Adilshahs continued to keep their courts open for practitioners from different faith groups.

Hence, while I certainly see evidence for a relationship between content and form of texts and other products of scholarly work in the domain of 'the sciences' in the post-classical period and the religious compositions of individual Islamic societies and their knowledge practices, I do not think that this relationship is simple or straightforward. With regard to the religious composition of communities of scholars practicing 'the sciences' I do not believe that the religious beliefs of these scholars are sufficient to explain the qualitative differences between products of the classical period and those of the post-classical period. Nor do I believe that the decreasing permeability of religious communal boundaries can be pinpointed as a cause for these qualitative differences. If anything at all from the religious realm can be seen as directly connected to the content and form of later scientific texts, images and instruments, it is the shift in institutional and normative contexts, the emergence of systematic and culturally negotiated and regulated bodies of knowledge and their institutionally and normatively sanctioned production and reproduction in *madrasa*s, mosques and related settings. With these brief general statements of my current views on the socio-cultural spaces of 'the sciences' in various post-classical Islamic societies in mind, let me now turn to the more specific issues of faith-neutral versus faith-based scholarly communities of 'the sciences' and cross-denominational 'cooperation'.

2 Cross-denominational 'cooperation'?

Goldstein described 'the sciences' in ninth- and tenth-century 'Abbasid Baghdad as a 'neutral zone for inter-religious cooperation'.[2] He argued that 'cooperation' was a social and cultural activity different from what Sabra had termed in 1987 'the appropriation of the ancient sciences' 'for it requires cooperation with

2 Bernard Goldstein, 'Science as a 'Neutral' Zone for Interreligious Cooperation,' in *Certainty, Doubt, Error: Aspects of the Practice of Pre- and Early Modern Science*, in Honour of David A. King, eds. Sonja Brentjes, Benno van Dalen, François Charette, *Early Science and Medicine* VII.3 (2002), 290–291, Special Issue, p. 290.

contemporary scholars, not just accepting the legacy of the ancients'.[3] Quoting Gutas, Goldstein emphasized that the patronage of translating Greek and other scientific works 'cut cross all lines of religious, sectarian, ethnic, tribal and lingustic demarcation. Patrons were Arabs and non-Arabs, Muslims and non-Muslims'.[4] Believing that 'early Islam was generally tolerant to diversity within Islam and within limits of contacts with members of the protected religious communities', Goldstein claims that things went beyond mere tolerance arguing that the Sabian scholar Thabit b. Qurra and prominent Muslim astronomers cooperated in 'joined research projects'.[5]

I believe that things are more complex on methodological, conceptual and evidential grounds. Conceptually I think cooperation describes a specific kind of social activity and without getting too pedantic I see it as an activity that to some extent is voluntary, contains an element of choice and takes place between partners of some kind of equality or autonomy. There are other activities that bring people together in work relationships such as salaried labor, patronage, servitude or enforced labor. In the case of Thabit b. Qurra and the Banu Musa to whom Goldstein referred, it seems to be clear from Ibn al-Nadim and other sources that the Banu Musa were patrons and Thabit b. Qurra was one of their clients. They paid him a monthly stipend for which he translated and perhaps wrote new texts for them. But because Thabit b. Qurra was a very bright and gifted person and because his patrons themselves were active and highly competent scholars their patron-client relationship seems to have evolved over time into a very special kind that furthered a shift from services rendered by a client to his patron to working together on translating a particularly difficult Greek mathematical text such as Apollonios' *Conics*. Thus, we can accept Goldstein's example indeed as a possible instance of cooperation between scholars of different creeds. But was it widespread, the norm of the day or rather only one and one of the rarer forms of social relationships that existed among scholars in 'the sciences' in the early Abbasid period?

This question is not at all easy to answer due to the sources we dispose of and their rhetoric. The scientific manuscripts of the period, if they are at all attributed to an author, are mostly ascribed to one author alone. Evidence for shared authorship of papers, let alone joined projects, is rare. The only projects where scholars did work together are the measurements of one degree of meridian and other astronomical observations under al-Ma'mun as well as under Buyid *umara'*. What do we learn from the descriptions of these projects in later sources? In the case of the projects carried out during al-Ma'mun's caliphate, those that involve

3 Abdelhamid I. Sabra, A.I. Sabra, 'The Appropriation and Subsequent Naturalization of Greek Science in Medieval Islam,' *History of Science*, 25: 1987, 223–243; Goldstein, Science as a 'Neutral' Zone, p. 290.

4 Dimitri Gutas, *Greek Thought, Arabic Culture: The Graeco-Arabic Translation Movement in Baghdad and Early 'Abbasid Society (2nd-4th/8th-10th centuries)*, London, 1998, p. 5.

5 Goldstein, Science as a 'Neutral' Zone, p. 291.

more than one participant are described by several ninth-century as well as later authors. The stories differ in their details already in the accounts by ninth-century writers. In one type of narrative, all projects are presented as headed by the caliph himself as an active and often only scientific decision maker. They are described as controlled and evaluated by a non-scientist, the Baghdadi head-judge Ibn al-Aktham. Everybody else allegedly occupied if not precisely the same, then a very similar status while contributing specific acts such as constructing specific instruments or carrying out specific observations or measurements. Another storyline presents the caliph in a mixed role. He is portrayed as an eager student of the ancient sciences who turns for advice about their practitioners to the head-judge Ibn al-Aktham and listens to the knowledge of the scholars recommended by the judge. On the other hand, he is depicted as the man who had the power, means and will to control results achieved by the scientists and to determine their validity.[6] Undeniably most of the project participants were working together in some, but not all moments of the project. Ibn al-Aktham was present in certain project phases. Apparently and perhaps luckily, he did not work in the project as a full participant, but rather as an advisor and rapporteur. In contrast al-Ma'mun himself was present only occasionally. He mainly, but importantly provided the questions to be studied and the money for the various activities. He ordered the control of the projects and evaluated their results. Where non-Muslims active in these projects? As far as I can tell, yes, at least one scholar participated in them who was not born to Muslim parents – Sanad b. ʿAli. Ibn al-Nadim in his *Kitab al-Fihrist* even describes him as having been "in charge of all the [astronomical] observations."[7] But was he still a member of his ancestral religious community? I do not know since the sources that provide biographical data about him stress that he had converted under the influence or should we better say 'pressure' (?) of al-Ma'mun.[8]

The scholars who particpated in astronomical projects carried out in the tenth century at the courts of various Buyid rulers in Iraq and Iran were almost all, to the best of my knowledge, Muslim. Their creeds may have differed nonetheless, but we do not know much about these beliefs if anything at all. The few exceptions may have been Abu l-Qasim Ghulam Zuhal, an astrologer in the service of ʿAdud al-Dawla (338–371/949–983), whose name points to a non-Muslim upbringing, and Abu Saʿd al-Fadl b. Bulis, nicknamed the 'Christian from Shiraz', who participated in the observations of 988 headed by Abu Sahl al-Kuhi under the patronage of ʿAdud al-Dawla's son Sharaf al-Dawla (376–79/987–990).[9] Did

6 David King, 'Too Many Cooks . . . A New Account of the Earliest Muslim Geodetic Measurements,' *SUHAYL* 1 (2000), 207–242.

7 *The Fihrist of al-Nadīm: A Tenth-Century Survey of Muslim Culture*, Bayard Doge (Editor and Translator), New York: Columbia University Press, 1970, 2 vols., vol. 2, p. 653.

8 *The Fihrist of al-Nadīm*, vol. 2, p. 652.

9 J. L. Berggren, *Patronage of the Mathematical Papers in the Buyid Courts*, unpublished. I thank Len Berggren for giving me access to his research results.

their being clients of the same Buyid patrons and in some instances being affiliated to the same observatory with a set of projects mean that they cooperated? Again, their products rarely claim joined production, but their involvement in the observatory is clearly described in biographical sources as of different degrees of responsibility as well as of specific content. Hence, cooperation may be seen as one of the social activites linked to astronomical group projects at some Abbasid and Buyid courts.

3 Narrative violence versus cultural tolerance?

Does the presence of scholars adhering to non-Muslim or non-majority Muslim religious beliefs at courts in Islamic societies signify tolerance on the side of the patrons, the courts or even urban society at large? Did the group projects in astronomy and astrology signify something higher than tolerance? Before I try to sketch my negative answer to these questions suggested by Goldstein's discussion let me emphasize a methodological issue. Neither scientific manuscripts nor biographical dictionaries or historical chronicles provide us with a direct access to social or cultural realities. All that they present to us are narratives and while I believe that narratives constitute human realities I also believe that they do not exhaust human reality. Hence, before we can try at all to reflect whether the one or the other statement in a historical text tells us anything about the conditions of scholars at a given place in time we need to analyze the narratives presented in our sources. In the rest of this paper, I will discuss a few of my observations of Arabic biographical dictionaries and their narratives about members of minority creeds.

In addition to my previous methodological and conceptual points, I disagree with Goldstein because his position does not include a reflection on the minority position of Muslims in the overall population of individual court towns during the early centuries of Muslim rule, the various scholarly skills needed for translating Greek, Middle Persian or Sanskrit scientific works, the link to Islamic religious purposes of at least parts of the astronomical group projects, the geographical limitation of such group projects to Baghdad and some Buyid strongholds in Iran and on the tenor of the narratives when non-Muslim as well as non-mainstream Muslim persuasions and convictions were concerned.

Examples for cross-denominational activities are rare and are mostly limited to reports about patron-client, servant-master, teacher-student and author-reader relationships. The first two kinds of relationship are not merely socially hierarchical, but also denominationally hierarchical. In the sources that I read and which were, except for one, all authored by Muslims, the patron as well as the master adhered all to the one or the other Muslim creed. The two latter kinds of relationship were not of the same type with regard to religious beliefs held by their members. In the classical period (until the fall of the Abbasid caliphate in 1258) more than one example of non-Muslim teachers of Muslim students and vice versa is known.[10] In the post-classical period the examples are difficult to come by. Those few that I found in the biographical dictionaries all have a Muslim in the role of the teacher.[11] On the other hand, in these few cases, the shared knowledge did not concern merely 'the sciences', but also the religious books of each other's beliefs.[12] Authors and readers in the classical period came from a variety of religious backgrounds. If we wish to emphasize a 'faith-neutral' zone in scientific activities during the classical period it is first and foremost here in the shared literary production and its usage that it was created.

The other large domain of scientific group activities – medicine – is equally difficult to penetrate. It is clear that many of those who wrote mathematical, astronomical, philosophical and other scientific texts were physicians, both in the classical and the post-classical period. The question is though, did they cooperate and did their joined presence in the entourage of rulers, princes and occasionally princesses signify that they lived together either in the palace or in the same quarters? Different Islamic societies offer different answers. In some cases such as the Ottoman court, the court physicians lived at court and were hierarchically structured as a group. In other cases such as the Ayyubid courts the physicians lived in their own houses in town. They met every day in a special part of the fortress, the ruler's house or the hospital. Their structure as a group was less formalized than the one found in the Ottoman Empire. It apparently depended more on their individual relationship to the patron and his recognition of their status. If we broaden the scope of inquiry, we find that Ayyubid physicians exchanged opinions about their patron's disease or teaching matters, exchanges that were not always free of conflict or even downright hostility. They also worked together at the Nuri hospital in Damascus and presumably at other hospitals built by their patrons, for example, in Cairo or Jerusalem.[13]

Other examples of cross-denominational activities such as friendship, intermarriage, shared festivities, shared burial rites or burial places are even harder to come by. Ibn Abi Uṣaybiʿa testifies to social relationships between Muslim,

10 The best-known examples are Thabit b. Qurra (Sabian) who studied with Ahmad b. Musa (Muslim); Abu Bishr Matta b. Yunus (Nestorian) who studied with Abu Ishaq Ibrahim al-Quwayri (Muslim); Abu Husayn Ishaq, known as Ibn Karnib (Muslim) and a certain Benjamin, perhaps Benjamin al-Nihawandi (Jew); al-Farabi (Muslim) who studied with Matta b. Yunus; and Yahya b. 'Adi (Jacobite Christian) who studied with Abu Bishr Matta b. Yunus (Nestorian) and al-Farabi (Muslim). See for instance L. E. Goodman, 'Translations of Greek Material into Arabic,' in *Religion, Learning and Science in the 'Abbasid Period*, eds. M. J. L. Young, John Derek Latham, Robert Bertram Serjeant, Cambridge: Cambridge University Press, 1990, pp. 477–449; p. 492.

11 A well-known example is that of Ibn al-Nafīs (d. 687/1288) and Ibn al-Quff (685/1286).

12 Ibn al-'Imad al-Katib, *Shadharat al-dhahab fi akhbar man dhahaba*, vol. 5, p. 300; www.islamport.com.

13 Sonja Brentjes, 'Ayyubid Princes and Their Scholarly Clients from the Ancient Sciences,' in *Court Cultures in the Muslim World: Seventh to Nineteenth Centuries*, eds. Jan-Peter Hartung, Albrecht Fuess, London: Routledge, 2011, 326–356.

Christian and less often Jewish court physicians. He tells the one or the other story of social advice and support provided by doctors for their professional fellows across denominational boundaries. The truly important boundary that he draws in his narrative for the Ayyubid period is the social status as a princely client or as the independent physician who proudly rejected the patronage offer.[14]

Most references to non-Muslim scholars in biographical dictionaries, except Ibn ʻIbri's who was a Christian writer, while straightforward and apparently neutral, concern conversion. This means they report that either the scholar himself or his father or grandfather had converted. In the classical period it is often under the pressure of the patron that such a conversion is reported to have taken place.[15] In the post-classical period it is often a spiritual or intellectual experience of the superiority of Islam in general or of a particular performance by a Muslim scholar that is presented as the cause.

Despite this apparent neutrality towards other religions and conversion, classical as well as post-classical narratives are replete with individual stories about violence against non-Muslims, including scholars, or holders of Muslim minority creeds. A systematic investigation indicates that conflicts are, in addition to conversion to the true faith, the second most important narrational focus for talking about adherents of other religious beliefs. Many of these conflicts arise from social strictures such as dress rules, rules for displaying social status such as riding or sexual relations between believers of different creeds, but they also arose from rebellions among some of the religious communities as well as from rulers' religious policies. Examples for the latter kind of conflicts are al-Ma'mun's order to the Sabians first to determine whether they were *dhimmis* and second to prove such a claim to a protected status or to die; alleged enforced mass conversions under the Abbasids in Hamadan and other Iranian towns after a defeated rebellion; or a mass conversion of Christian subjects ordered by Mamluk sultans. There can be no doubt. The overall atmosphere that the biographical dictionaries conjure in their narratives is one of Muslim superiority and non-Muslim precariousness.

The enmity against non-Sunni Muslim groups and converts in these narratives is, however, even more important than the animosity against non-Muslim communities, whether *dhimmi*s or not. So-called *zindiq*s or *rafidis* among scholars are portrayed as the most dangerous species and often, albeit not always, as punished with death for their deviating beliefs.[16] Shiʻi or Sunni scholars, whenever part of the minority, also can be disparaged for their religious creeds as can Sufis. Adherents of particular

14 Ibn a. Uṣaybiʿa, *'Uyūn al-anbā' fī ṭabaqāt al-aṭibbā'*., ed. Nizār Riḍā, Bayrūt: Dār Maktabat al-Ḥayat, n.d., 584–586, 589–591, 598–601, 635–637 *et passim*.

15 On average, the conversion is described as having taken place 'under the hand of . . .' a caliph or some other patron. A clearer expression of the threat involved in caliphal desires for conversion is Ibn al-Nadim's description of Sinan b. Thabit b. Qurra's story: "He was a distinguished surgeon, who fled when [the Caliph] *al-Qāhir* desired him to become a Muslim. But [later] he did embrace Islām. Fearing al-Qāhir, however, he went to Khurāsān.", *The Fihrist of al-Nadīm*, vol. 2, p. 709.

16 See, for instance, Ibn al-ʻImad al-Katib, *Shadharat al-dhahab*, vol. 5, p. 300.

teachings such as *kalām*, philosophy or the 'ancient sciences' in general or authors such as Ibn al-Rawandi, Muḥammad b. Zakariyya' al-Razi, al-Farabi, Ibn Sina, Sayf al-Din al-Amidi, Shihab al-Din Suhravardi, Kamal al-Din b. Yunus, Ibn 'Arabi or Nasir al-Din Tusi were designated in a variety of manners as 'spoilers of religion' and 'traitors'. While being acknowledged as important members of the scholarly world and as dissident members of Islam, they were talked about also as apostates, servants of apostates, infidels or as simply lacking in religion due to their intellectual preferences.

Given all this verbal and verbose violence, tolerance is not a value I consider as part of scholarly communities in classical or post-classical Islamic societies and their broader contexts. Such an answer depends of course on how we want to understand the term 'tolerance'. To me it means at least the acceptance of the right of any member of a society to adhere to the specific religious beliefs held by his or her parents without persecution or pressure to convert or to renounce such beliefs publicly. The forms in which such a right and its acceptance are expressed and made practicable will differ of course over time and space. Hence, when I say that I do not consider tolerance a value that was held and perhaps even practiced by scholarly communities in various Islamic societies, I mean to say that the stories told in the biographical dictionaries reflect a belief that an Islamic society should ideally consists of Muslims only. This belief did not mean to reject the reality of the existing religious minorities. It highlighted the advantages of the majority belief and put the minorities into a hierarchical sequence. The ranks that are ascribed to individual communities in dictionaries until the thirteenth century deviate from what one might expect – Christians, Sabians, Jews, Zoroastrians, other Muslim creeds, heretics and apostates.

Did the discursive violence in the biographical dictionaries prevent encounters, shared spaces or shared intellectual practices? Exceptions seem to suggest that while mainstream attitudes hardened apparently over time and opportunities for trespassing shrank, there also were localities of coexistence and moments of cooperation. In the BnF, there is an astronomical manuscript (Persan 173) that contains a few Hebrew notes on its cover pages and a repeated Persian statement about the cooperation in observations between a Jewish and a Muslim astrologer in town. A certain Husayn b. Muhammad b. Ahmad, a philosopher labelled by Ibn al-'Imad in his biographical dictionary *Shadharat al-dhahab* as a *rafidi*, taught in Irbil, in northern Iraq, according to a quote from Shams al-Din al-Dhahabi, the 'ancient sciences' to everybody who wanted to listen, 'Muslims, Innovators, Shi'is, Jews, Christians'.[17] He was acknowledged not only by the two conservative Muslim biographers as an emminent representative of the so-called 'rational' sciences which include disciplines that are based on reason such as *usul al-din*, *usul al-fiqh*, metaphysics, logic or grammar. He also enjoyed the appreciation and patronage of the local ruler al-Malik al-Nasir.[18]

17 Ibn al-'Imad al-Katib, *Shadharat al-dhahab*, vol. 5, p. 300.
18 Ibn al-'Imad al-Katib, *Shadharat al-dhahab*, vol. 5, p. 300.

Another of such rare examples is 'Ala' al-Din Abu l-Hasan 'Alī b. Taybugha al-Halabi al-Hanafi al-Muwaqqit (d. Aleppo 793/1391) who wrote a commentary on a part of Maimonides' legal text *Mishneh Torah* in which he included an extensive paraphrase of a part of Ibn Kammuna's commentary on Shihab al-Din Suhravardi's *al-Talwihat*.[19]

There are several important aspects to this last example. The most striking aspects in our context concern the relationship that existed between a great-great-grandson of Maimonides, Rabbi David ben Joshu'a, and 'Ali b. Taybugha and the results that this relationship had for the intellectual activities as well as the reputation of the Muslim scholar. According to Schwarb, Rabbi David most likely considered the first four chapters of that part of his ancestor's work upon which 'Ali b. Taybugha wrote his commentary as "a straightforward, 'exoteric' digest of all theoretical sciences, in such a way that a well-conceived commentary upon these chapters could serve as a 'Guide to the theoretical sciences for the knowledgeable' (*dalalat al-'alim fi l-'ulum al-naẓariyya*). According to David's and indeed Maimonides' own understanding such a commentary should, of course, not be limited to reorganizing the contents of the *Guide*, but rather build upon the most advanced scientific knowledge available at a given time".[20] Schwarb concluded:

> Hence, a 14th century commentary on *Hilkhot Yesodei ha-Torah* I – IV was meant to take the form of a scientific-philosophical 'encyclopaedia', including all theoretical sciences according to the standards of its time. In David's view, the perfect commentator of *Hilkhot Yesodei ha-Torah* would be a leading scientist of his time, no matter what his religious affiliation. David had himself a broad scientific knowledge, and evidently considered himself apt to carry out such a project, but he may well have thought that a savant of Ibn Taybughā's standing would even better match the profile of the perfect commentator . . . Ibn Taybughā indeed conceived of his commentary as a philosophical 'encyclopaedia' of all theoretical sciences and composed it to fit this generic framework.[21]

Summarizing previous research, Schwarb highlights that Ibn Taybugha 'based his commentary on the Arabic translation handed to him by David, but moreover relied on David's own *Sharḥ* on *Hilkhot Yesodei ha-Torah* I – IV and finally presented his commentary to David as he could not think of other potential readers'.[22]

Thus, David ben Yoshu'a and 'Ali b. Taybugha seem to have entertained an exceptionally high regard for each other, maybe even an exceptionally close

19 Information by Sabine Schmidtke.

20 Gregor Schwarb, '"Ali Ibn Taybugha's Commentary on Maimonides" Mishneh Torah, Sefer Ha-Madda, Hilkot Yesodei Ha-Torah I-IV: A Philosophical "Encyclopaedia" of the 14th Century,' Part I, *Medieval Encounters* 16 (2010), 7–8.

21 Schwarb, '"Ali Ibn Taybugha",' pp. 7–8, 9.

22 Schwarb, '"Ali Ibn Taybugha",' p. 4.

relationship involving not only discussions of philosophy, theology, and other scholarly matters, but apparently also acts of immediate cooperation. David ben Yoshu'a may have translated Maimonides' text into Arabic for 'Ali b. Taybugha and 'Ali had studied David's own commentary.[23] But the Muslim author, it seems, was not very familiar with the Jewish community in Aleppo and its members' education nor did he think another Muslim might have been interested in his commentary on Maimonides' work.

The work 'Ali b. Ṭaybugha invested in his commentary was, according to Schwarb, substantial and profound. In addition to the personal contacts between two scholars of different religious persuasion, the author-readership chain from Suhravardi to Ibn Taybugha links a Muslim scholar of the twelfth century with a Jewish scholar of the thirteenth century and the son of a Turkic Mamluk of the fourteenth century. Suhravardi was an innovator with a broad spectrum of intellectual interests who paid for his innovative curiosity and confrontation with more traditional scholars in Aleppo with his life. Ibn Kammuna, the Jewish scholar, contributed to the revival and survival of Suhravardi's teachings, although he did not comment on Suhravardi's writings on *Ishraq* and was himself rather a follower of Ibn Sina's philosophy.[24] 'Ali, the son of a Mamluk, was a Sunni who worked on timekeeping and related astronomical issues, something only very few relatives of Mamluks did. He commented on a philosophical text in Aleppo, the city whose Muslim scholars had pushed the Ayyubid sultan to execute Suhravardi. He did so in a time and region which is believed by many current historians of having been void of any kind of interest in philosophical themes. But more so, in addition to *'ilm al-miqat* he also studied *'ilm al-hay'a* (mathematical cosmography) and the two *asl*, two of the religious disciplines among the rational sciences.[25] Apparently, he also counseled occasionally a Mamluk *amir* in rebellion about his fortunes predicted by the stars.[26] None of these activities seem to have been standard for *muwaqqit*s in Mamluk Syria. Thus, if we would study biographical dictionaries more often and take their narratives more seriously, we would know that our views were not shared by those who wrote such dictionaries, independent of whether they liked philosophy and the other 'ancient sciences' or not.

But the story of Muslim-Jewish author-reader chains and shared literary activities is not closed with 'Ali b. Taybugha. His commentary of a part of Maimonides' *Mishneh Torah* survived thanks to the travels and intellectual interests of a Jewish scholar from fifteenth-century Yemen, Sa'id b. Da'ud al-'Adani. This traveler came to Aleppo, found a manuscript of 'Ali b. Taybugha's commentary written by David b. Yoshu'a in Arabic and copied it in Hebrew letters.[27]

23 Schwarb, '"Ali Ibn Taybugha",' p. 5.
24 I thank Sabine Schmidtke for this clarification.
25 Schwarb, '"Ali Ibn Taybugha",' p. 10.
26 Schwarb, '"Ali Ibn Taybugha",' p. 10, fn 61.
27 Schwarb, '"Ali Ibn Taybugha",' p. 2, fn and p. 4, fn 16.

'Ali b. Ṭaybugha was accosted by a Muslim compatriot in Aleppo who doubted his firmness in religious belief. He was described by two biographers as lax in praying and a wine drinker, who did not care for purity of religion and knowledge. The reason for this negative reputation is located beyond the standard acts of religious disorder in his 'decay of creed'.[28] His contacts and cooperation with a Jewish scholar may have contributed to such a verdict, but were not mentioned. His focus on the mathematical as well as parts of the rational sciences alone appears to have been sufficient for raising the hackles among the more conservative Muslims in town.

4 *Umma*s and their properties

A third domain, where religious communities are represented with regard to their relevance in 'the sciences', is the discussion of '*umma*s'. The contribution of different '*umma*s' to 'the sciences' was a theme that according to Gutas was formed as part of a legitimizing strategy for al-Ma'mun's new cultural politics of philhellenism and anti-Byzantinism.[29] Choosing the term '*umma*' set the debate *nolens volens* into a religious context, albeit not every later contributor to this narrative stressed this aspect for each chosen group equally. According to Gutas' reading of some of the early testimonies to this narrative written by al-Jahiz, the Byzantines were portrayed as having broken with their ethnic scientific ancestors on religious grounds, while the Muslims, in particular those of a Mu'tazili stance, had taken them up due to their subscription to reason as the only means open to humans for dealing with any matter of intellectual bearing, whether religious or otherwise. While some writers of the eleventh and twelfth centuries, in particular in al-Andalus, took up this view of the ennobling power of 'the sciences' with regard to the comparative status of '*umma*s', Abu l-Fida', despite his reliance on Sa'id al-Andalusi as well as one not well-known Ibn 'Isa al-Maghribi, rather followed al-Shahrastani and focused on the religious beliefs, not the scientific knowledge of the '*umma*s' he discussed. Intermingled with the summaries of religious beliefs is information on contemporary or near contemporary holders of them. Abu l-Fida', for instance, lists several ethnic groups outside the Islamic world as Christians – Armenians, Russians, Bulghars, Germans, French, Genovese, Venetians, the people of Rome, and a group proposed by Maribel Fierro to be read as Bashqunish (Basques), whom Abu l-Fida' locates between the Germans and the French.[30] All of them are identified first and foremost by an ethnic name and their supposed geographical location. Several of them are also related to Muslims either by being described as having converted to Islam or as having Muslims living among them

28 Schwarb, '"Ali Ibn Taybugha",' p. 10.

29 Dimitri Gutas, *Greek Thought, Arabic Culture: The Graeco-Arabic Translation Movement in Baghdad and Early 'Abbasid Society (2n-4th/8th-10th centuries)*, Abingdon, New York: Routledge, 1998, pp. 84–95.

30 Abu l-Fida', *al-Mukhtasar fi akhbar al-bashar*, vol. 2, pp. 59–60, www.islamport.com.

or as having fought against the ancestors of Abu l-Fida'. In the latter case, the description of the wars is either matter of fact like the conquest of Damietta by the crusaders under the banner of the French king and the subsequent retaking of the town, as Abu l-Fida' claims, after the death of al-Salih Ayyub b. al-Malik al-Kamil in 648 h.[31] That means he talks about the Sixth Crusade led by Louis IX against Egypt from 1248–1251. In the case of the Third Crusade (1189–1192), Abu l-Fida' emphasizes the military might of the German king, that is, Frederick Barbarossa (r. 1152–1190), who marched allegedly with 100,000 men towards al-Sham to defeat Salah al-Din. Abu l-Fida' merely reports that most of his men perished before they arrived, without mentioning any of the other details such as the death of the king himself at Antioch, the continuation of this crusade by a small force under his son or the parallel campaigns by Philipp Augustus and Richard Lionheart listed by other Muslim historians. Historical precision obviously was not Abu l-Fida's purpose in his narrative about the '*umma*s'. His remarks can rather be regarded as principally similar to matter-of-fact statements about religious affiliations of individual scholars in the Islamic world, which I propose to understand as elements of a salvation history. In contrast, his comments on the expanse of territorial possessions coupled with statements about the large quantity of the people belonging to the one or the other group can be seen as expressions if not of admiration, so at least of acknowledgement that peoples outside the Islamic world holding other religious beliefs could achieve power and wealth. Thus, groups living outside the Islamic world did not have to be portrayed as enemies or as necessarily inferior. Salvation history had not to be applied to them. Here Abu l-Fida's narrative can perhaps best be understood as part of a representation of the inhabited world (i.e., as part of geography, as a narrative about terrestrial features such as location, direction or distance; urban properties such as size and quantity of households; and noteworthy properties and phenomena such as buildings, natural wonders and beyond). Only in one case at the end of his list of Christian groups beyond the borders, Abu l-Fida's turns to the issue of knowledge, but it is a negative one and again related through its choice of words like *jahl* (ignorant; a major term used to describe the pre-Islamic Arabs) and *jufa'* (vain, futile) more to religion than to 'the sciences'.[32]

Hence, for Abu l-Fida' 'the sciences' and religious beliefs seem to be widely separate themes. This changed perspective might be an expression of a different kind of shift than the one I had meant to investigate. Rather than reflecting a transformation of 'the sciences' from 'a neutral zone' to a 'faith-based' communal fragmentation, Abu l-Fida' when discussing the '*umma*s' seems to be a voice similar to that constituted by Ibn 'Ibri's *Mukhtasar Kitab al-duwal* or to that of the books collected by the Jacobite patriarch of Antioch, Ni'matallah, and his predecessors which he brought to Rome in 1578. Ibn 'Ibri wrote his history

31 Abu l-Fida', *al-Mukhtasar*, vol. 2, p. 59.
32 Abu l-Fida', *al-Mukhtasar*, vol. 2, p. 60.

not only in Arabic, but very much in the format and rhetoric of the historical chronicles and biographical dictionaries composed by Muslim authors, although he was a subject of the Ilkhanids before the latter's conversion to Islam. In the late sixteenth century, the Patriarchal library of Antioch contained, in addition to manuscripts on Christian topics, books on 'the sciences' which were almost exclusively written by Muslim authors and covered topics not overly different from a library held at a *madrasa*, a mosque, a Muslim court or in the household of a Muslim scholar. While the library surely could not compete with those of major Muslim courts, the books that Ni'matallah carried to Rome could have been also held by a Muslim scholar in any of the three other spaces of 'the sciences'. 'The sciences' as well as basic forms of historical narratives concerned with them and their practitioners apparently had become 'universally' accepted across the various religious communities in at least major parts of the central lands of Islam. Thus, while cooperation remained exceptional and other forms of working together may have also found less and less space, reading and writing mathematical, astronomical, medical and other scientific texts had become a shared cultural practice that transcended denominational boundaries at specific times and places. Whether the limited evidence that speaks for a 'universal status' of 'the sciences' in Islamic societies can or should be taken to mean that all religious groups looked upon 'the sciences' as practiced by the scholars of the majority creed as a supra-religious commodity or rather as one facet of religious domination and superiority remains an open question that needs much more research than the one I have done for this paper.

5 Convivencia?

Where do all the examples and observations that I assembled here as well as those that I left out for lack of opportunity lead us in regard to the question of *convivencia*? As I have stressed, at times scholars of different creeds populated the same space and practiced the same kinds of sciences. But this does not mean that they lived together qua scholars, let alone as inhabitants of the same city. They met sometimes physically as doctors and astrologers, teachers and students or more often metaphorically as readers and authors. They coexisted in differently regulated spaces that were marked by easily recognizable colors even at times that were congenial for physical encounters as the 160 years of Ayyubid rule. They lived in hierarchically structured spaces where members of the religious minorities could find themselves occasionally at top places of power, even able to reject demands of their princely patrons and at times willing to mock them and being allowed to do so. But for all their power they were always in a precarious position. Foreign visitors could throw stones at them without being persecuted because the court physician rode a horse forbidden to the minorities. Patrons could be fickle and demand conversion in exchange for continued patronage. Muslims and, above all, recent converts could write elaborate and biting apologies against their religious beliefs. Over time, even the *dhimmi*s became to be seen as unclean, unfit for company, and could be declared infidels. But since all these elements are part and parcel of stories and the greater narratives they served it is difficult to gauge to what extent they reflect lived realities. Other sources such as court documents, letters or diaries are needed to get a firmer grasp of the relationships between narration and lived experiences. For practitioners of the mathematical sciences at least such a demand seems to be very difficult to satisfy.

INDEX

ʿAbbās I 90–91, 93, 126
ʿAbbās II 7n12
ʿAbd Allāh b. Muḥammad b. ʿAbd al-Razzāq Ḥāsib 110, 117
ʿAbd al-ʿAzīz, Jalāl al-Dawla 65–67, 118
ʿAbd al-Bāqī Muḥammad Akbar 109
ʿAbd al-Fattāḥ b. Maṣʻūd Šarīf 104
al-ʿAbd al-Ġarīq 91
ʿAbd al-Laṭīf ibn Uluġ Beg 67
ʿAbd al-Majīd ibn Muḥammad Quṭb al-Dīn 77
ʿAbd al-Raḥīm 103
ʿAbd al-Razzāq b. Muḥammad Muʻīn Kāšānī 67, 118
ʿAbd al-Waḥḥāb-Ḫān Bahādu 77
al-Abharī, Aṯīr al-Dīn 106
Abū l-ʿAbbās Aḥmad ibn Zāġū 145n47
Abū ʿAbd Allāh Muḥammad ibn an-Naǧǧār 145n47
Abū Bishr Mattā b. Yūnus 219n10
Abū l-Faḍl ibn al-Imām 145n47
Abū l-Fidāʾ (Abu l-Fidaʾ) 147, 224–225
Abū Ḥanīfa 46
Abū l-Ḥasan ʿAlī ibn Qāsim 145n47
Abū l-Ḥasan b. ʿAbd al-Qādir al-Šarīf al-Ṭabīb Muḥammad ʿAlī al-Šarīf al-Ṭabīb 123
Abū Ḥusayn Isḥāq, known as Ibn Karnib 219n10
Abū Isḥāq 68
Abū Isḥāq ibn ʿAbd Allāh 69, 79
Abū Isḥāq Ibrāhīm ibn Aḥmad ibn Abī Bakr 145n47
Abū Kāmil al-Miṣrī, Shujāʿ b. Aslam 20–21
Abū Maʿšar 82, 111, 116, 119
Abū Naṣr Aḥmad b. Ibrāhīm b. Fāris b. Ḥasan [Ḥusayn] 116
Abu Saʿd al-Fadl b. Bulis 217
Abū Saʿīd Gürägān 67–68
Abū l-Wafāʾ Būzjānī 20–21, 43–45, 51, 137n20, 190, 198–199, 204–206, 205n32, 206n35
Abū Yaʿqūb Yūsuf ibn Ismāʿīl 145n47
Adelard of Bath 157
ʿAḍud al-Dawla 217
al-Afghani, Jamal al-Din 188n3, 191–192
Aḥmad 46–47, 50
Ahmad ibn Bayezid II 72
Aḥmad b. Muḥammad Naṣr Allāh 102–103
Aḥmad b. Mūsā 219n10
Aḥmad b. Rusta 14
Aḥmad ibn Yūsuf 164
Ahmed Dede 73
Akbar, ʿAbd al-Bāqī Muḥammad 109
Akbar, Jalāl al-Dīn 75–78, 78n81, 87
Albertus Magnus 160
Algazel 190, 190n5, 191n6
ʿAlī b. ʿAbd al-Wahhāb 119
ʿAlī b. Abī Ṭālib 9
ʿAlī b. Ḫiẓr 120
ʿAlī b. Riḍwān, Abū l-Ḥasan 23, 116, 118
ʿAlī Qušjī (Qušçī, Qushjī) 66–73, 78, 82, 85, 87, 92–94, 102–106, 108–109, 115–119, 121, 137n20, 140, 147
ʿAlī Rīżāvi 82, 88
ʿAlīšīr Navāʾī 66
al-ʿAllāma al-Ḥillī 53
Allard, André 155–156
ʿAlqama al-Faḥl 145n47
Alverny, Marie-Thérès d' 155, 157, 166
al-Āmidī, Sayf al-Dīn 53, 221
al-ʿĀmilī, Bahāʾ al-Dīn (ad-Dīn; Behâ Eddìn) 64, 65n5, 72, 76–79, 81–82, 86, 88–90, 92, 101, 104–105, 108, 110–111,

114–115, 117, 120–126, 137n20, 147, 190, 196, 202–203, 203n28, 209
ʿĀmilī, Muḥammad Ḥasan Ḥurr 112
ʿĀmilī, ʿAbd al-Munʿim 125
Amīr ʿAbd al-Karīm 79
Amīr Abū l-Ḥasan 126
Āmulī, Rukn al-Dīn ibn Šaraf al-Dīn Ḥusaynī 66–67, 119
Āmulī, Šams ad-Dīn (Shams al-Dīn) Muḥammad ibn Maḥmūd 68, 137n20
Andijānī, Muḥammad Taqī b. Muḥammad Aqdaş Šarīf ʿĀmilī 123
Apollonius (Apollonios) 16, 92, 137n20, 147, 203, 203n28, 216
Āqā Jalāl Muḥammad 128
Archimedes (Archimède) 18, 91–92, 105–106, 113, 203, 203n28
Ardakānī, Mīrzā Qāżī Kāšif al-Dīn Muḥammad Yazdī 108, 115, 126
Aristarchos 91–92
Aristotle (Aristote) 11–2, 116, 119, 190, 191n7 and 9, 192n11
Arneth, Arthur 206–207
Arnold of Saxony 159–160
Arużī, Aḥmad b. ʿUmar Niẓāmī 61
Aṣmaʿī, Aḥmad Taqī b. Faqīh Muḥammad 108
Astarābādī, Abū al-Qāsim Yaḥyā 91, 105
Astarābādī, Ḥasan b. Ġiyāṯ al-Dīn 101
al-Asṭurlābī, ʿAlī ibn ʿĪsā al-Ḥarrānī 82, 110
ʿAṭā Allāh Qārī 76
ʿAṭā Allāh Rašīd ibn Aḥmad Nādir 75
Autolykos 91–92, 112, 114, 127–128
al-Avārī, Muḥyī al-Dīn b. Badr al-Dīn 125
Averroes 158, 191n9
Avicenna (Avicenne) 158, 190
Awjānī, Muḥammad Zamān b. Kuttāb Allāh 119
Awrangzīb 75, 87
Ayas Paša 72
Āzādānī, Muḥammad Ṣādiq ibn Muḥammad Ṣāliḥ Iṣfahānī 76

Bābur Ḫān Bahādur, Abū l-Qāsim 66, 68
Bachet de Mezériac 202, 202n23
Bacon, Francis 9
Badaḫšānī, Rustam Beg Ḥārithī 78
Baġdādī, ʿAbd Allāh ibn Muḥammad al-Ḫuddāmī 78
al-Baghdādī, ʿAbd al-Laṭīf 12
Balḫī, Abū l-Qāsim ʿAlī b. Aḥmad 116, 118
Banū Mūsā 93, 106, 137n20, 216
Barceló, Carmen 169
Barsbay 143
Bartholomew the Englishman 160
al-Baṣrī, Ayyūb b. Sulaymān 20
al-Baṣrī, Muḥammad b. ʿAbdallāh, called Ibn Labbān 20
al-Bayḍāwī 53, 145n47
Bayezid II 69, 71, 74n54
Berlinghieri, Francesco 74n54
Bernardo, Lorenzo 146
Bessarion 191n7
Bhāskara 20n32, 75, 204–205
Bihištī, Abū ʿAlāʾ Isfarāʾaynī 102, 106
Bihištī, Abū Muḥammad b. Aḥmad Isfarāʾaynī 101
Biot, Edouard 190–191
Biot, Jean-Baptiste 188n3, 190, 197, 209
Birjandī, Niẓām al-Dīn 72, 77–78, 82, 86–89, 92–93, 101–105, 114–116, 118, 124–125
al-Bīrūnī, Abū l-Rayḥān 12, 14–15, 30–31, 45, 61, 71, 82, 92, 112, 132
Bolad Ch'eng-Hsiang 57
Boncompagni, Baldassare 188–189, 189n4, 200–201, 204, 211
Brahe, Tycho 74, 190, 198
Brentjes, Sonja 158, 161–162, 168
Buchner, C.F. 193
Buḫārī, Abū l-Maḥāmid Muḥammad b. Maṣʿūd b. Muḥammad Ġaznavī 122
Buḫārī, ʿImād al-Dīn ibn Jamāl al-Dīn 67
al-Bukhārī, Muḥammad b. Ismāʿīl 50
al-Būrī, Abū ʿAbd Allāh Muḥammad 145n47
Burnett, Charles 155, 157–158, 161, 163–164
Busard, Hubertus L. L. 155
al-Būzīdī, Abū Rabīʿ Sulaymān 145n47

Čaġmīnī, Sharaf al-Dīn Maḥmūd ibn Muḥammad 66–67, 71, 73, 82, 92–93, 102, 107, 109, 112, 125, 147
Cantor, Moritz 193, 196, 206–208
Cassini, Jean-Dominic 189
Caussin de Perceval, Jean-Jaques Antoine 190
Cervantes 198
Chasles, Michel 197, 205
Charles X 187
Clagett, Marshall 155
Clavius, Christopher 78

Colebrooke, Henrry Thomas 203, 203n28
Comte, Auguste 189
Condorcet, Jean Antoine Caritat de 189
Constantinus Africanus 159–160
Copernicus 5

Damāvandī, Favāris 122
Damġānī, ʿAlī b. Āḫūnd Mullā Babāyān 118
Dārā Šikōh 76
Daštakī, Ġiyāṯ al-Dīn Manṣūr Šīrāzī 75, 77, 87–89, 101–102, 109, 124
Daštakī, Ṣadr al-Dīn 88
David ben Joshuʿa 222–223
Davvānī, Jamāl al-Dīn (ad-Dawwānī, Ǧalāl ad-Dīn) 75, 87, 104, 129n2, 139n29
Delambre, Jean-Baptiste 190
Delisle, Joseph-Nicolas 189
al-Dhahabī, Shams al-Dīn (aḏ-Ḏahabī, Šams ad-Dīn) 135, 221
Dhārma Narāyāṇ 75
Dihlavī, Farīd al-Dīn Masʿūd 76–77
Dihlavī, Ṭayyib ibn Ibrāhīm 77
Diophant 203
Draelants, Isabelle 159–160
Duhem, Pierre 154, 160
al-Dunayṣarī, ʿImād al-Dīn 33

Empedocles 49
Euclid (Euclide) 11, 14, 20, 23, 31, 53–54, 61, 66, 68–71, 73, 76–78, 80, 84–85, 91, 91n133, 104, 113, 119, 124, 127–128, 137n20, 168n79, 191n7, 203, 203n28

al-Fārābī (Alfarabi) 12, 121, 131–132, 164, 190, 190n5, 191n6, 219n10, 220
al-Farghānī, Aḥmad b. Muḥammad 14
al-Fārisī, Abū al-Ḫayr Taqī al-Dīn Muḥammad 89, 103
Fārisī, Abū Zayn al-Ḥasan 103, 119
al-Fārisī, Badr al-Dīn 12
al-Fārisī, Kamāl al-Dīn 78
Fayżī 75
Fayż Kāšānī, Muḥsin ibn Murtaḍā 82
Federico da Montefeltro 74n54
Fidora, Alexander 163–166
Firdawsī 81
al-Fīrūzābādī 7
Folkerts, Menso 155–156, 158, 187, 189, 193, 206
Frederick Barbarossa 225
Freudenthal, Gad 163–165
Fück, Johannes 162

Gabriel du Chinon 6
Galen 11, 159
Galilei, Galileo 74
Ganjaʾī, ʿAlī Taqī b. Mullā Valī 125
Gerard of Cremona 164
Gerbert of Aurillac 156
al-Ghazālī, Abū Ḥamīd 4, 12, 28n1, 34, 53, 60
al-Ghaznawī, Abū l-Maḥāmid Muḥammad 19
Ghulam Zuhal, Abu l-Qasim 217
Gīlānī, al-ʿĀlim ibn ʿAbd al-Ġanī ʿAbd al-Karīm Raštī 91
Goldstein, Bernard 2, 213, 215–216, 218
Goldziher, Ignaz 29n1, 135
Grant, Edward 155
Grunebaum, Gustav E. von 29n1, 135
Gunābādī, Muẓaffar ibn Muḥammad Qāsim 82, 88, 90, 91, 107, 119, 124, 126–127
Gundisalvi, Dominicus 164–165
Gutas, Dimitri 29n1, 46, 51–52, 150, 163, 167, 216, 223

Ḥabaš al-Ḥāsib 62
Hadī b. Muḥammad ʿAlī b. Ḥājjī Malik Qāsim 122
Ḥāfiẓ-i Abrū 58, 113
Ḫafrī, Šams al-Dīn, *see* al-Khafrī
al-Ḥajjāj b. Yūsuf b. Maṭar 53, 55n27
al-Ḥalabī, ʿAlāʾ al-Dīn Abū l-Ḥasan ʿAlī b. Ṭaybughā al-Ḥanafī al-Muwaqqit 222–223
Ḫalḫālī, Šams al-Dīn ʿAlī Ḥusaynī 123
Ḫalḫālī, Šams al-Dīn Muḥammad ibn ʿAlī 78
al-Ḫalīlī, Šams al-Dīn 72
Halley, Edmund 202, 202n23
Ḥamdallāh Mustawfī 58
Hames, Harvey J. 164, 166
Hammer-Purgstall, Joseph von 207
Ḥamza b. Abī Bakr 115
Hankel, Hermann 207, 209
al-Ḫāqānī, al-Ġaffūr b. Maṣʿūd 106
Haravī, ʿAlī Jān b. Ḥaydar ʿAlī 105
Haroun Alrachid 202, 202n23
Ḥasan 143
Ḥasanī, Muḥammad Rašīd 112
Ḥasan Muḥammad Taqī 109
al-Hāshimī, ʿAlī 11

Haskins, Charles Homer 151–163, 166
al-Ḥasnāwī, Abū l-Ḥasan ʿAlī ibn Ibrāhīm 145n47
al-Ḥasnāwī, Mūsā ibn Ibrāhīm 145n47
Hasse, Dag Nikolaus 161, 163
Ḥātim Beg, Ḫwāja Naṣīr al-Dīn 90, 114, 126
Ḫātūnābādī, Muḥammad Bāqir b. Mīr Ismā'īl Ḥusaynī 127
Ḫātūnābādī, Muḥammad Zamān b. Ḥusayn 105
al-Ḫayyām, ʿUmar, *see* al-Khayyām
Ḫāzinī, Muḥammad 109
Herbelot, Barthélémy d' 189
Herder, Johann Gottfried von 193, 195
Hermes 116, 119
Hibat Allāh al-Ḥusayn 13
Hippocrates 137n20
al-Ḫirāzī 145n47
Hochheim, Adolf 207
Høyrup, Jens 158
Ḫudābanda, Mīrzā Ḥamza ibn Muḥammad 90, 126
Hultsch, Friedrich 207
Humāyūn 69, 78–9, 115
Humboldt, Alexander von 200, 204
Humboldt, Wilhelm von 187
Ḥusayn 90, 108, 125, 127
Ḥusaynī, ʿAbd al-Ġanī 116, 118
Ḥusaynī, ʿAlī b. Niʿmat Allāh 1056
Ḥusaynī, Amīr ʿAbd al-Karīm Muḥammad Muḥsin 119
Ḥusaynī, Ḥasan b. ʿAbd al-Ġaffūr b. Jāmī 125
Ḥusaynī, Luṭf Allāh 123
Ḥusaynī, Muḥammad Bāqir b. ʿAbd al-Qādir b. Hibat Allāh 103
Ḥusaynī, Muḥammad Taqī b. Ḥaydar 110–111
Ḥusaynī, Nūrallāh b. Muḥammad 104
Ḥusaynī, Sayyid Muḥammad Masīḥ Šīrāzī 86, 116, 119
Ḥusayn b. Muḥammad b. Aḥmad 221
Ḫwāja Niẓām al-Dīn Mīrijān 104, 125
Hypsikles 91, 91n133, 127

Ibn Abī Bakr, Abū Isḥāq Ibrāhīm ibn Aḥmad 145n47
Ibn Abī Jumhūr 53
Ibn Abī l-Qāsim 145n47
Ibn Abī Šukr al-Maġribī 82, 92, 111–113, 124, 141
Ibn Abī Sulmā, Zuhayr 145n47
Ibn Abi Usaybiʿa 37–38, 219
Ibn Afšūš 145n47
Ibn al-Aktham 217
Ibn al-ʿAmīd 44
Ibn ʿArabī 49, 221
Ibn al-ʿAṭṭār, Muḥibb al-Dīn Muḥammad b. Muḥammad 30
Ibn al-Bannāʾ 145n47
Ibn al-Ǧawzī 135
Ibn Ḥāǧib 145n47
Ibn al-Hāʾim 72, 137n20
Ibn Ḫaldūn (Khaldoun) 144, 191n8
Ibn Ḥanbal, Aḥmad 55
Ibn Ḫātūn, Šayḫ (Shaykh) Asad Allāh Muḥammad 86
Ibn al-Haytham (Hayṯam) 12, 17n34, 19, 21–24, 22n42, 30–31, 132, 137n20
Ibn ʿIbri 225
Ibn al-ʿImād 221
Ibn Kammūna 12, 12n25, 222–223
Ibn al-Majdī (al-Maǧdī) 72, 137n20, 143
Ibn Mālik 145n47
Ibn Muḥammad Raḥim Muḥammad Ṣāliḥ 122
Ibn al-Muqaffaʿ 54–55
Ibn al-Nadīm 14, 45, 216–217, 220n15
Ibn al-Nafīs (an-Nafīs) 137n20, 219n11
Ibn al-Quff 219n11
Ibn Qurqmas al-Ḥanafī, Naṣīr ad-Dīn 143
Ibn al-Rawandī 221
Ibn Rušd 129, 129n2, 132, 191
Ibn Saʿdān 44, 52
Ibn aṣ-Ṣalāḥ aš-Šahrazūrī 135
Ibn al-Sarī, Aḥmad b. Muḥammad 12, 19, 23–24, 71
Ibn al-Sarrāj 13, 140
Ibn al-Shāṭir 12–13, 72, 140
Ibn Sīnā (Sina) 30–31, 42, 46–47, 51–52, 59, 73, 88, 124, 129n2, 132, 137n20, 147, 158, 160–161, 220n15, 221, 223
Ibn Ṭaymīya 135, 140, 141n31
Ibn Yūnus 18n36, 190, 199
Ibn Zayn al-ʿĀbidīn, Muḥammad Bāqir Yazdī 82, 86, 91, 107, 114–115, 126–127
Ibrāhīm b. Sinān 16–17
al-Ījī, ʿAḍud al-Dīn 30
al-Ījī, Muḥammad b. Ibrāḥīm 30
Ikhwān al-Ṣafāʾ 50
Ilḥāq ibn Abī Isḥāq 69
Imrūʾ al-Qays 145n47

al-Iṣfahānī, ʿAbd Allāh al-Laṭīf Muḥammad Šarīf b. Ḥājjī Maqṣūd ʿAlī 112
Iṣfahānī, Aḥmad 101
Iṣfahānī, Āqā Zayn al-ʿĀbidīn Ḫādim Šarīf 86, 89, 103
Iṣfahānī, ʿInāyat Allāh Ḥusayn 104
Iṣfahānī, Muḥammad Ašraf ibn Muḥammad Jaʿfar 89, 104
Iṣfahānī, Muḥammad Hadī b. Šayḫ Bahāʾ al-Dīn Muḥammad Miʿmār 125
Iṣfahānī, Muḥammad Qāsim b. Muḥammad Ṭāhir 114
Iṣfahānī, Muḥammad Ṣādiq 124
Iṣfahānī, Saʿd al-Dīn Muḥammad b. Kamāl al-Dīn 102
al-Isfizārī, al-Muẓaffar 141
Isḥāq b. Ḥunayn 55–56, 55n27
Iskandar Sulṭān 58, 66–68, 66n7, 72, 81
Ismāʿīl 89, 104, 125
ʿIṣmat (or: ʿĀṣim) Allāh ibn ʿAẓīm ibn ʿAbd al-Rasūl 78
Istajlū, Abū l-Qāsim 93, 106
ʿIzz al-Dawla 44

Jābir b. Ibrāhīm al-Ṣābīʾ 12
Jacobi, Carl Gustav Jacob 193
Jacquart, Danielle 159
Jaʿfar b. Muḥammad Muʾmin 117
Jahāndār Šāh 76
Jahāngīr 77
al-Jahz 224
Jāmāsb 110
al-Jawharī, al-ʿAbbās b. Saʿīd 14
al-Jawharī, Abū Naṣr 7
Jawnpuri, Rawshan ʿAli 196
Jourdain, Amable 190
Jurjānī, al-Sayyid al-Šarīf 72, 79, 88, 112
Juvaynī, Šams al-Dīn Muḥammad 87
Jūzjānī, ʿUbaydallāh 73

Kamāl al-Dīn b. Yūnus 221
Kara Mustafa Paša 72
al-Karajī, Muḥammad 15
Kāšānī, Muḥammad b. Muḥsin b. Murtażā Fayż 121
Kāšānī, Muḥammad Saʻīd ibn Muḥammad Amīn 88, 114
Kāšānī, Muḥammad b. Ṭāhir Fāżil 106–107
Kāšānī, Muḥsin b. Murtażā Fayż 122
Kašġārī, Muḥammad b. Muḥammad 116
al-Kāshī (Kāšī), Ghiyāth (Ġiyāṯ) al-Dīn (ad-Dīn) 58, 66–68, 66n7, 72, 82, 209
al-Kāshī, Maḥmūd 58
Kāšī (Kāshī), Muʻīn al-Dīn 66–67
Kāšifī, Ḥusayn 106
al-Kātibī, Najm al-Dīn 71
Kāẓimī, Fāżil Javād 122
al-Kāẓimī, Ḥusayn b. Ḥājj Muḥammad 111
Kāẓimī, Kalb ʿAlī b. Mullā Javād 86, 112
al-Kāẓimī, Muḥammad Javād ibn Saʿd 78, 86
al-Khafrī, Shams al-Dīn 30, 62, 82, 84, 86–88, 93, 101–102, 104, 114–115, 117, 124, 139n29
al-Khayyām (Alkhayyāmī, Khayyam), ʿUmar 132, 141, 202–203, 203n24, 205, 205n31
al-Khāzin, Abū Jaʿfar 13–14, 43–45, 51
al-Khwārizmī, Muḥammad b. Mūsā (Mohammed Ben Moûçâ) 13, 16–17, 20–22, 20n40, 45–46, 196, 202–203, 203n28, 208
al-Kindī, Yaʿqūb ibn Isḥāq 30–31, 45, 55n27, 93, 127, 132, 137n20
King, David A. 18n36, 59–60, 62, 69, 81, 132, 169
Kirmānī, Ġiyāṯ al-Dīn Abū Isḥāq Muḥammad ʿĀšiqī 126
Kiyādehī, Ṣadiq b. Maḥmūd 94, 121
Komnenos, Manuel 167
Koningsveld, Pieter Sjoerd van 160
Köprülü Fazıl Ahmed Paša 72
Korkut ibn Bayezid II 72
Kremer, Alfred von 207
al-Kūhī, Abū Sahl Wayjan b. Rustam 17–18, 18n36, 82, 91n133, 93, 127, 217
Kūhistānī, Fasīḥ al-Dīn Muḥammad ibn ʿAbd al-Karīm Niẓāmī 67
Kunitzsch, Paul 156–157, 162
Kūšyār ibn Labbān 94, 107–108, 113

Labarta, Ana 169
Lafīnī, Ġulām 111
Lāhījānī, Āqā Muḥammad Masīḥ 122
Lahījī, ʿAbd al-Qādir Ruyānī 118
Lāhūrī, Aḥmad 77
Lange, Nicholas de 166–167
Langermann, Y. Tzvi 167
Laplace, Pierre-Simon 190
Lārī, Muṣliḥ al-Dīn 77, 88, 115, 121
Lārī, Quṭb al-Dīn ʿAbd al-Ḥayy ibn ʿIzz al-Dīn Ḥusaynī 94, 106, 108, 123–124, 126
Lascaris 191n7
Lasker, Daniel J. 167

Lemay, Richard 158
Leurquin, Régine 156
Libera, Alain de 159
Libri, Guillaume 197–198
Lindberg, David Charles 155
Lorch, Richard 155
Lorenzo de' Medici 74n54
Louis UX 225
Louis Philippe 187
Lucas, Paul 75n57
Luṭf Allāh Muhandis 77

al-Maghribi, Ibn ʿIsa 223
Maḥābat Ḫān 79
al-Māhānī 16
Maḥmūd or Muḥammad Sirāj 106
Maimonides 11n24, 222–223
Majlisī, Muḥammad Bāqir ibn Muḥammad Taqī 81, 90, 119, 127
Makdisi, Heorge 29n1, 34n15, 135–136
Malet, Antoni 163, 168
al-Malik al-Nāṣir 221
al-Maʾmūn (Almamoun) 18n36, 22, 202, 202n23, 216, 216–217, 220, 224
Manūchihr Ḫān 83, 109
Marcianus Capella 203, 203n28
al-Māridīnī, Ismāʿīl b. Ibrāhīm, known as Ibn Fallūs 35
al-Māridīnī, Jamāl al-Dīn (Ğamāl ad-Dīn al-Maridānī) 31, 137n20
Marre, Aristide 207
Martin, Craig 161
Martínez Gázquez, José 166
Marwazī, Šams al-Dīn Abū Bakr Muḥammad b. Aḥmad 123
Mašhadī, Ġulām ʿAlī b. Darvīš ʿAlī 103
Mašhadī Ḥusaynī, Muḥammad Taqī 93, 107, 109, 113
Mašhadī, Zamān ibn Šaraf al-Dīn Ḥusayn 89
Maṣʿūd b. Ḥabīb Allāh 107
al-Mawālī, Ṣadr al-Islām Ibn ʿAlī 103
Maybudī, Mīr V al-Dīn 102, 123
Māzandārānī, Ḥabīb Allāh 118
Māzandarānī, Muḥammad ʿAlī 86
Maẓhar al-Dīn Muḥammad ibn Bahāʾ al-Dīn ʿAlī Qārī 89
McVaugh, Michael Rogers 156
Mehmed Fātih 69–70, 74n54
Menelaos 91, 105, 113, 137n20
Mercier, Raymond 156
Micheaud, Françoise 159
Millàs Vallicrosa, Josep Maria 155, 169
Mīr Dāmād 89
Mīr Ḫwānd 147
Mīr Sayyid Muḥammad Saʿīd Mīr Muḥammad Yaḥyā 77
Mīrzā ʿAbd Allāh Efendī Iṣfahānī 110
Mīrzā ʿAzīz Koka 78n81
Mīrzā Ḥakīm 103
Mīrzā Ibrāhīm 30, 88
Mīrzā Ẓahīr al-Dīn Muḥammad Ibrāhīm 81, 83, 90
Montucla, Jean-Étienne 195
al-Muʾayyad al-Manṣūr Fakhr al-Mulk Abū Ghālib 22n42
Muḥammad 9, 50, 86, 169
Muḥammad ʿAlī b. Muḥammad Ḥusayn 124, 126
Muḥammad Amīn 101
Muḥammad Ašraf Munajjim b. Muḥammad Ṣādiq 127
Muḥammad Bāqir ibn ʿImād al-Dīn Maḥmūd 73
Muḥammad ibn Hacı Atmaca al-Kātib 71
Muḥammad Hādī 86
Muḥammad b. Ḥājjī 108
Muḥammad Ḥusayn 101
Muḥammad Ḥusayn b. ʿAlī 102, 127
Muḥammad ibn Ḥusayn 86
Muḥammad ibn Kātib Sinān 71
Muḥammad b. Muḥammad b. Lakhmī al-Muhandis 21
Muḥammad ibn Marzūq ibn Ḥafīd 145n47
Muḥammad ibn Yūsuf 71
Muḥammad Muʾmin 124
Muḥammad Muʾmin Aḥmad 103
Muḥammad Muʾmin b. Niẓām al-Dīn ʿAlī 117
Muḥammad Qulī 106
Muḥammad Raḥīm Munajjim 108, 127
Muḥammad Rašīd al-Dīn 76
Muḥammad Rīżā b. Muḥammad Taqī 88, 112
Muḥammad Rustam ibn Muḥammad Ḫalīfa 77
Muḥammad Ṣādiq 102
Muḥammad Saʿīd b. Faḫr al-Dīn 113
Muḥammad Ṣāliḥ b. Pīrzāde 117
Muḥammad Ṣanīʿ 82, 88
Muḥammad Taqī Ḥusaynī ibn Muḥammad Bāqir 80, 112
Muḥyī al-Dīn Muḥammad Kātib 121
Mullā Chand 77
Mullā Ṣadrā 82, 86
Mullā Tarzī 77
Murad II 71–72

Murad III 73–74
Murdoch, John Emery 155–156
Muṣṭafā ibn ʿAlī 72

Nabāṭī, ʿAlī b. Aḥmad 105
an-Nābiġa aḏ-Ḏ ubyānī 145n47
Nādir Šāh 91n133
Najafī, Maḥmūd b. Ibrāhīm b. ʿAbd Allāh 121–122
Najm al-Din ʿAli Khan 196
al-Nawbakhtī, al-Ḥasan b. Mūsā 11
Needham, Joseph 134, 134n16
Nesselmann, Georg Heinrich Ferdinand 187, 193–197, 200, 209
Niʿamatallah 225–226
Nicolaidis, Efthymios 163, 166–167
Nicomachus 196
Niebuhr, Barthold Georg 193
al-Nihāwandī, Benjamin 219n10
al-Nīsābūrī, Niẓām al-Dīn 12, 72–73, 77, 79, 87–89, 104, 108–109, 114, 117–118, 120, 139n29
Niẓām al-Mulk 28, 30, 60–61, 137n20
al-Nuʿaymī, ʿAbd al-Qādir b. Muḥammad 28–29, 32–33, 36–37, 39–40
Nūḥ b. Naṣr 44

Pascal, Blaise 9
Petrus Alfonsi 164
Petrus Venerabilis 157
Philipp Augustus 225
Plato 49
Plessner, Martin 20n1
Proclus 195
Ptolemy (Ptolémée) 11–12, 77, 92, 101–102, 107–108, 110, 116, 118, 121, 127, 137n20, 152, 156, 191n7
Pythagoras 49

Qāḍī Aḥmad b. Mīr-Munshī 8
Qāḍīzāda (Qāżīzāda; Qadizade) al-Rūmī (ar-Rūmī, Rūmī) 66–68, 73, 82, 85, 92–93, 102–103, 107, 109, 111, 118, 125, 137n20, 147, 209
al-Qāhir 220n15
Qāʾinī, Abū Jaʿfar Kāfī b. Muḥtašam b. ʿAmīd b. Muḥammad Šāhinšāh 107, 110
Qāʾinī, Ḥasan ibn Sāʿd 83, 109
Qāʾinī, Muḥammad Kāfī b. Abī l-Ḥasan 127
Qāqshāl, Muḥammad Barānī Ummī 8
Qāsim b. Ḥusayn b. Zāl 119
Qāżī Muḥammad Šarīf b. Mullā Muṣṭafā 112
Qazvīnī, Mīr Ṣadr al-Dīn Muḥammad b. MuḥammadṢādiq Ḥusaynī 126
Qazvīnī, Mīrzā Muḥammad Ṭahir b. Ḥusayn Ḫān Vāḥid 109
al-Qazwīnī (Qazvīnī), Zakariyyāʾ b. Muḥammad 15, 84
Qiwām (Qivām) al-Dīn Ḥusayn b. Šams al-Dīn Ḫafrī 86, 124
Qummī, ʿAlī b. Ḥājjī Muḥammad Gurgānī 115
al-Qunawī, Ṣadr al-Dīn 53
al-Qurashī, ʿAlī b. al-Khiḍr b. al-Ḥasan 19–22
Qusṭā b. Lūqā 11, 72, 82, 93
Quṭb Šāh 69
al-Quwayrī, Abū Isḥāq Ibrāhīm 219n10

Rāḍī al-Dīn ʻAlī 76
Ranke, Leopold von 193
Raphael du Mans 6
Rashīd al-Dīn 57, 79
al-Rāzī, Fakhr al-Dīn (Faḫr ad-Dīn; Fakhr-eddin) 11n20, 76, 129n2, 190, 190n5
Rāzī, Muḥammad Taqī b. Muḥammad Rīżā 108
al-Rāzī, Muḥammad b. Zakariyyāʾ 14, 219, 221
Reif, Stefan 167
Renan, Ernest 188, 188n3, 191–193, 191n7–9
Richard Lionheart 225
Ricklin, Thomas 163–164
Riet, Simone van 158
ar-Rīfī, Abū Yaʿqūb Yūsuf 145n47
Rius, Mònica 169
Rīżā ʿAbbāsī 80
Rosen, Friedrich August 196, 207
Rosenthal, Franz 29n1, 31
Rukn al-Dawla 44
Rustam b. Šāh Vīrdī 121
Rustam ibn ʿUmar Šayḫ (Shaykh) 66

Šabākī, Ḥusayn 123
Sabra, A. I. 1–2, 10, 29n1, 131, 134, 139, 149, 215
Saʿdī 81, 147
al-Ṣafhānī 7
Ṣafī 81, 114, 126
al-Ṣaghānī, Abū ʿAlī b. Muḥtāj 44
Šāh ʻĀlam Bahādur Šāh 79
Šāh (Shāh) Jahān 75–76

Salah al-Din 225
al-Salih Ayyub b. al-Malik al-Kamil 225
Samarqandī, Muḥammad Fāżil ibn ʿAlī ibn Muḥammad Miskīnī Qāżī 76
Samarqandī, Shams (Šams) al-Dīn 67–68, 85, 91, 123
al-Samawʾal al-Maghribī 12, 15
Samsó, Julio 169
Sanad b. ʿAli 217
Šaraf al-Dīn Ḥusayn 118
Sāravī, ʿAlī Taqī b. Ḥājjī Muḥammad Amīn Qārī 116–117
al-Sardafī 60
Šarīf Ḥusaynī, Muḥammad Rafīʻ b. Muḥammad 123
Šarīf Qāʾinī, Abū al-V 103
Sarton, George 154
Sayf al-Dīn Maḥmūd b. Ḥājjī Ibrāhīm 118
Sayyid Muḥammad Manṣūr 125
Sayyid Saʿīd 104
Šams al-Dīn Muḥammad b. Bahāʾ al-Dīn 119
Scheuren, Claudy 156
Schmidtke, Sabine 51, 166, 222n19, 225m24
Schwartz, Yossef schwartz 164
Sédillot, Jean-Jacques Emmanuel 190, 207
Sédillot, Louis-Amélie 188n3, 190–191, 193, 197–200, 198n20, 202, 204, 207–209
Shāhmardān b. a. l-Khayr 58
al-Shahrastani, Tāj al-Dīn Abū al-Fath Muhammad ibn ʿAbd al-Karīm 224
Shāhrukh 58
al-Shakkāz, ʿAlī b. Khalaf 13
Sharaf al-Dawla 217
Shīrāzī (Šīrāzī), Muḥammad Riḍā (Rīżā) 30, 117
al-Shīrāzī, Quṭb al-Dīn 15, 30–31, 46–47, 49–51, 59, 67–68, 78–80, 83–84, 92, 102, 107, 122, 137n20, 141
Sibṭ al-Maridānī 72, 137n20
al-Sijistānī, Abū Sulaymān 43–44, 50
Sijistānī, Šams al-Dīn Muḥammad b. Ḥusayn 123
al-Sijzī, Aḥmad ibn Muḥammad ibn ʿAbd al-Jalīl 30, 82, 93, 137n20
Silvestre de Sacy, Antoine-Isaac 196
Sinān Paşa 70
Sinān b. Thābit b. Qurra 220n15
Sirais, Nancy 161
Šīrāzī (Shīrāzī), ʿAlāʾ al-Dīn Manṣūr 73
Šīrāzī, Amīr Muʿīn al-Dīn Muḥammad Ašraf b. Ḥabīb Allāh 108
Šīrāzī (Shīrāzī), Fatḥ Allāh 75
Šīrāzī, Šāh-Mīr 87
Sokullu Mehmed Paša 72
Spiesser, Maryvonne 158
Steinschneider, Moritz 207
al-Ṣūfī, ʿAbd al-Raḥmān 12, 68, 74, 79–80, 82–84, 90, 104, 109, 119
Ṣūfī, Ḥasan b. Aḥmad b. ʿAbd Allāh 119
Suhrawardī (Suhravardī), Shihāb al-Dīn 49–51, 221–223
Sulaymān 81, 89, 126–127
Süleyman 71
Sultan Cem 74n54
Šuštarī, Nūr Allāh 78
Suʿudi, Mehmed 73
Sylla, Edith Dudley 155–156

al-Ṭabarī, Jarīr 61
al-Ṭabarī, Muḥammad ibn Ayyūb Ḥāsib 94, 107
al-Ṭabarī, ʿUmar b. Farrukhān 14
Ṭabasī, Malik Ḥusayn b. Muḥammad 103
Ṭabasī, Qurbān ʿAlī ibn Ramaḍān Šams al-Dīn 86, 118, 121
Ṭabasī, Muḥammad Ḥusaynī 104, 124
Ṭabātabāʾī, ʿAzīz Allāh b. Yūsuf 114
Ṭabāṭabāʾī, Muḥammad Ašraf ibn Ḥabīb Allāh al-Ḥasanī al-Ḥusaynī 78
Tabrīzī, Muḥammad Ibrāhīm b. Muḥammad Taqī 122
Tabrīzī, Muḥammad Rīżā b. Ḥajjī Sulṭān Muḥammad 107
Tabrīzī, Muḥammad Saʿīd b. Muḥammad Muʾmin 123
Tafrīšī, ʿAlī ibn Masʿūd Ḥusaynī Qummī 88, 109
al-Taftazānī, Saʿd al-Dīn 71
Ṭahmāsb 90, 125–126
Taine, Hippolyte-Adolphe 201
Ṭāliqānī, Muḥammad Ḥusayn b. Muḥammad Yūsuf 112–113
Taqī al-Dīn b. Maʿrūf 73
Ṭarafa ibn al-ʿAbd 145n47
Taršīzī, Muḥammad Ṣādiq b. ʿAbd al-ʿAlī 110
Ṭāshköprüzāde 54
al-Tawḥīdī, Abū Ḥayyān 44, 50, 52
Thābit (Ṯābit) b. Qurra 12, 16–18, 53, 55, 55n27, 91n133, 127, 196, 216, 219n10
Theodosios 91–92, 105–106, 111, 114, 117, 122, 127–128
Thorndike, Lynn 154
Tihon, Anne 156, 166
at-Tilimsānī, Abū Bakr 145n47

at-Tīrūnī, Yaʿqūb 145n47
Tischler, Matthias M. 166
Titus Livius 147
al-Tosyavī, Muṣṭafā ibn Maḥmūd 71
Tūnī, Ḥāfiz Ḥasan b. Šujāʿ b. Muḥammad b. Ḥasan 113
Tūnī, Muḥammad Rīżā b. ʿAzīz Allāh 123
al-Tunikabānī, Muḥammad Ṣādiq 126
Turkumānī, ʿAlī b. Šāhīd ʿAlī 107
al-Ṭūsī, Muḥammad b. Maḥmūd 15
al-Ṭūsī, Naṣīr al-Dīn (Nassir eddin al Thusi,) 12, 12n25, 30, 45, 66–69, 72, 77, 79, 82, 85, 87–92, 93, 101–102, 104–107, 109–122, 124, 129n2, 132, 137n20, 139n29, 141, 147, 202, 202n23, 209, 221

Ulugh (Uluġ) Beg 12, 58, 64–73, 85, 89, 94, 104, 112, 137n20, 141, 147, 200, 209
al-ʿUqbānī, Abū l-Qāsim ibn Saʿīd 145n47
al-Ūqlīdisī, Abū Yūsuf 20–21
al-ʿUrḍī, Muḥyī l-Dīn (d-Dīn) 141
ʿŪsjānī, Muḥammad Qāsim b. Bāqir 124
Ūzūn Ḥasan 69

Valle, Pietro della 94
Vernet Ginés, Juan 155
Viladrich, Mercé 169
Vincent of Beauvais 160

al-Wanšarīsī, Abū Bakr ibn ʿĪsā 145n47
Washmgīr 44
Weil, Gustav 207
Wenrich, Johann G. 207
Woepcke, Franz 189, 193, 200–207, 205n32, 207–209
Wüstenfeld, Ferdinand 207

Yaḥya b. ʿAdī 219
Yaḥya Kiyā 118
al-Yaldāvī, Mūsā ibn Ibrāhīm 73
Yaʿqūb Bahādur Ḫān 69
Yaʿqūb b. Ṭarīq 14–15
Yazdī, Jalāl al-Dīn Muḥammad b. ʿAbd Allāh 112
Yazdī, Muḥammad b. Amr Allāh 115
Yazdī, Muḥammad Ašraf b. Ḥājjī Muḥammad 119–120
Yazdī, Muḥammad Bāqir 88

al-Zawāwī, Muḥammad ibn Muḥammad Abū l-Faḍl al-Mašdāllī al-Biǧāyī 145, 145n47
Zonta, Mauro 164–165
Zuhayr ibn Abī Sulmā 145n47
Zunīl 126

www.ingramcontent.com/pod-product-compliance
Lightning Source LLC
Chambersburg PA
CBHW060253220326
41598CB00027B/4083
9781032445069